Biology of Lichenized Fungi

James D. Lawrey

QK 583 .L38 1984
Lawrey, James D., 1949-
Biology of lichenized fungi

PRAEGER

PRAEGER SPECIAL STUDIES • PRAEGER SCIENTIFIC

New York • Philadelphia • Eastbourne, UK
Toronto • Hong Kong • Tokyo • Sydney

Library of Congress Cataloging in Publication Data

Lawrey, James D., 1949—
Biology of lichenized fungi.

Bibliography: p.
Includes index.
1. Lichens. 2. Fungi. I. Title. II. Title:
Lichenized fungi.
QK583.L38 1984 589.1 84-9908
ISBN 0-03-060047-2

Published and Distributed by the
Praeger Publishers Division
(ISBN Prefix 0-275)
of Greenwood Press, Inc.,
Westport, Connecticut

Published in 1984 by Praeger Publishers
CBS Educational and Professional Publishing
A Division of CBS, Inc.
521 Fifth Avenue, New York, NY 10175 USA

456789 052 987654321

Printed in the United States of America
on acid-free paper

Contents

Preface

Lichenized fungi make up about half of the 50,000 species of cup fungi (Ascomycetes). Their unique symbiotic relation with algae has enabled these fungi to colonize and flourish in a wide range of habitats from the Antarctic continent to the rain forests of the tropics. Their ecological success in so many types of habitats depends on a number of unique structural and functional adaptations that are only now beginning to be generally appreciated by mycologists and microbiologists.

Why have lichens not been studied more extensively? Biologists have complained that lichenology is an esoteric science and that lichens are difficult research organisms. In reality, one of the main reasons is the lack of suitable reference works that present topics relevant to other workers. The objective of this text is to present information on lichen biology from a diverse set of sometimes obscure journals and reference works and to make this information relevant to the research interests of mycologists and microbiologists.

Throughout the text, I have attempted to foster an appreciation of the various ways lichens solve basic biological problems: the need to reproduce, disperse, acquire resources, and tolerate competitors, predators, and physical stress. Although I have attempted to be as comprehensive as possible, I have chosen to emphasize the unusual attributes that make these fungi unique. Toward this end, I have included chapters on the specific and complex requirements for symbiont resynthesis, the production of bizarre lichen secondary substances, and the unusual ability of some lichens to colonize the most unfavorable physical environments on earth.

Unfortunately, in a text of this length, it is not possible to adequately discuss topics of interest to all readers. There is practically no discussion of lichen ontogeny or phylogeny. Fungal morphology is only briefly discussed, and this discussion is merely an introduction to structural solutions to biological problems; it is not meant to be a diagnostic guide. Also, throughout the book, emphasis is placed on the lichen fungus. Except for certain selected topics, the biology of the algal host is not discussed.

Biology of Lichenized Fungi provides a framework for experimental investigation of lichens. It is not meant to be an introduction to lichen biology; there are other texts much more suitable for the beginning student of lichens. Rather, it attempts to direct graduate students and established investigators interested in fungi and microorganisms to research areas involving lichens that relate in a meaningful way to their own investigations. It is my hope that **Biology of Lichenized Fungi** will attract mycologists and microbiologists to the study of lichen biology, and that experimental lichenologists will find valuable suggestions for future concentrated effort.

I gratefully acknowledge a number of my colleagues who provided previously published and sometimes unpublished information and/or figures. Doctors C. Culberson, M. Galun, E. Nieboer, G. Chilvers, R. Showman, J. Armesto, G. Vobis, W. Jordan, E.I. Friedmann, R. Tapper, H.M. Jahns, and E. Peveling kindly provided previously published figures. Doctors V. Ahmadjian, R. Slocum, G. Hoffman, F. Slansky, L. Sigal, D. Porter, M. Lechowicz, and M. Groulx donated unpublished figures and/or data. Mr. Jan Endlich graciously provided technical assistance in some of the ultrastructural work reported in the book. My inexpressible thanks are extended to Dr. Mason E. Hale, Jr., who provided numerous published and unpublished scanning electron micrographs and who has been a constant source of advice and support, not only during the development of this book, but for all the years I have known him. Finally, I thank my wife, Sara, for her patience, understanding, and encouragement.

James D. Lawrey

1 Vegetative Structure

The lichen thallus is a remarkably complex structure considering the low level of organization achieved by isolated symbionts. To describe this complexity, lichenologists have developed a sometimes confusing terminology that is in large part unique to lichenology. Only the broader details of lichen descriptive morphology, however, are presented here since there are already a number of excellent references that can be consulted by students interested in this subject (Smith, 1921; des Abbayes, 1951; Ozenda, 1963; Henssen & Jahns, 1974; Hale, 1983). The objective of this chapter is to provide the nonspecialist with an understanding of lichen structure necessary to appreciate the chapters on lichen reproduction, growth, physiology, and ecology that follow.

GENERAL THALLUS ORGANIZATION

Lichen thalli are typically composed of several distinct layers (Figure 1.1). While there are a number of variations on this pattern, foliose lichens are surprisingly constant in general design. The upper and lower cortical layers provide both protection and structural support. Crustose lichens also have an upper cortex, often only weakly organized, but lack a lower cortex.

The algal layer is generally found just below the upper cortex in a distinct zone, an arrangement that is called stratified or heteromerous. Some groups have an unstratified or homoiomerous organization with algal cells scattered throughout the medulla, with or without cortices.

The medulla makes up the bulk of most lichen thalli and is weakly

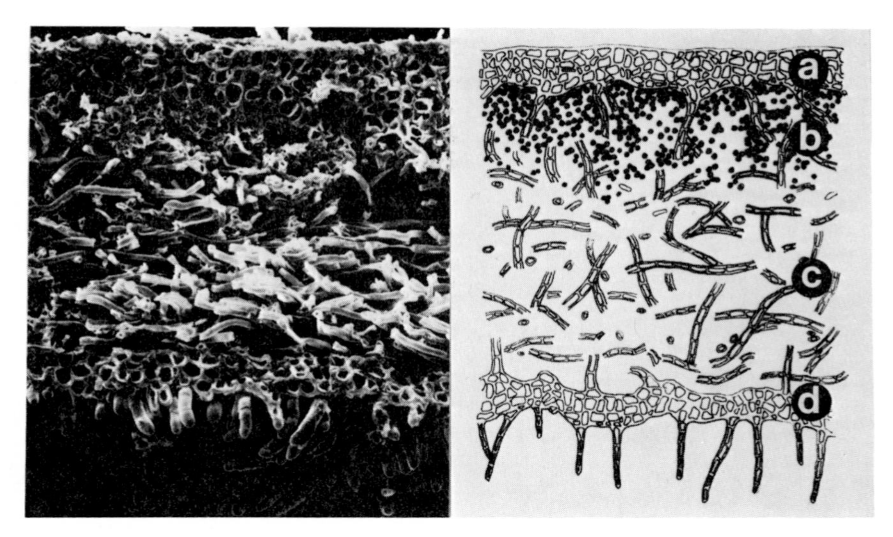

FIGURE 1.1. Typical thallus structure. SEM of *Sticta weigelii* CS (×400) with schematic diagram. (a) Upper cortex; (b) Algal layer; (c) Medulla; (d) Lower cortex with rhizines. SEM from M.E. Hale, Jr. 1976. "Lichen structure viewed with the scanning electron microscope." In *Lichenology: Progress and Problems* (D.H. Brown, D.L. Hawksworth, & R.H. Bailey, eds.), pp. 1–15. London: Academic Press. Reprinted with permission of Academic Press.

organized if it shows any organization at all. In crustose lichens, the medullary hyphae attach the thallus directly to its substrate. In species that produce secondary compounds, the medulla is generally the site of the greatest chemical diversity, although some compounds are produced only by cortical hyphae.

TISSUES

In most lichens, the fungal partner (mycobiont) dominates over the algal partner (phycobiont), at least in bulk, and determines thallus organization. During thallus development, when a suitable phycobiont is available, fungal hyphae become aggregated into *tissues* in various forms that distinguish one species from another. Although they are not true tissues comparable to the cellular structure of higher plants, the term will be used in this discussion to refer to organized and identifiable groups of hyphae.

Various types of lichen tissues have been assigned names based on the direction of hyphal growth, degree of branching and crowding of filaments, frequency of septation, and thickening of cell walls (Smith, 1921). Numerous lichen tissue terms have been proposed (Hale, 1976)—some of which are seldom used by lichenologists. This section will use Hale's (1976) terminology.

Paraplectenchyma

This tissue is formed from short, randomly oriented hyphae packed closely together so that they appear cellular in both cross and long sections (Figure 1.2). Cortical layers that provide support and protection are frequently made up of this tissue. In most lichens, the upper and lower cortices are multilayered regions that vary in thickness and degree of conglutination of the hyphae. In *Leptogium*, however, the upper cortex is usually a single hyphal layer thick, and the short, densely packed hyphae are easily recognized, both in cross section (Figure 1.3a) and from the surface (Figure 1.3b).

Prosoplectenchyma

This tissue is formed from long, parallel hyphae with heavily conglutinated cell walls, giving it a much different appearance in cross section than in long

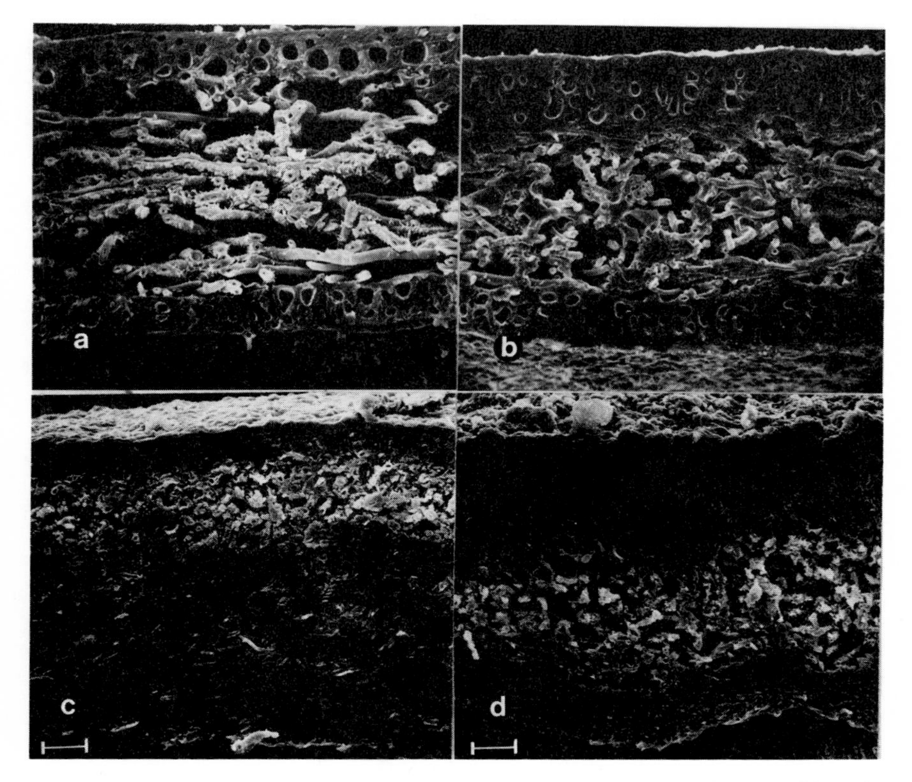

FIGURE 1.2. Paraplectenchyma. (a) *Cetraria pinastri* CS (×850); (b) *Cetraria andrejevii* LS (×600); (c) *Physcia hypomela* CS, bar = 20 μm; (d) *Physcia aipolia* LS, bar = 20 μm. Provided by M.E. Hale, Jr.

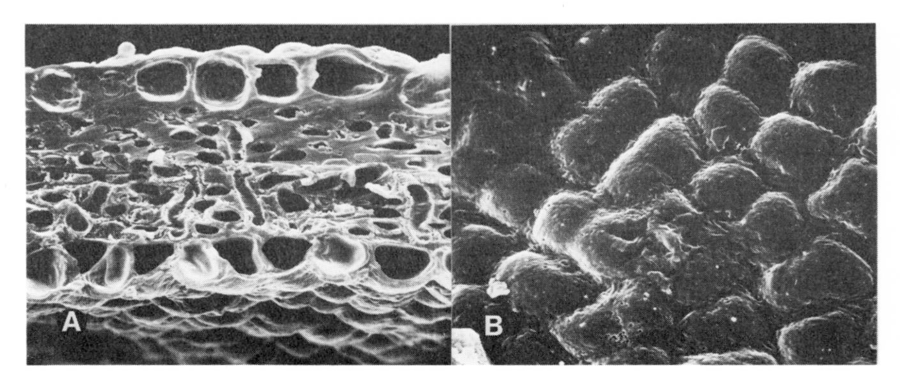

FIGURE 1.3. *Leptogium azureum*. (a) CS (×1500); (b) Dorsal view (×2000). Provided by M.E. Hale, Jr.

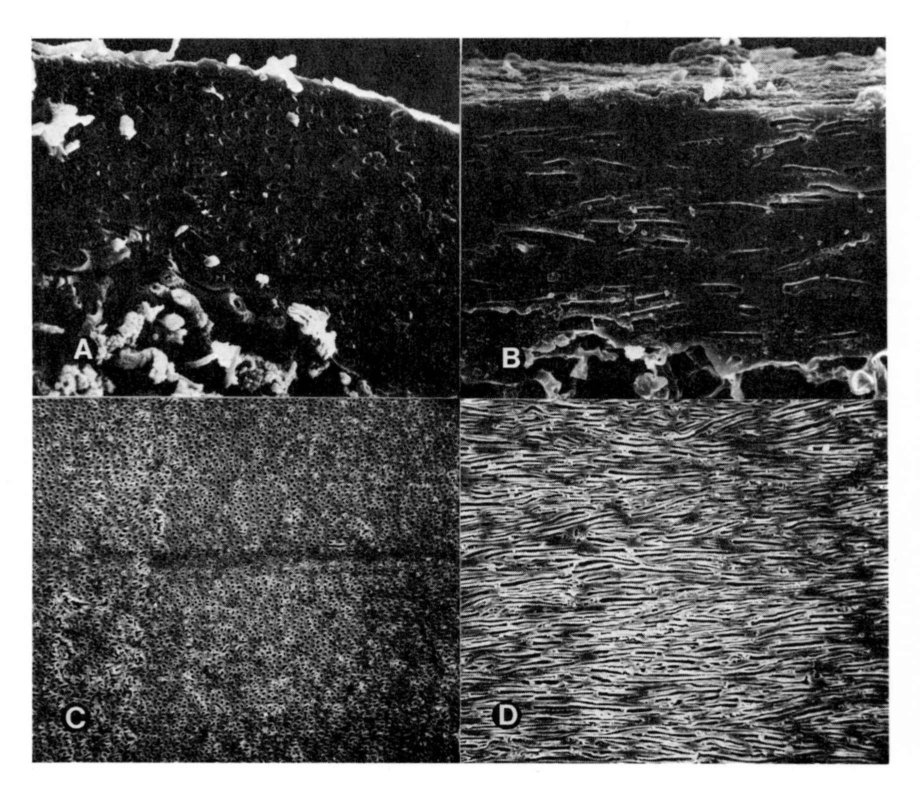

FIGURE 1.4. Prosoplectenchyma. (a) *Alectoria cornicularioides* CS (×1100); (b) *A. cornicularioides* LS (×1000); (c) *Usnea strigosa* cord CS (×400); (d) *U. strigosa* cord LS (×400). Provided by M.E. Hale, Jr.

section (Figure 1.4). Like paraplectenchyma, it is a protective and supportive tissue frequently found in cortical layers. Some groups have paraplectenchyma in the upper cortex and prosoplectenchyma in the lower cortex.

Prosoplectenchyma is also found in fruticose lichens where it forms medullary cords in the Usneaceae (Figure 1.4) and podetial stalks in the Cladoniaceae. The cortex of some lichens (e.g., some *Cornicularia* and *Ramalina* species) is unique because it appears to be formed of paraplectenchyma in the upper part and prosoplectenchyma in the lower layers (Figure 1.5a). Finally, all lichen rhizines and cilia appear to be prosoplectenchymatous (Figure 1.5b).

Palisade Plectenchyma

This tissue is composed of more or less vertically oriented short hyphae that are similar in appearance to the palisade mesophyll of vascular plant leaves (Figure 1.6). It is associated with the pored epicortex (discussed below) in numerous groups of lichens. The loose columnar arrangement of hyphae may facilitate gas exchange between the algal layer and the environment (Hale, 1973b; 1981), but there is no experimental evidence for this. The cortices of several nonepicortical fruticose groups (e.g., *Roccella*) also have palisade (fastigiate) plectenchyma.

Medullary Plectenchyma

The medullary region of most lichens is composed of loosely interwoven, randomly oriented hyphae situated between the upper and lower cortex (Figure 1.1). An exception is *Cornicularia*, in which medullary hyphae form compressed, flattened strands (Figure 1.5a). Tissues similar to the medullary plectenchyma also compose the tomentum produced by many species and the hypothallus of both *Anzia*, where it arises from the medulla, and *Pannoparmelia* (Figure 1.7), where it arises from lower cortical tissue (Hale, 1976).

NONREPRODUCTIVE THALLUS STRUCTURES

Lichens produce a number of distinctive vegetative structures—some of which are also found in nonlichenized fungi (e.g., rhizines, tomentum, and cilia); others are unique to lichens (e.g., cyphellae, pseudocyphellae, and cephalodia). All are important diagnostic characters and are assumed to be biologically important as well, although almost no experimental evidence is available to suggest how they function.

Only nonreproductive structures will be reviewed here. Reproductive

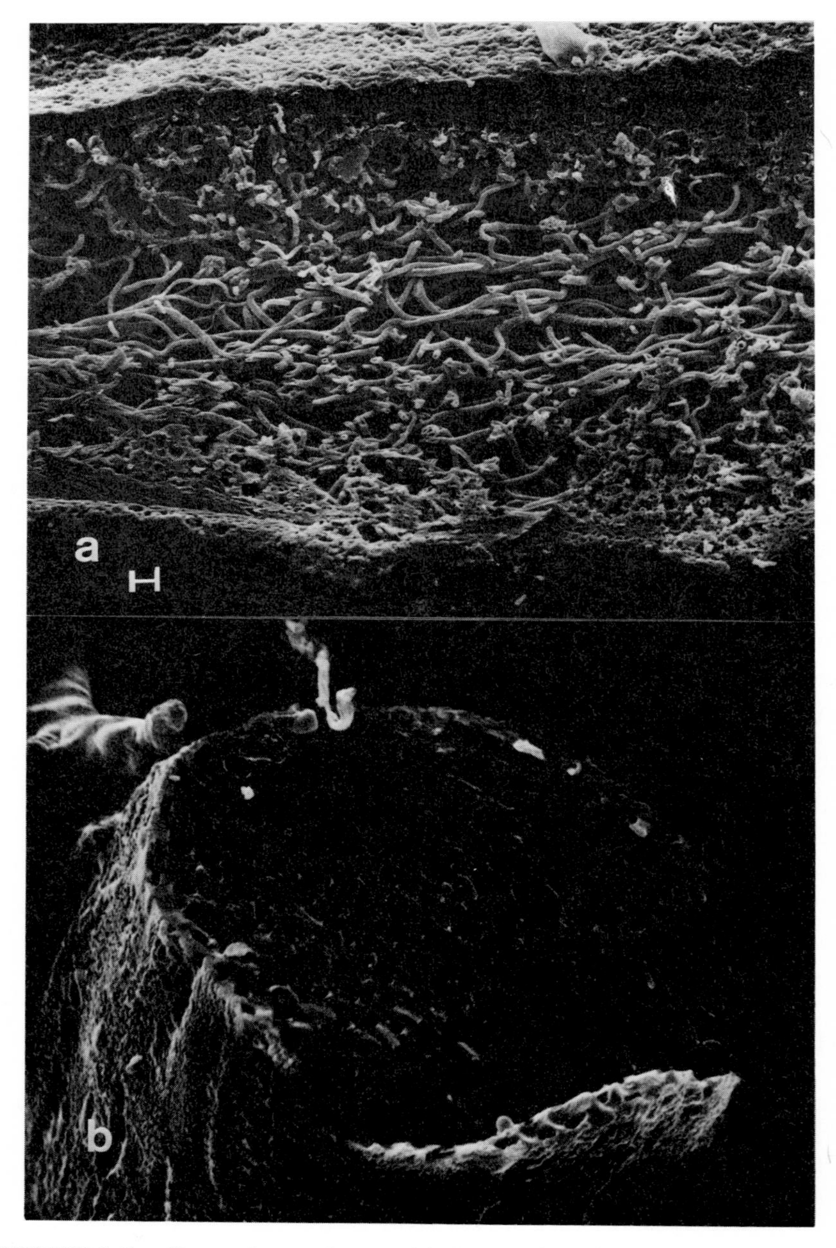

FIGURE 1.5. Prosoplectenchyma. (a) *Cornicularia californica,* CS, bar = 10 μm; (b) *Parmelia densirhizinata* rhizine CS (×2000). Provided by M.E. Hale, Jr.

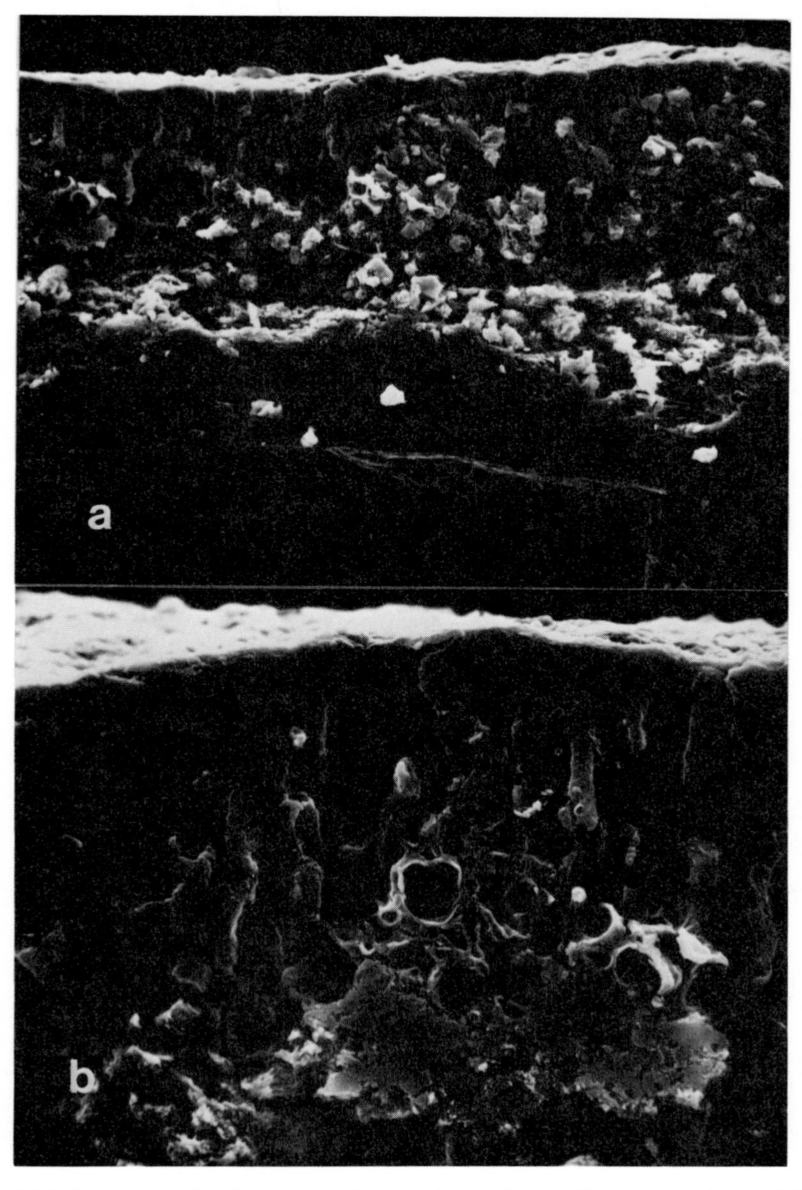

FIGURE 1.6. Palisade plectenchyma. (a) *Relicina fluorescens* LS (×350); (b) *R. fluorescens* LS (×1000). Provided by M.E. Hale, Jr.

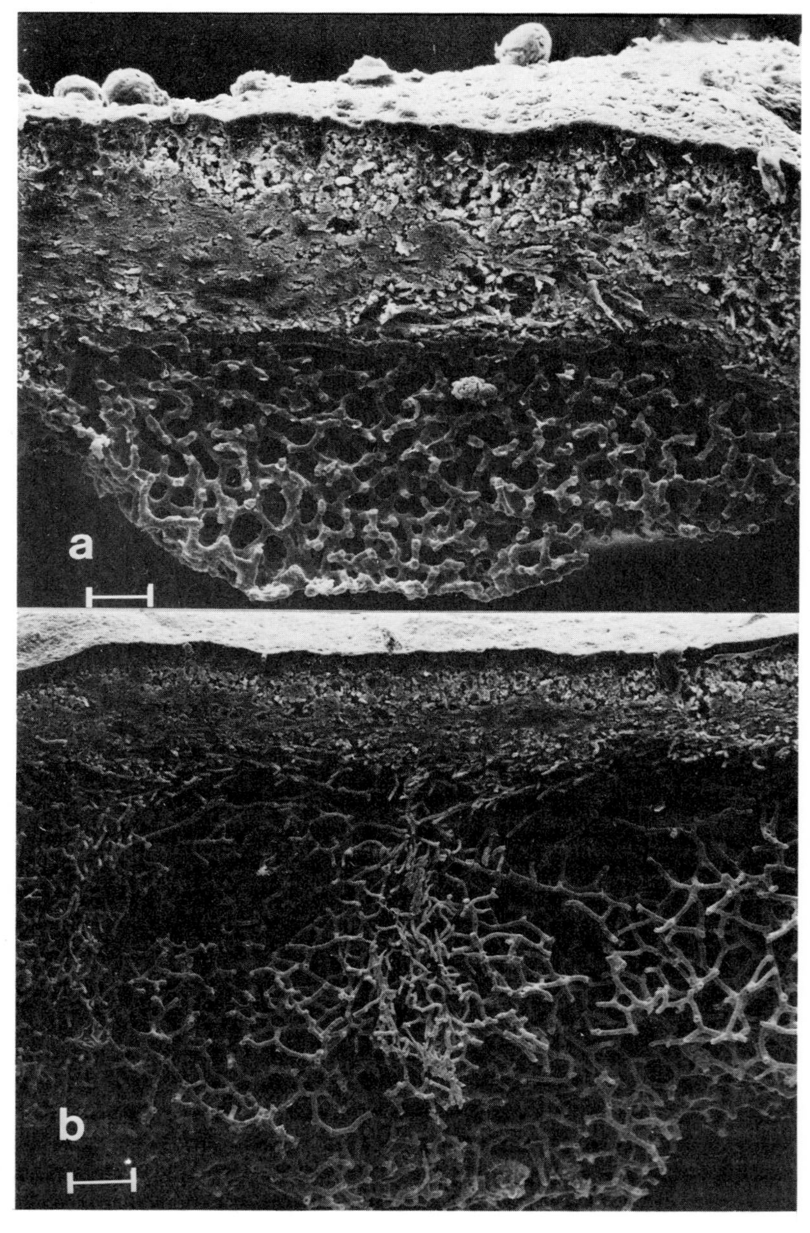

FIGURE 1.7. Tomentum. (a) *Pannoparmelia wilsonii,* CS, bar = 50 μm; (b) *Anzia colpodes,* CS, bar = 100 μm. Provided by M.E. Hale, Jr.

characters, including vegetative structures that have propagative or dispersal functions (i.e., soredia, isidia, and hormocysts) will be discussed in Chapter 3. Cephalodia, unique structures that apparently involve synthesis with "foreign" algae, are discussed in Chapter 5 on synthesis.

Cyphellae

Cyphellae are open depressions formed on the undersurface of lichens in the genus *Sticta* (Figure 1.8). They are 0.5–2.0 millimeters in diameter with a distinctive margin in the lower cortex and a base in the medulla. The function of these depressions is not completely understood. It has long been thought to involve gas exchange, and indeed a recent study (Green, Snelgar, & Brown, 1981) of the cyphellate *Sticta latifrons* has provided strong indirect evidence for this. In this study, cyphellae were found to act as air pores, providing passageways for CO_2 to enter the thallus. Even at thallus water contents allowing maximal photosynthetic rates, CO_2 exchange rates measured through the upper cortex were much lower than those through the lower cortex; lower cortex exchange rates approximated those of the whole thallus. The authors suggested that cyphellae, found only on the lower cortex, were responsible for these differences.

Pseudocyphellae

Pseudocyphellae are pores in the cortical tissue (Figure 1.9) that are usually smaller, 0.1–1.0 millimeters in diameter, and less differentiated than cyphellae. They can be found on both the upper and lower cortices of numerous lichens—particularly in the Parmeliaceae and Lobariaceae. They appear to develop by a progressive disintegration of the relatively thick, tightly compressed cortical tissues characteristic of pseudocyphellate species; this process leaves a cavity that fills with medullary hyphae from below (Figure 1.9b)

The mechanism responsible for triggering the development of pseudocyphellae is not known, but Hale (1981) has suggested that it may be related to thallus lobe surface-to-volume ratios. He reasoned that gas exchange through the cortex is impeded as lobes grow larger. Some species are able to respond by producing pseudocyphellae; nonpseudocyphellate species are either small-lobed or they produce other structures to promote gas exchange (e.g., breathing pores or dactyls of *Hypotrachyna*, cortical cracks, etc.). The fact that small-lobed foliose lichens (e.g., Physciaceae and *Cetraria*) rarely have pseudocyphellae and related large-lobed groups (e.g., *Cetrelia* and *Plastismatia*) frequently do have them tends to support Hale's (1981) hypothesis, but experimental evidence is needed.

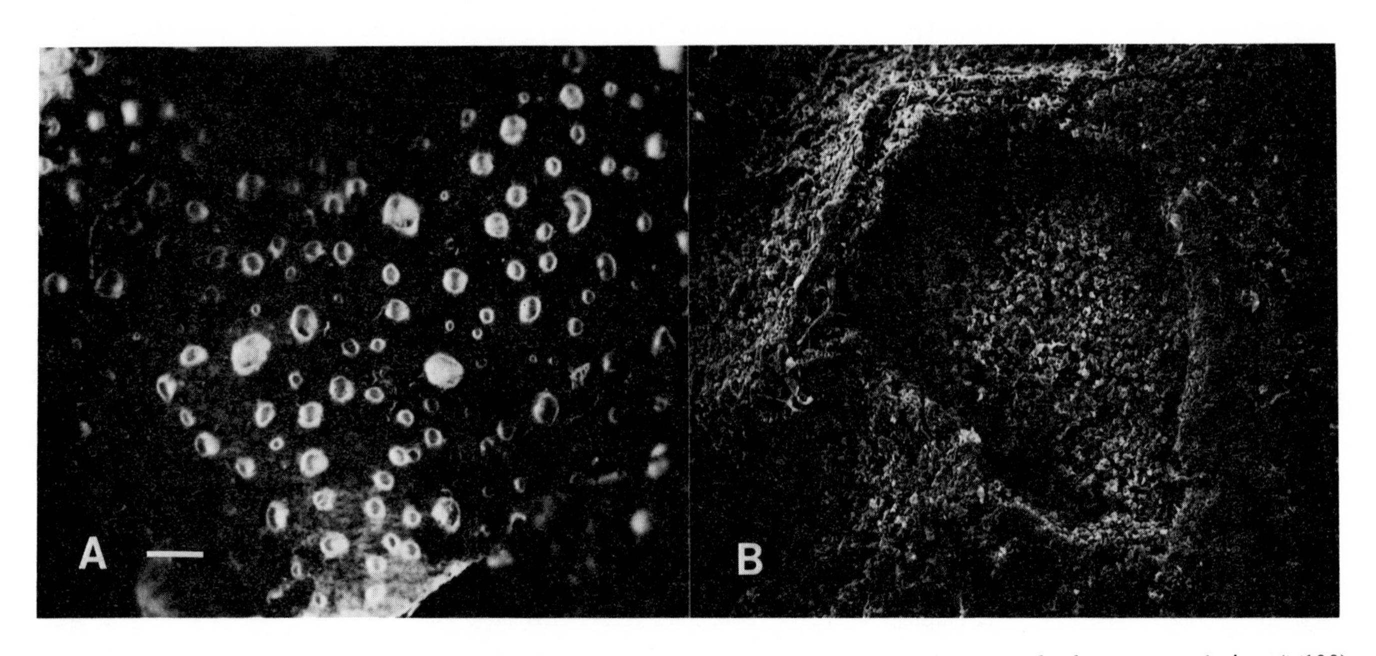

FIGURE 1.8. Cyphellae. (a) *Sticta damaecornis,* ventral view, bar = 0.1 mm; (b) *Sticta limbata,* ventral view (×100). Figure 1.8b from M.E. Hale, Jr. 1976. "Lichen Structure viewed with the scanning electron microscope." In *Lichenology: Progress and Problems* (D.H. Brown, D.L. Hawksworth, & R.H. Bailey, eds.), pp. 1–15. London: Academic Press. Reprinted with permission of Academic Press.

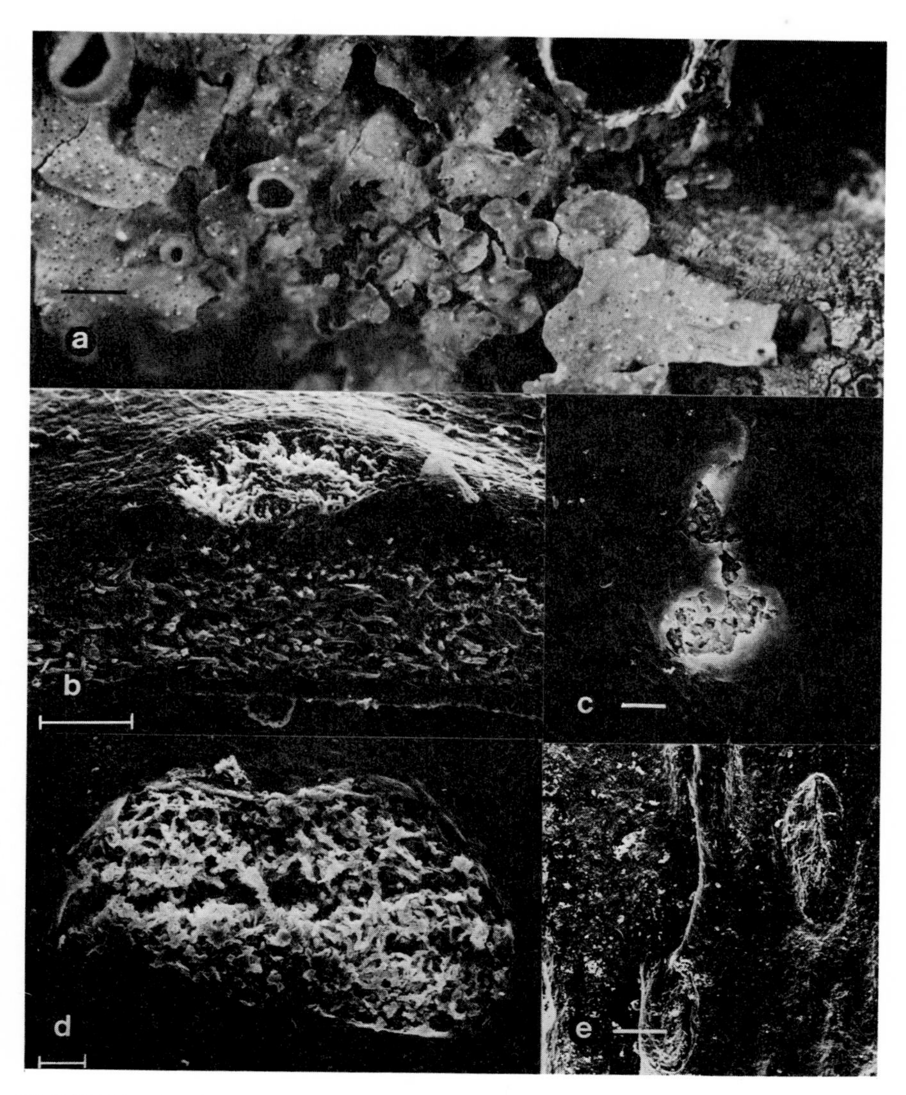

FIGURE 1.9. Pseudocyphellae. (a) *Parmelia subpraesignis,* dorsal surface, bar = 2 mm; (b) *Cetraria straminea* CS, bar = 50 μm; (c) *Parmelia praesignis,* dorsal surface, bar = 20 μm; (d) *P. praesignis,* dorsal surface, mature pseudocyphella, bar = 20 μm; (e) *Ramalina americana,* dorsal surface, bar = 100 μm. Figures 1.9c and 1.9d from M.E. Hale, Jr. 1981. "Pseudocyphellae and pored epicortex in the Parmeliaceae: Their delimitation and evolutionary significance." *Lichenologist* 13:1–10. Reprinted with permission of Academic Press. Figure 1.9e from M.E. Hale, Jr. 1978. "A new species of *Ramalina* from North America (Lichenes: Ramalinaceae)." *Bryologist* 81:599–602. Reprinted with permission of the American Bryological and Lichenological Society.

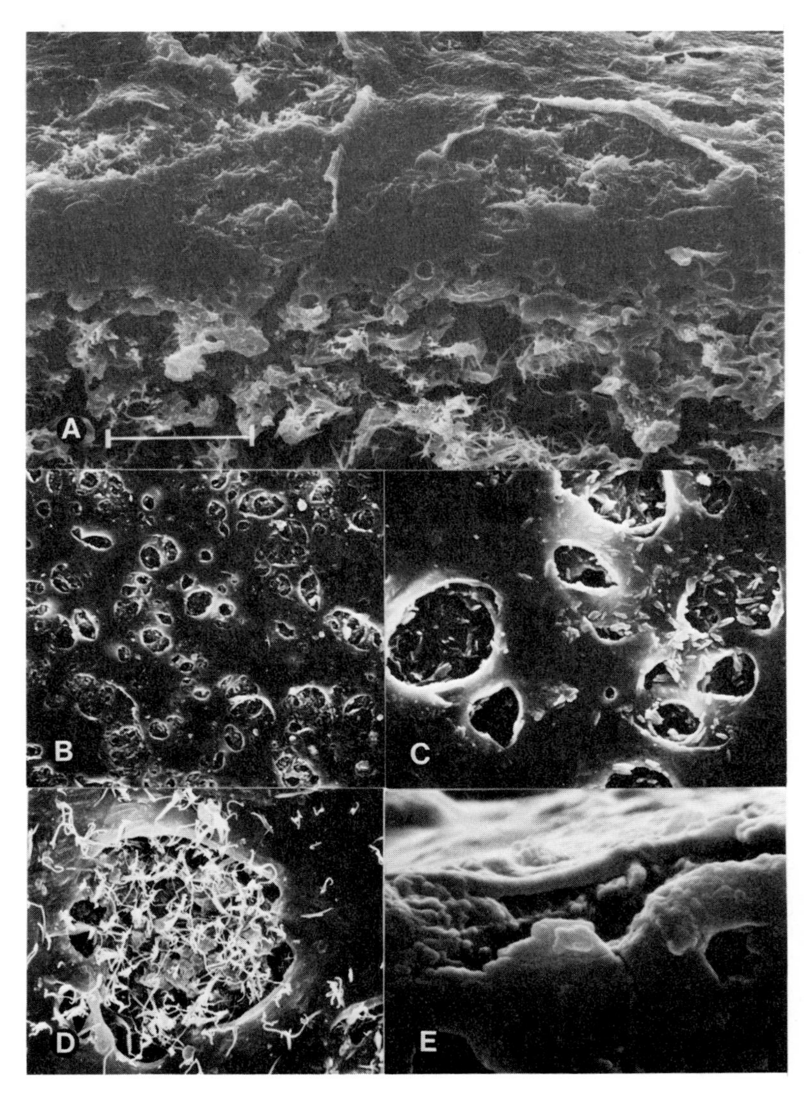

FIGURE 1.10. The pored epicortex. (a) Dorsal surface of epicortex and pores of *Parmeliopsis ambigua* showing cortical accretions, bar = 20 μm; (b) *Parmelia enormis,* dorsal surface (×500); (c) *P. enormis,* dorsal surface (× 2000); (d) *Parmelina minarum* with curly artifacts on dorsal surface (×2000); (e) *Bulbothrix laevigata* CS (×7500). Figures 10a and 10e from M.E. Hale, Jr. 1981. "Pseudocyphellae and pored epicortex in the Parmeliaceae: Their delimitation and evolutionary significance." *Lichenologist* 13:1–10. By permission of Academic Press. Figure 10d by permission of the Smithsonian Institution Press from *Smithsonian Contributions to Botany,* No. 10, Fig. 1b, p. 13; Smithsonian Institution, Washington, D.C.

The Pored Epicortex

The pored epicortex (Figure 1.10) is a thin polysaccharide sheet about 0.6 micrometers thick that covers the upper cortex and is presumably produced by cortical hyphae, although nothing is known of its ontogeny. It is perforated by numerous holes 15–25 micrometers in diameter; cortical secondary compounds may sometimes be apparent through the pores (Figure 1.10c). There may also be various artifacts, as in *Parmelina minarum* (Figure 1.10d).

Although Peveling (1970) initially observed this layer, Hale (1973b) was the first to recognize it as unique. He has recently found a number of differences between lichens with pseudocyphellae and those with a pored epicortex (Table 1.1) that suggest a similar biological role for these structures, namely, the facilitation of gas exchange. Lichens with the two morphologies show distributional and structural patterns that Hale (1981) considers *prima facie* evidence for separate and distinct biological lines. Pseudocyphellate genera (and closely related, narrow-lobed, nonpseudocyphellate genera) are:

1. temperate, boreal, or arctic
2. limited in morphological and chemical diversity
3. rather conservative with small species assemblages

In contrast, genera with a pored epicortex are:

1. tropical to subtemperate
2. rich in morphological and chemical diversity
3. often highly speciated

TABLE 1.1. Comparison between Pseudocyphellae and the Pored Epicortex

Pseudocyphellae	*Pored Epicortex*
Formed by pore in cortex	Separate from cortex
Pierced cortex	Entire cortex
Contain intrusions of medullary hyphae	No medullary hyphae present
Associated acids derived from medullary hyphae	Associated acids entirely cortical in origin
Diameter 200–2000 μm	Diameter 15–40 μm
Density 1–2 pores mm^{-2}	Density 100–400 pores mm^{-2}

M.E. Hale, Jr. 1981. "Pseudocyphellae and pored epicortex in the Parmeliaceae: Their delimitation and evolutionary significance." *Lichenologist* 13:1–10. Reprinted with permission of Academic Press.

Rhizines

Rhizines are compacted strands of hyphae that originate from the lower cortex and anchor the thallus to the substrate. They are prosoplectenchymatous (Figure 1.11) and vary in color, degree of hyphal conglutination, and branching complexity. Rhizine variability can best be illustrated by considering the genus *Relicina* (Hale, 1975). *Relicina* species produce rhizines of various shapes and sizes ranging from relatively short, simple threads to highly branched, bushy masses of hyphae (Figure 1.11).

Aside from anchoring the thallus to its substrate, rhizines are thought by some investigators to facilitate water and essential element uptake. There is some experimental evidence to support this idea. Larson (1981) compared water saturation rates of thalli with rhizines removed to those obtained using

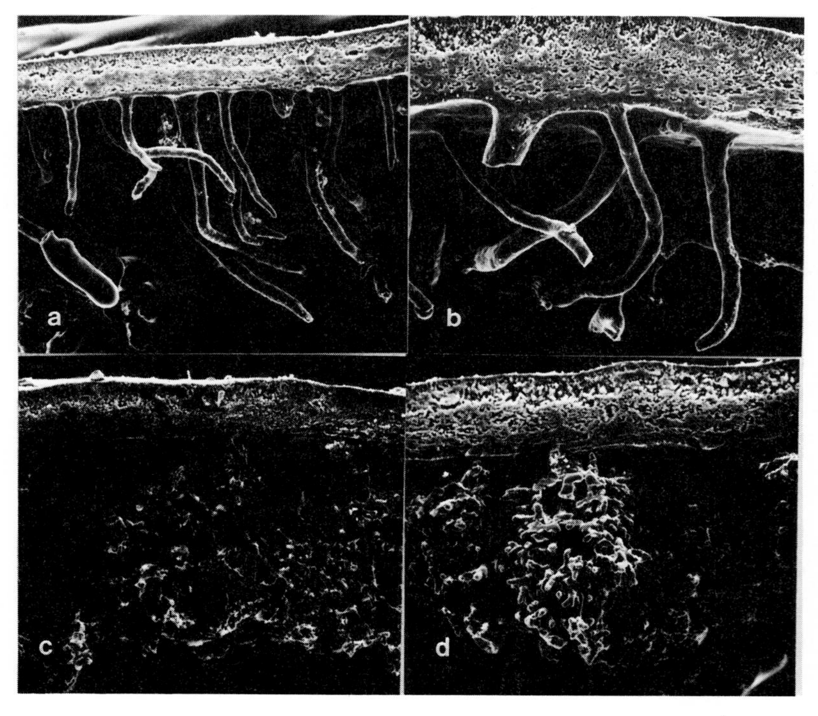

FIGURE 1.11. Rhizines. (a) *Relicina abstrusa* CS (×75); (b) *R. limbata* CS (×100); (c) *R. ramosissima* CS (×65); (d) *R. sublanea* CS (×100). From M.E. Hale, Jr., 1975. By permission of the Smithsonian Institution Press from *Smithsonian Contributions to Botany,* No. 26, Fig. 8a,b,f,g, p. 10; Smithsonian Institution, Washington, D.C.

normal thalli and demonstrated that rhizines play a major role in the water relations of some species (e.g., *Umbilicaria* species), but not others (e.g., *Peltigera* species). Other investigators have done similar tests (Smyth, 1934; Jahns, 1973). Goyal and Seaward (1981) found that the rhizines of various terricolous lichens often exhibited high concentrations of certain metal elements, presumably obtained from the substrate; in addition, thalli with rhizines removed were invariably found to have lower concentrations of metals. These observations indicated that rhizines play an important role in lichen element absorption and storage.

Cilia and Cortical Hairs

Cilia are rhizinelike strands of hyphae (Figure 1.12) that are found in numerous foliose genera, particularly those that are structurally advanced (Hale, 1983). They originate from lobe margins and also occur on the exciples of lecanorine apothecia in *Usnea*. In *Bulbothrix* and *Relicina*, bizzare, bulbate cilia with enlarged, fluid-filled bases are produced (Figures 1.12c, d). Nothing is presently known about the origin or significance of these strange structures.

In some genera (e.g., *Phaeophyscia, Melanelia, Collema, Peltigera, Teloschistes*), minute, dorsal, hairlike structures, *Glaszilien*, are produced that have been called glass cilia by some workers (Peveling & Poelt, 1974) and cortical hairs by others (Esslinger, 1977; 1978). These structures are much smaller than true cilia and practically transparent (Figure 1.12b). They are thought to aid in water uptake from humid atmospheres. Since species from warm, xeric regions (Peveling & Poelt, 1974) and fog zones frequently have them, this hypothesis appears to be worth testing.

Tomentum

Tomentum is a cottony mat of loosely organized hyphal strands similar in structure to rhizines. It is most frequently produced on the underside of the thallus, either from the lower cortex (Figure 1.7a) or the medulla (Figure 1.7b); it may also be found on the upper surface of some species (Hale, 1983). In *Anzia*, the tomentum forms a great hyphal net many times greater in thickness than the thallus itself (Figure 1.7b).

If there is a functional role for tomentum, it is not known. The similarity between tomentum and rhizines suggests they had the same origin and function similarly, probably in water absorption. Snelgar, Green, & Wilkins (1981) have suggested that species with a thick, tomentose lower cortex are able to hold water longer than nude species; however, there is an associated increase in CO_2 resistance at higher water contents, a common characteristic of whole lichen thalli (Bewley, 1979). Heavily tomentose species in the genus *Sticta* may increase gas exchange through cyphellae (Green, Snelgar, & Brown, 1981).

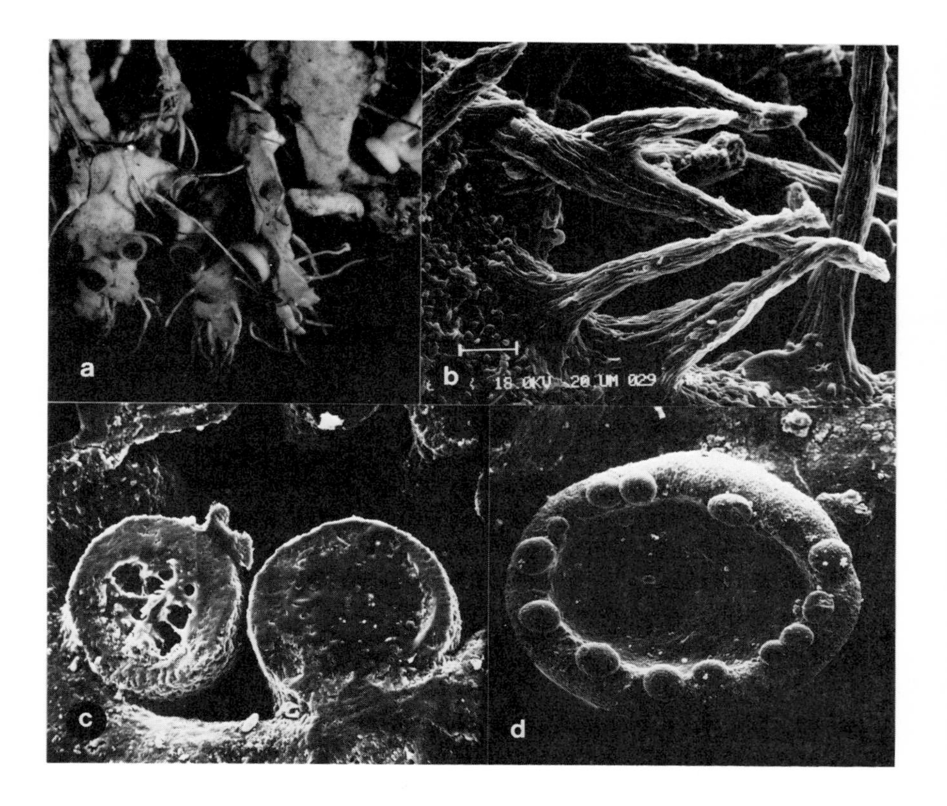

FIGURE 1.12. Cilia. (a) *Heterodermia erinacea* with cilia, dorsal surface (×6); (b) *Phaeophyscia cernohorskyi* glass cilia, dorsal surface, bar = 20 μm; (c) *Relicina sublanea* bulbae (bulbate cilia) CS (×200); (d) *Bulbothrix subcoronata* bulbae, dorsal surface (×75). Provided by M.E. Hale, Jr.

Pruina

This term refers to a whitish covering on the cortex and apothecia of certain lichens that appears to originate several different ways. In some groups, the outermost cells become necrotic, giving the thallus a "scurfy" appearance (Jahns, 1973). In others, the pruina is crystalline, apparently consisting of carbonates and oxalates (Figure 1.13). Although there is no experimental evidence, these crystals are probably formed from excretory products of the cortical hyphae.

Calcium oxalate and calcium carbonate are both found in many lichens

FIGURE 1.13. Pruina. (a) Dorsal surface of *Parmelia testacea,* bar = 20 μm; (b) Dorsal surface of *Parmelia saxatilis,* bar = 4 μm. Provided by M.E. Hale, Jr.

(e.g., Schade, 1970; Schade & Seitz, 1970). Concentrations of calcium oxalate up to 60 percent (*Rhizocarpon umbilicatum*) have been reported (Syers, Birnie, & Mitchell, 1967; Mitchell, Birnie, & Syers, 1966). This is apparently related to the fungal production of oxalic acid, which forms an extracellular calcium oxalate deposit. The significance of this in lichen element accumulation processes is discussed in greater detail in Chapter 6.

2 Cellular Structure

Since the first ultrastructural details of lichen symbionts were published by Moore and McAlear (1960) and Ahmadjian (1960a), numerous transmission electron microscope (TEM) and scanning electron microscope (SEM) investigations of lichens have appeared. Interest in this aspect of lichenology has increased steadily as new techniques develop.

These approaches were initially used to understand symbiont interactions at the cellular level, and they provided details not available from light microscope investigations. Since these early studies, TEM and SEM approaches have also been used to investigate a number of other problems in lichenology, including the following:

1. subcellular comparisons of isolated and lichenized symbionts
2. comparisons of thalli from differing environments
3. aspects of symbiont cell wall structure and synthesis
4. various processes involved in lichen synthesis
5. carbohydrate and essential element flow between symbionts

This section will review some of the available information on these subjects, attempting whenever possible to reduce what is rapidly becoming a large and complicated body of observations to a limited number of generalizations. Excellent reviews by Peveling (1973; 1976) are also recommended.

CELLULAR STRUCTURE OF LICHEN ALGAE

The most common blue-green and green lichen phycobionts have been investigated using TEM. By far, the largest number of investigations have involved *Trebouxia* and *Pseudotrebouxia* species.

The Algal Cell Wall

The cell walls of *Trebouxia* and *Pseudotrebouxia* species consist of two layers differing in electron density (Brown & Wilson, 1968; Jacobs & Ahmadjian, 1969). Walls of *Coccomyxa* species apparently have three layers (Peveling, 1973; Honegger & Brunner, 1981). Blue-green algal cells have a sheath surrounding a cell wall that may have as many as four distinct layers (Peat, 1968; Peveling, 1969; Paran, Ben-Shaul, & Galun, 1971).

Until recently, practically nothing was known about the chemistry of phycobiont cell walls. Recent ultrastructural and microchemical studies of *Coccomyxa* and *Myrmecia* phycobionts by Honegger and Brunner (1981) have revealed a number of characteristics of the cell walls. Ultrathin sections of these phycobionts showed that the cell walls have a tripartite structure. They consist of:

1. An outermost membrane-like trilaminar wall layer about 13 micrometers thick with particles embedded in an amorphous, carbohydrate-containing matrix on the inner and outer surfaces—in between these surfaces is a smooth middle layer that appears to contain sporopollenin, although this is still being debated.

2. An underlying electron-dense layer of uniform thickness containing short, probably cellulosic fibrils embedded in an amorphous matrix.

3. An inner wall layer of variable thickness containing hemicelluloses and possibly also Golgi-derived vesicular inclusions.

The apparent discovery of *sporopollenin* in the outermost trilaminar layer is significant if true. The term sporopollenin refers to the highly resistant exine material of pollen and spores. It has also been found in cell walls of myxobacteria and algae and sexual spore walls of various phycomycetes and ascomycetes (Honegger & Brunner, 1981). Its presence in cell walls is thought to inhibit microbial degradation in general and fungal parasitism in particular. In lichens, the significance of this is obvious because the presence of sporopollenin in algal walls may determine algal resistance to fungal penetration, a characteristic Ahmadjian (1982b) believes is strongly correlated with successful lichenization. Honegger and Brunner suggested that a microchemical investigation of the cell walls of related phycobionts exhibiting different degrees of mycobiont penetration will demonstrate even more clearly the involvement of sporopollenin in regulating lichen symbiont interactions.

The Algal Protoplast

In *Trebouxia* the protoplast is dominated by a large central chloroplast surrounded by a small area of cytoplasm (Figure 2.1). The chloroplast is characterized by a complex thylakoid network with a large central pyrenoid. In cultured cells, the chloroplast is generally smaller (Ahmadjian & Jacobs, 1983).

Associated with the pyrenoid are unique osmiophilic globules that range in diameter from 40 to 100 nanometers. These structures, called *pyrenoglobuli* by Peveling (1968), are lipid-containing globules that have been found in

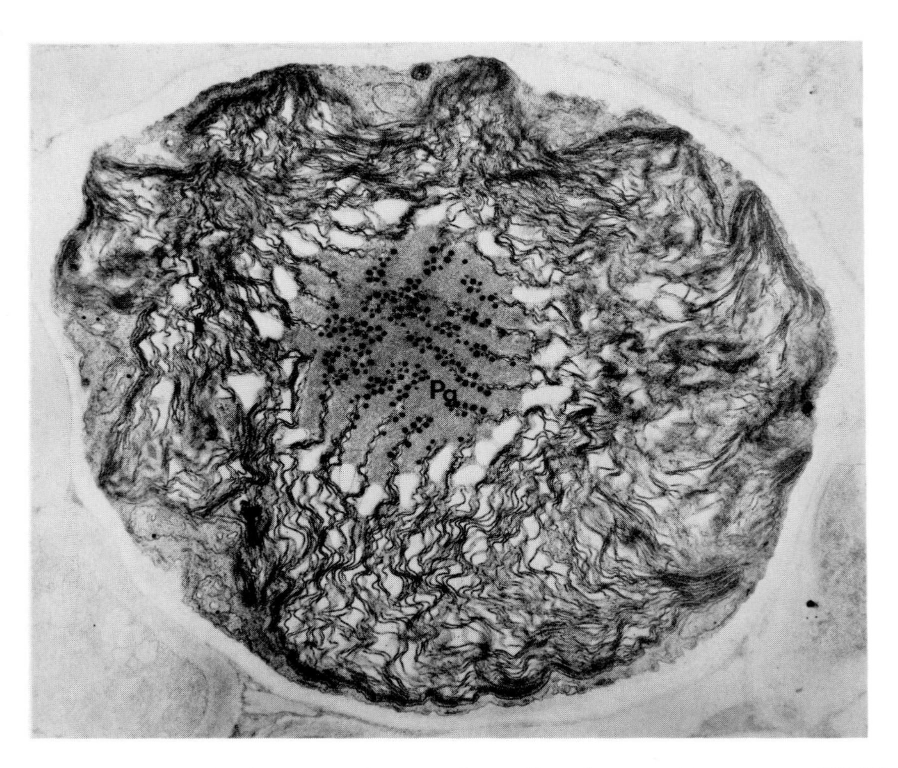

FIGURE 2.1. *Trebouxia gelatinosa* from *Pseudoparmelia caperata* CS (× 30,000). Note large chloroplast with pyrenoglobuli (Pg) in the central pyrenoid. Provided by R. Slocum.

almost all lichens and a few nonlichenized algae. The only exceptions among lichens are species of *Usnea* (Chervin, Baker, & Hohl, 1968; Jacobs & Ahmadjian, 1969); however, Ahmadjian (1982a) believes these observations are due to the effects of various fixation procedures.

The biological importance of pyrenoglobuli is not clear as yet. Jacobs and Ahmadjian (1971a) believe they may be respired under dry conditions, thereby providing the symbionts energy and water. They are apparently more common in dry lichens; starch is formed when the phycobiont is hydrated, either in the lichenized or isolated state (Brown & Wilson, 1968). Hessler and Peveling (1978) consider starch to be the primary product of photosynthesis and the pyrenoglobuli to be synthesized later.

Ahmadjian (1982a) has suggested that pyrenoglobuli formation in lichens is a response to physiological stress. He cited a study by Smith and Morris (1980) of Antarctic phytoplankton that apparently incorporate large amounts of fixed carbon into lipid under stressful conditions. When removed to more favorable habitats, these algae synthesized more polysaccharide and protein and less lipid. For lichens, there is a limited amount of evidence in support of this hypothesis:

1. When lichens are hydrated, pyrenoglobuli generally decrease in number and starch granules increase.

2. When phycobionts are isolated from lichen thalli and maintained under optimum growing conditions, fewer pyrenoglobuli are observed (Ahmadjian, 1982a).

3. When transferred from an organic to an inorganic growth medium, isolated phycobionts exhibit an increase in the number of pyrenoglobuli (Jacobs & Ahmadjian, 1971a).

4. In the aquatic lichen *Hydrothyria venosa*, pyrenoglobuli are not found in the summer but are found in the fall, perhaps reflecting a buildup of storage materials prior to overwintering (Jacobs & Ahmadjian, 1973).

The *Trebouxia* nucleus and mitochondria are observed in the small volume of cytoplasm between the chloroplast and plasmalemma. *Dictyosomes* have been observed in isolated *Trebouxia* species (Jacobs & Ahmadjian, 1971a; Fisher & Lang, 1971) and appear to occur most frequently during cell division. This is probably why they are not frequently observed in lichen phycobionts. In their study of zoosporogenesis in lichenized *Trebouxia gelatinosa*, Slocum, Ahmadjian, and Hildreth (1980) found that dictyosomes were much more prevalent in zoospores than in surrounding vegetative cells. Ellis (1975) reported the presence of dictyosomes in the lichenized *Trente-*

pohlia species of *Chiodecton sanguineum*, a tropical lichen. She suggested this may reflect the loosely defined fungal-algal association in *C. sanguineum* or the phycobiont ontogeny.

CELLULAR STRUCTURE OF LICHEN FUNGI

The Fungal Cell Wall

Lichenized fungal cell walls vary in thickness depending upon their age and position in the thallus (Chervin, Baker, & Hohl, 1968; Jacobs & Ahmadjian, 1969; 1971a; Peveling, 1969; 1973; Malachowski, Baker, & Hooper, 1980). In most micrographs, they appear to be layered, although as Peveling (1973) has pointed out, it is often difficult to determine whether the various layers are part of the cell wall proper or are extracellular deposits.

Bednar and Juniper (1964) observed an extracellular microfibrillar substance around cortical hyphal cell walls of *Xanthoria parietina* that they suggested was secreted by the fungus and composed of polysaccharides. On the basis of different staining properties they concluded that this material is not the same as in the cell wall. Similar observations have been reported by many (Brown & Wilson, 1968; Fujita, 1968; Handley, Overstreet, & Grossenbacher, 1969; Jacobs & Ahmadjian, 1969; 1971a; Peveling, 1973; and Malachowski, Baker, & Hooper, 1980).

An extracellular fibrillar matrix is also commonly produced in culture by isolated mycobionts (Figure 2.2). Ahmadjian (1982a; also Jacobs & Ahmadjian, 1971a) identified three distinct layers of wall material in isolated *Cladonia cristatella* hyphae including an extensive gelatinous layer. Jacobs and Ahmadjian (1971a) noted that this gelatinous material lost its amorphous matrix and became fibrillar in areas where there were bacteria attached to the wall. Ahmadjian (1982a) has also noted that hyphae with extensive extracellular gelatinous layers generally do not bind to phycobiont cells during thallus resynthesis. Hyphae most likely to bind to algal cells have thin walls and probably attach by means of a gelatinous substance around the algae.

Crystals of lichen secondary compounds, frequently observed in SEM on hyphal surfaces (Figure 2.3), are also observed in and on hyphal cell walls viewed with TEM (Peveling, 1973; Ellis, 1975). These substances are usually extracted during specimen preparation; Ellis (1975) was able to view them clearly in *Chiodecton sanguineum* with freeze-fracture techniques. Brown and Wilson (1968) observed unit membrane-bound crystals near the fungal plasmalemma of *Physcia aipolia* that suggested a cellular source of these substances.

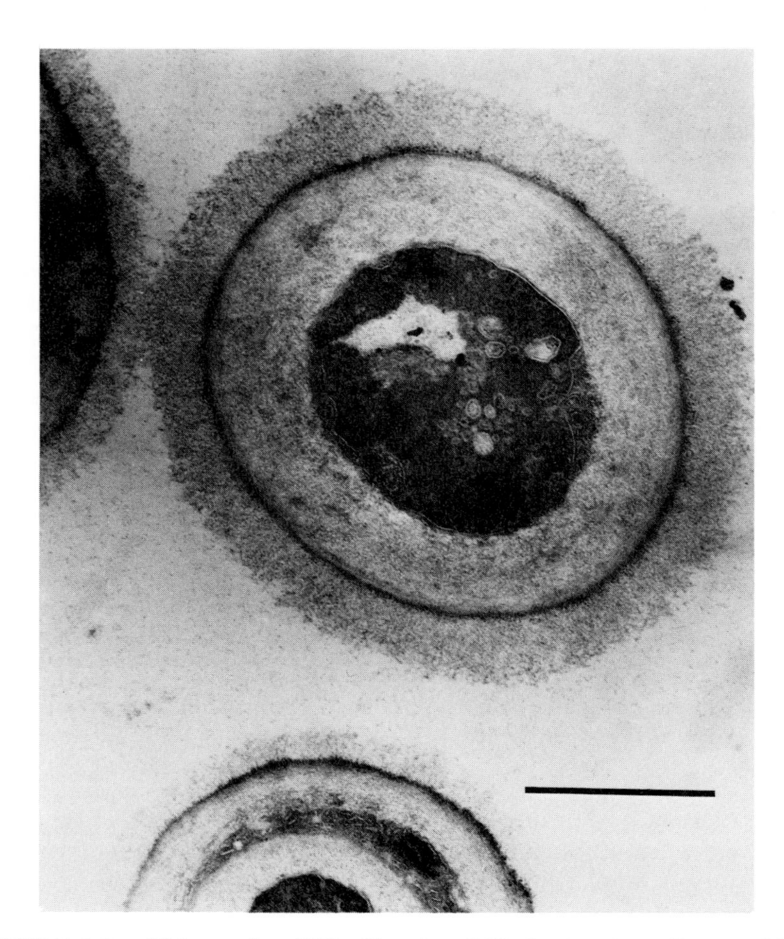

FIGURE 2.2. Thin section (CS) of hypha of *Cladonia cristatella* mycobiont cultured in liquid malt-yeast extract medium. Note extensive gelatinous layers outside and within the hyphal wall; this type of hypha does not bind to algal cells. From V. Ahmadjian. 1982a. "Algal/fungal symbioses." In *Progress in Phycological Research* (F.C. Pound & D.J. Chapman, eds.), Vol. 1, pp. 179–233. Amsterdam: Elsevier Biomedical Press. Reprinted with permission of Elsevier.

FIGURE 2.3. Salazinic acid crystals on hyphal surfaces of *Parmotrema latissima* (×5000). Provided by M.E. Hale, Jr.

Until recently, the chemical nature of lichen hyphal cell walls was almost totally unknown, due mainly to problems in obtaining isolated cell wall preparations (Galun et al., 1976). The most important problems are the following:

1. Little hyphal wall material can be obtained in culture because of the slow growth of lichen mycobionts.
2. The dual nature of lichenized thalli complicates isolation of only the fungal cell walls.

Galun et al. (1976) circumvented these problems somewhat through the use of two novel experimental approaches, *microautoradiography* and *lectin binding*, in an investigation of hyphal wall chemistry in three isolated mycobionts: *Xanthoria parietina, Tornabenia intricata*, and *Sarcogyne* sp.

They exposed newly germinated mycobionts to radioactively labelled N-acetylglucosamine, a chitin precursor. Following exposure, autoradiographs showed that hyphal tips had incorporated the label. This suggested that chitin is an important component of hyphal walls. Further support for this idea came from lectin-binding studies with the same isolated mycobionts. Fluorescein-conjugated wheat germ agglutinin, which is known to interact specifically with chitin oligomers, was used as a probe to locate chitin-rich walls in newly germinated mycobionts. These tests further confirmed that chitin is a mycobiont hyphal wall component. Other lectin-binding experiments demonstrated that the specific chemistry of the hyphal wall may vary considerably from species to species.

Another result of this study was the observation that labelled mannitol and mannose were actively incorporated by hyphae of the three mycobionts but not by hyphae of two free-living fungi, *Trichoderma viride* and *Phytophthora citrophthora*. This indicated that these compounds serve as precursors for carbohydrate polymer synthesis in the mycobionts but not in the hyphae of free-living fungi.

The Fungal Protoplast

At the ultrastructural level, lichenized ascomycetes and basidiomycetes are similar to those that are nonlichenized. The cytoplasm contains numerous ribosomes, nuclei, mitochondria, extensive endoplasmic reticulum, dictyosomes, and various storage granules (Jacobs & Ahmadjian, 1971a; Peveling, 1973).

Lomasomes can frequently be observed near the plasmalemma. These structures, apparently found only in the Eumycota, were initially characterized by Moore and McAlear (1961) and first studied in lichens by Jacobs and Ahmadjian (1969). They are apparently formed by endoplasmic reticulum encapsulation of small vesicles with subsequent movement of the vesicle to the plasmalemma. The vesicle membrane then fuses with the plasmalemma and the fusion point dissolves, releasing the vesicle contents outside the protoplast. The vesicle contents have not yet been determined; they may be either secreted ergastic substances trapped within the plasmalemma or cell wall precursors (Moore & McAlear, 1961). The biological significance of lomasomes obviously depends on a clearer understanding of their contents.

Figure 2.4 shows lichenized fungal hyphae frequently have a highly invaginated plasmalemma (Brown & Wilson, 1968), particularly those close to algal cells (Peveling, 1973). For this reason, it is assumed that increased fungal protoplast surface area promotes chemical exchange between the symbionts. Jacobs and Ahmadjian (1971a) observed this characteristic in hyphae of both lichenized and isolated *Cladonia cristatella* mycobionts. Peveling (1973) suggested it is less frequently observed in isolated mycobionts.

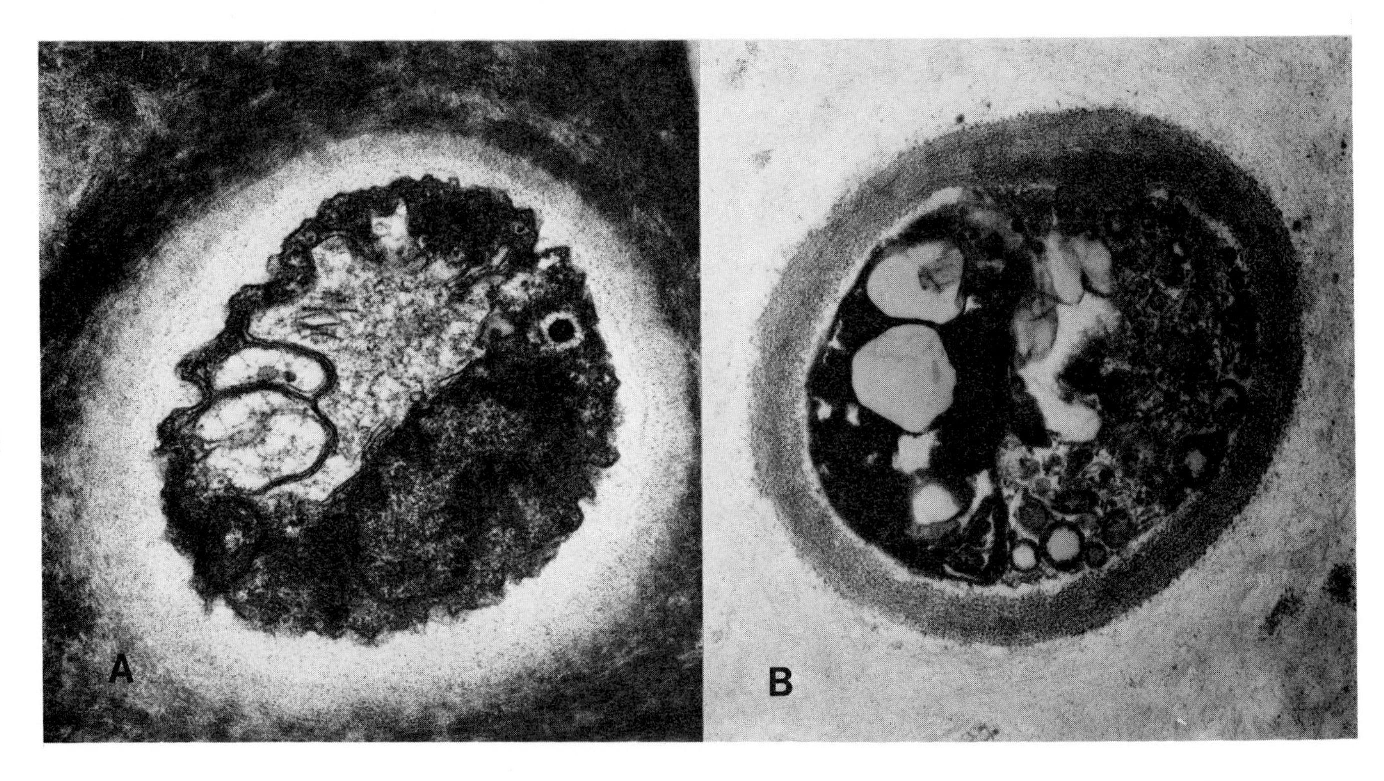

FIGURE 2.4. *Graphis scripta* mycobionts. (a) Lichenized CS (×50,000); (b) Isolated and maintained on malt-yeast extract agar CS (×48,000).

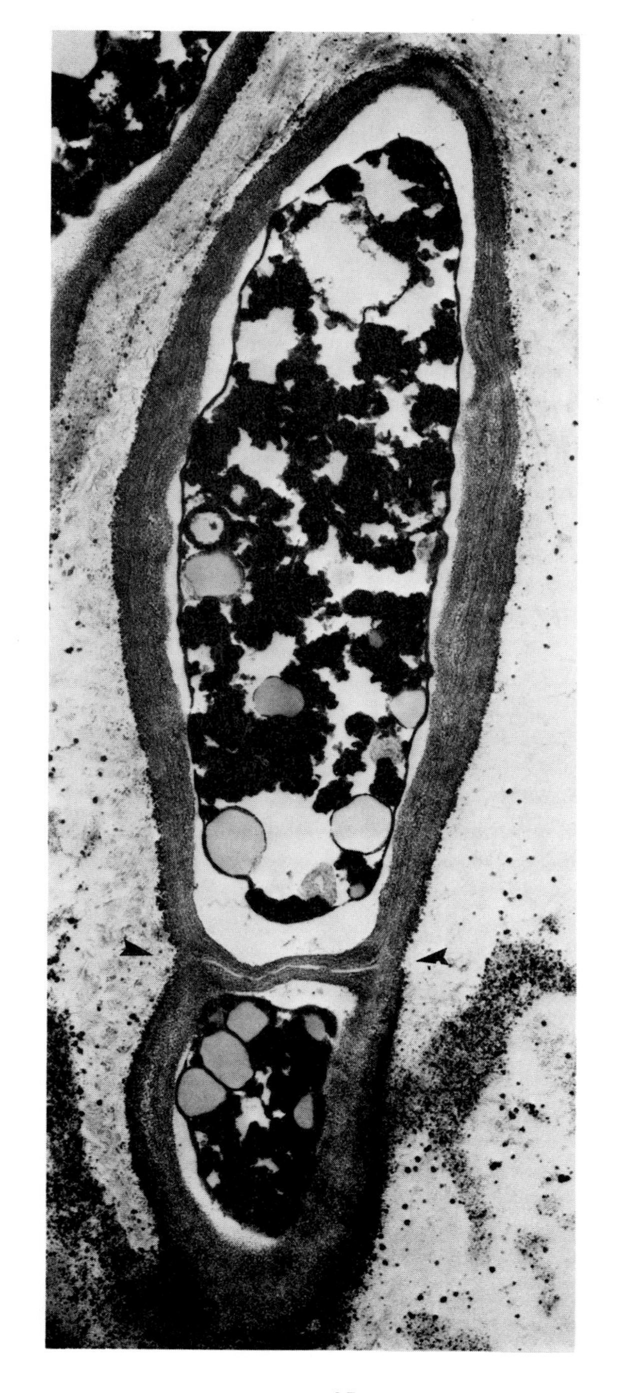

FIGURE 2.5. Hyphal septum formation (arrows) in the isolated mycobiont of *Graphis scripta* LS (×15,000).

Hyphal septa are apparently formed in ascolichens as in nonlichenized ascomycetes (Figure 2.5). Roskin (1970), Slocum & Floyd (1977), and Slocum (1980) have also observed dolipore septa in basidiomycetes (Figure 2.7a).

Concentric Bodies

Concentric bodies are round or ellipsoidal subcellular particles (Figure 2.6) that are most frequently observed in lichenized ascomycetes, although they have also been found in nonlichenized pathogenic ascomycetes (Griffiths & Greenwood, 1972; Bellemère, 1973; Tu & Colotelo, 1973; Granett, 1974; Samuelson & Bezerra, 1977). The structure of these bodies is apparently the same in all fungi that produce them. They are about 300 nanometers in diameter and have a transparent core surrounded by concentric bands of varying density. An additional halo may surround the body. Peveling (1969) and Griffiths and Greenwood (1972) have discussed the fine structure of these bodies in detail.

It is not presently known precisely what concentric bodies are, where they come from, or what they do. Brown and Wilson (1968) originally conceived of concentric bodies as centers of membrane synthesis or material transport. They noted that these structures are frequently found in close association with membranes and with the nucleus. Jacobs and Ahmadjian (1971a) also considered them to be connected to the internal membrane system of the fungus. This is supported by occasional observations of a tubular system apparently connecting concentric bodies near nuclei (Boissière, 1982). Ellis and Brown (1972) thought of concentric bodies as functionally equivalent to the Golgi apparatus. They appear in greatest numbers in thallus zones of maximum cell division and are perhaps passively distributed in differentiated hyphae (Boissière, 1982). The only cytochemical study of concentric bodies indicated they are proteinaceous; they also apparently have a high heat resistance (Galun, Behr, & Ben-Shaul, 1974).

Concentric bodies have been observed in the hyphae of cultured mycobionts in only one instance. This was a colony of *Cladonia cristatella* that formed pycnidia during a period of slow drying (Jacobs & Ahmadjian, 1971a; Ahmadjian, 1980a). They are almost always found in lichenized thalli, either natural or resynthesized. For this reason, presence of concentric bodies has been used as an indicator of the completeness of thallus resynthesis (Ahmadjian & Jacobs, 1970; Marton & Galun, 1976). Concentric bodies have also been observed in ascospores of lichenized *Rhizocarpon geographicum* (Ascaso & Galvan, 1975). They have not been observed in basidiolichens; in addition, there have been occasional reports of their absence in ascolichens. They have not been found in the aquatic lichen, *Hydrothyria venosa* (Jacobs & Ahmadjian, 1973), or two *Gonohymenia* species from the Middle East (Paran, Ben-Shaul, & Galun, 1971). These species all have blue-green

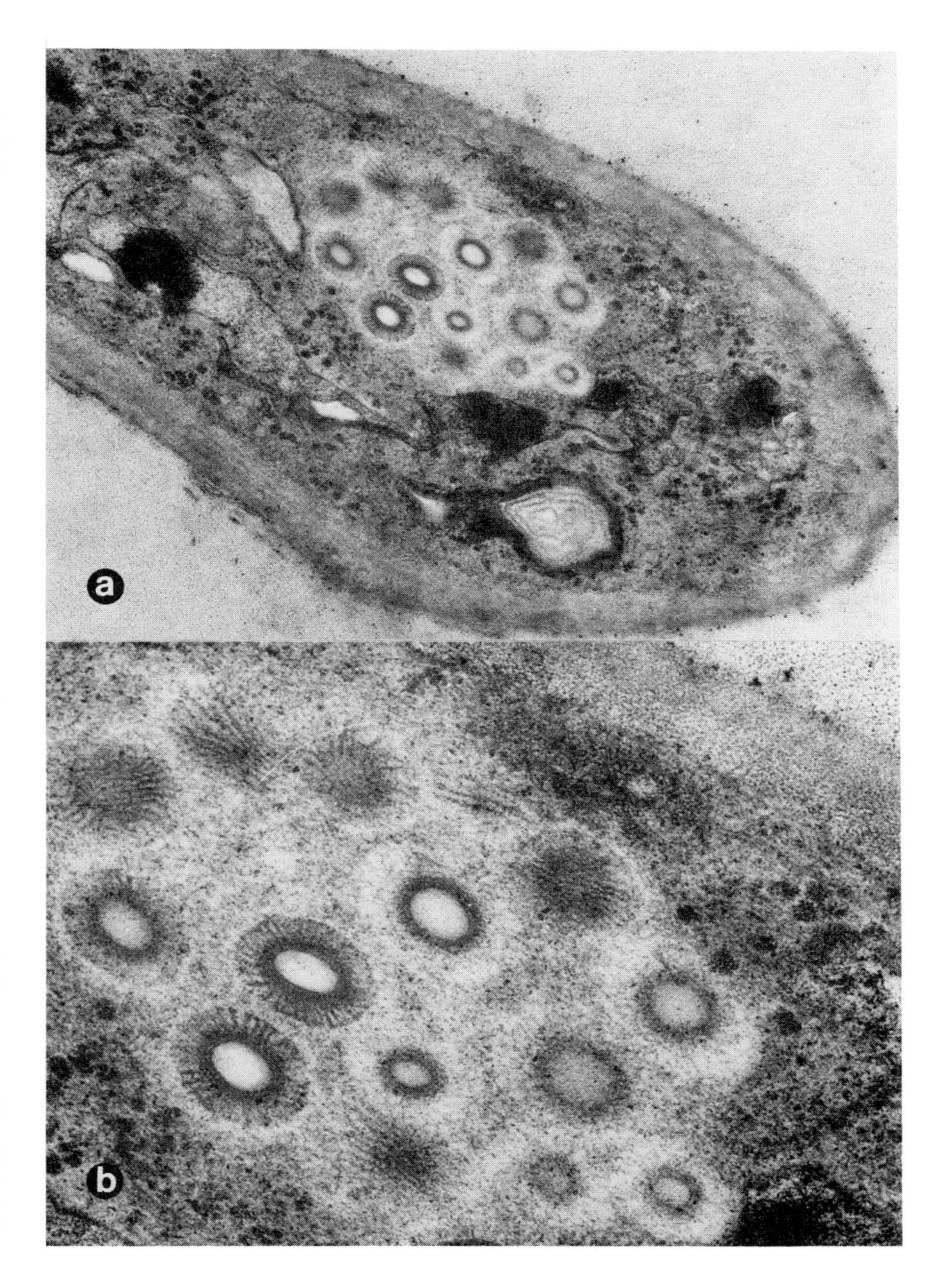

FIGURE 2.6. Concentric bodies in hyphal sections of *Collema cristatum*. (a) CS (×37,000); (b) CS (×93,000). Provided by E. Peveling.

phycobionts; however, this is apparently not the explanation for the absence of concentric bodies, because concentric bodies are produced by many other lichens with blue-green hosts (Peveling, 1973).

FUNGUS-ALGA CONTACTS

Ever since Moore and McAlear (1960) first studied the fine structure of lichen haustoria, electron microscopy has been used to investigate the cellular relationships between symbionts in lichenized thalli. Although an early study by Durrell (1967) suggested that haustorial contacts were rare in lichen phycobionts, the accumulated evidence since then has demonstrated not only their presence but their relatively high frequency in many lichenized thalli. Ahmadjian and Jacobs (1981) found that in natural squamules of *Cladonia cristatella*, well over half the algal cells were penetrated by haustoria. According to Ahmadjian (1982a), the reasons haustoria are infrequently seen are:

1. They are sometimes small and do not penetrate very far beyond the algal cell wall.
2. The large chloroplasts of *Trebouxia* and *Pseudotrebouxia* sometimes obscure haustorial penetrations. He suggested that the squash method of Ahmadjian and Jacobs (1981) is the most accurate one to use to quantify haustorial contacts in lichens.

By far, the greatest amount of TEM work on symbiont interactions has been done by Galun and colleagues in Israel and Peveling and colleagues in Germany. Ahmadjian (1982a) has discussed this entire body of work, and provided a number of generalizations about symbiont interactions at the ultrastructural level, including the following:

1. As Plessl (1963) first concluded, structurally primitive lichens have a relatively large percentage of their algae penetrated by fungal haustoria. In contrast, advanced lichens rarely show haustorial penetration of healthy cells; symbiont contact is limited for the most part to wall-to-wall apposition.

2. In pyrenocarpous lichens, plasmalemma-bound haustoria are frequently observed, whereas in discocarpous species, the haustoria are apparently always wall-bound (Galun, Paran, & Ben-Shaul, 1971a; Galun, Ben-Shaul, & Paran, 1971b; Galun et al., 1973; Jacobs & Ahmadjian, 1969; Peveling, 1973).

3. The plasmalemma of the alga is never ruptured except in dead

cells. It invaginates as the haustorium penetrates (Galun, Paran, & Ben-Shaul, 1970b).

4. Different degrees of haustorial penetration can be observed in the same thallus; older portions show intracellular haustoria most frequently (Galun, Paran, & Ben-Shaul, 1970a; 1970b). This is perhaps responsible for the slow growth rates characteristic of most lichens (Jacobs & Ahmadjian, 1971a).

5. Haustorial penetrations of phycobiont cells are most frequent in extreme environments (Galun, Paran, & Ben-Shaul, 1970c)—perhaps a result of increased algal susceptibility to fungal attack in these environments (Ahmadjian, 1982a).

Haustorial penetration of algal cells is apparently accomplished by both chemical and mechanical processes. In an ultrastructural study of *Parmelia sulcata*, Webber and Webber (1970) noted the presence of mesosomes and "lysosomelike organelles" in the advancing edge of haustoria, suggesting that penetration results from enzymatic digestion of the algal cell wall. No cytochemical tests were done to confirm this, however. Boissière (1982) considers the mesosomelike bodies to be storage organelles that contain polyglucosides from the algal cells.

Other ultrastructural features characterize the zone of contact between mycobiont haustoria and the phycobiont (Figure 2.7). Fungal cell walls are frequently much thinner in hyphae close to phycobiont cells (Peveling, 1973). Also, fungal mitochondria are more numerous (Chervin, Baker, & Hohl, 1968; Webber & Webber, 1970).

In an ultrastructural study of the arctic-alpine *Cornicularia normoerica*, Walker (1968) described unusual channels that originated in the algal chloroplast and passed through the algal cell wall to the open space between the symbionts. She considered these channels, not seen in any other lichens investigated using TEM, to be a means of transporting photosynthetic products from the alga to the fungus. Jacobs and Ahmadjian (1969) rejected this interpretation and suggested instead that the channels were cytoplasmic remnants resulting from partial cellular decomposition during four weeks of unrefrigerated storage.

Those lichens that do not have haustorial connections between symbionts generally have an appressorial contact in which there is a thinning of both fungal and algal cell walls at the point of contact (Jacobs & Ahmadjian, 1969; Ellis, 1975; Lambright & Tucker, 1980; Meier & Chapman, 1983). Ahmadjian (1970) considers appressorial lichens to be more primitive than haustorial forms, although there may also be environmental correlates. For example, xeric environments may promote haustoria formation (Galun, Paran, & Ben-Shaul, 1970a; Kushnir & Galun, 1977). Some apparently advanced

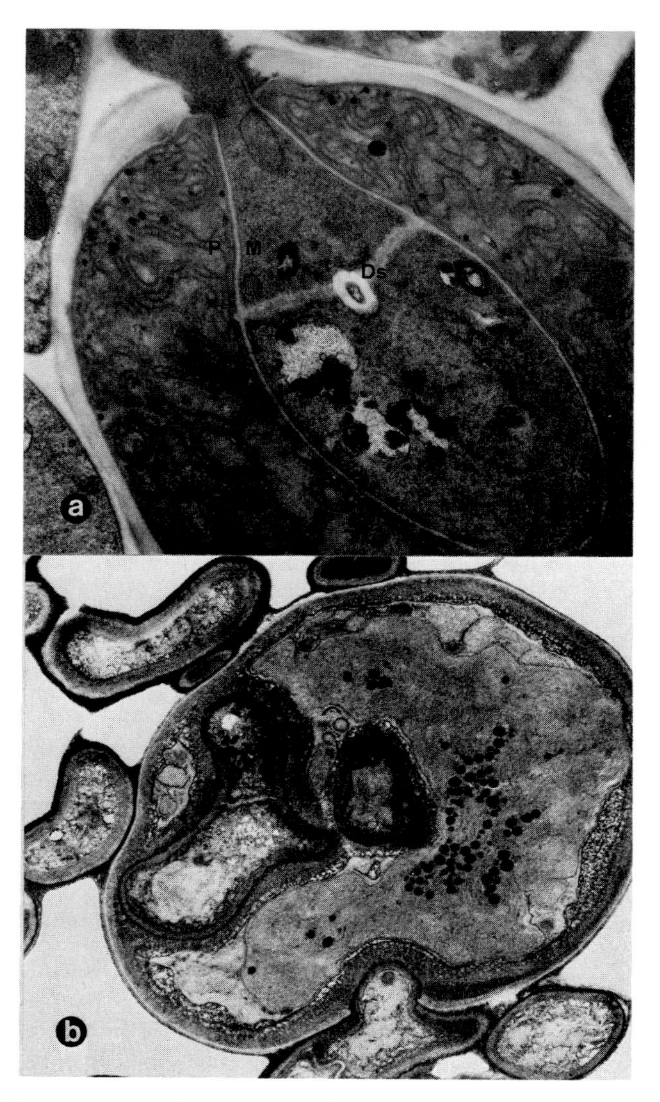

FIGURE 2.7. Haustorial contacts in lichenized thalli. (a) *Dictyonema irpicinum;* establishment of intracellular haustorium of mycobiont (M) by penetration of *Scytonema* phycobiont cell (P) by "shell" hypha. Dolipore septum (Ds). CS (×16,600). Reproduced by permission of the National Research Council of Canada from the *Canadian Journal of Botany,* Volume 58, 1980 and R.D. Slocum. (b) *Cladonia cristatella* synthesized with *Trebouxia erici* phycobiont CS (×14,575). From V. Ahmadjian. 1982a. "Algal/fungal symbioses." In *Progress in Phycological Research* (F.C. Pound & D.J. Chapman, eds.), Vol. 1, pp. 179–233. Amsterdam: Elsevier Biomedical Press. Reprinted with permission of Elsevier.

lichens, most notably *Physcia aipolia*, have also been found never to produce haustoria (Rudolph & Giesy, 1966; Brown & Wilson, 1968; Jacobs & Ahmadjian, 1969).

An unusual type of symbiont interaction was described recently by Malachowski, Baker, and Hooper (1980) for *Usnea cavernosa*. In addition to intracellular haustoria, *U. cavernosa* produces an extracellular hyphal sheath with distinctive protuberances that penetrate into the algal cell walls. The protuberances were only observed in older regions of the thallus. Malachowski and colleagues suggested that the sheath with its protuberances may facilitate photosynthate absorption by the fungus a number of ways:

1. by increasing surface area of the fungus

2. by acting as an anchor to hold algal cells close to absorptive fungal surfaces

3. by partially enclosing the algal cell and channeling substances released by the alga during wetting or drying.

CORRELATIONS BETWEEN ENVIRONMENT AND ULTRASTRUCTURE

Lichens show a number of interesting ultrastructural changes in response to changing environments. Although the evidence is still relatively equivocal and contradictory, a number of tentative generalizations can be made.

1. The production of pyrenoglobuli in phycobiont pyrenoids depends on the physiological and nutritional status of the thallus.

2. Generally, pyrenoglobuli are observed in the phycobiont pyrenoid of dry thalli and starch is found in thalli kept wet for considerable periods of time (Brown & Wilson, 1968; Jacobs & Ahmadjian, 1971a).

3. If thalli are desiccated for long periods of time, the number of pyrenoglobuli decreases from the center to the periphery of the pyrenoid (Jacobs & Ahmadjian, 1971a; Ascaso & Galvan, 1976a).

4. The extent of haustorial penetration is sometimes correlated with the amount of thallus drying—perhaps reflecting an increased algal susceptibility (Ben-Shaul, Paran, & Galun, 1969; Peveling, 1973; Galun, Paran, & Ben-Shaul, 1970c).

5. Environmentally induced ultrastructural changes to lichen fungi are seldom observed (Brown & Wilson, 1968; Jacobs & Ahmadjian, 1971a).

These statements are not true in all cases, and more research is needed to explain deviations from the normal pattern. For example, the absence of pyrenoglobuli in *Cladonia cristatella* ecotypes occupying fully exposed rock substrates (Jacobs & Ahmadjian, 1971a) is consistent with the hypothesis that these bodies store lipids that can be assimilated in stressful environments. However, the absence of these structures in some *Usnea* species (Chervin, Baker, & Hohl, 1968; Jacobs & Ahmadjian, 1969) and their presence in others (Peveling, 1968; Malachowski, Baker, & Hooper, 1980) cannot be explained entirely on the basis of environmental stress.

There have been some recent studies of environmentally induced ultrastructural changes in lichens. Peveling and Robenek (1980) observed changes in the plasmalemma structure of *Trebouxia* that apparently relate to thallus hydration. Using a freeze-fracture technique, they observed thalli of *Hypogymnia physodes* maintained under different moisture conditions. Freeze-etching revealed an algal plasmalemma with unique grooves on the outer membrane face (the P fracture face) and corresponding ridges on the inner membrane face (the E fracture face). Moreover, the size and frequency of these grooves and ridges changed rapidly with changing humidities. These structures were much more pronounced in dry thalli than in moist thalli. Other plasmalemma particles were also observed, but these responded more slowly to changing moisture conditions.

Structures similar to the grooves and ridges of *Trebouxia* have been observed in the plasmalemmas of yeast (Moor & Mühlethaler, 1963; Bauer & Sigarlakie, 1973) and red algae (Peveling & Robenek, 1980). They appear to be correlated with cell aging in yeast and environmental stress in red algae. Peveling and Robenek suggested that the presence of these structures in lichen phycobionts may also be stress related, but their function remains unknown.

3 Reproduction and Dispersal

Dispersal in lichens may involve both sexual and vegetative processes. Although neither process is well known, sexual reproduction is probably the least understood aspect of lichen biology. Hale (1983) has suggested that the main reasons for this are:

1. slow thallus growth and an inability to maintain whole thalli under laboratory conditions;
2. our present inability to induce production of sexual structures in pure mycobiont cultures.

This chapter reviews what little is known about lichen reproduction and dispersal, emphasizing the relationships between form and function rather than developmental morphology or taxonomy. These latter areas are discussed in a number of comprehensive reviews (Poelt, 1973; Letrouit-Galinou, 1968; 1973; Henssen & Jahns, 1974; Bellemère & Letrouit-Galinou, 1981; Henssen, 1981).

REPRODUCTION

Sexual Structures—Ascomycetes

Most lichenized fungi belong to the Ascomycetes, a large class of higher fungi in which eight spores are typically produced in a sac (ascus). Although yeasts and related ascomycetes (Endomycetales) produce naked asci, most ascomycetes produce a fruiting structure called an *ascocarp* that surrounds the asci.

The structure and ontogeny of the ascocarp have frequently been used to delimit various ascomycete groups. Two distinct morphogenetic lines can be recognized in ascocarp-forming ascomycetes. These are termed ascohymenial and ascolocular (Nannfeldt, 1932), and there are at least two major characteristics that distinguish them (Luttrell, 1951).

The first characteristic is the nature of the ascocarp wall. In ascohymenial groups, the ascogenous system or ascogonium gives rise to a true hymenium, which is the spore-bearing region composed of asci and sterile, unbranched interascal hyphae called paraphyses. The ascocarp wall then forms from adjacent sterile hyphae in response to the developing ascogonium. In ascolocular groups, asci are produced in cavities or locules in a preformed layer of vegetative hyphae called a stroma. Asci without true paraphyses develop in the locules; branched anastomosing paraphysislike hyphae called paraphysoid tissue or pseudoparaphyses, depending on their origin, typically separate the locules from each other.

A second major characteristic that distinguishes these two lines is the structure of the ascus. Ascolocular groups produce a bitunicate ascus (Figure 3.1) that consists of two layers: a thin, nonextendable outer layer and a thick, extendable inner layer. At dehiscence, the outer layer splits at the apex and the inner layer expands as a tubular extension at the tip of the ascus. Ascospores are shot out an elastic pore at the tip of the extension. Ascohymenial groups generally produce unitunicate asci that do not have this jack-in-the-box construction.

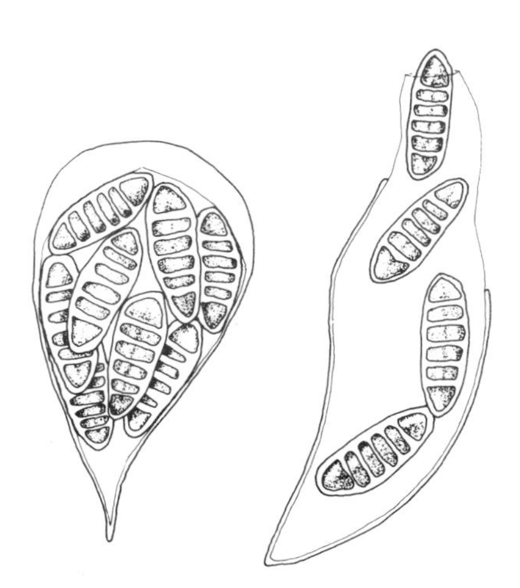

FIGURE 3.1. Bitunicate ascus of *Arthonia* sp.

There are also characteristics of the apical apparatus of asci that distinguish the Loculoascomycetes from other groups (Chadefaud, 1960). The nassasceous type of ascus, which possesses an apical cylinder *(nasse)* with no surrounding ring, is characteristic of the Loculoascomycetes. Other groups are either annellasceous—in which an apical ring but no cylinder is formed—or archetypal—in which both an apical ring and a cylinder are present. Members of the Lecanorales typically are thought to possess archetypal asci, although there are numerous subdivisions (Bellemère & Letrouit-Galinou, 1981).

Types of Ascocarps

Apothecia. The discomycetous lichens produce apothecia, which are disc-, saucer-, or bowl-shaped ascocarps (Figure 3.2a–c) with a thin hymenial

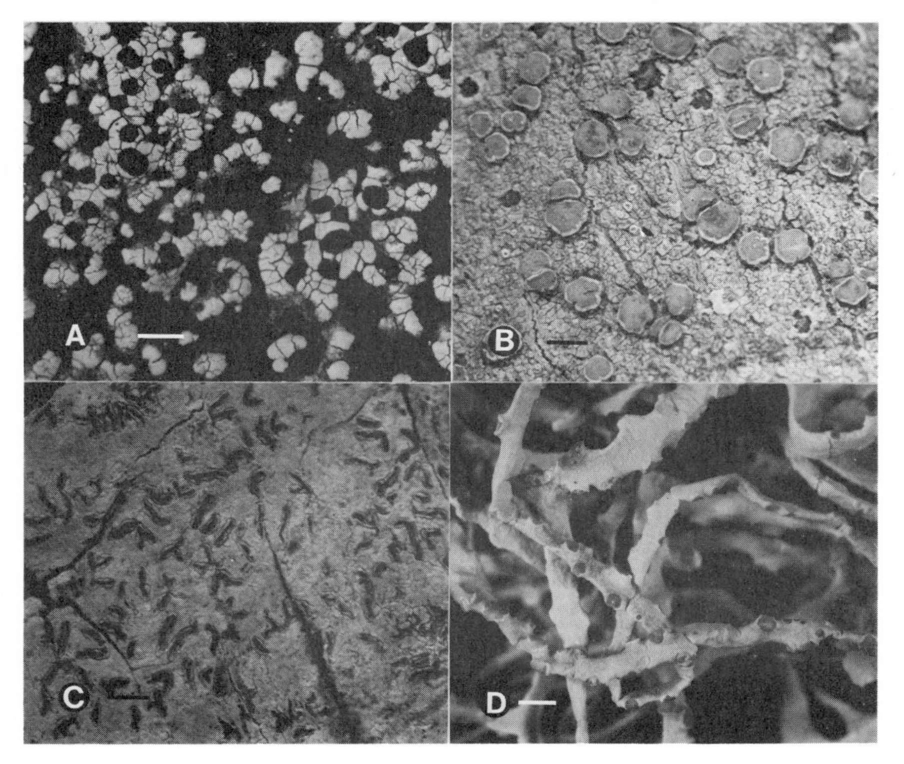

FIGURE 3.2. Lichen reproductive structures. (a) Lecideine apothecia of *Rhizocarpon alpicola*, bar = 1 mm; (b) Lecanorine apothecia of *Lecanora pallida*, bar = 2 mm; (c) Lirellae of *Graphis scripta*, bar = 1 mm; (d) Pseudothecia of *Dendographa* sp., bar = 2 mm.

layer of asci and paraphyses lining the inner surface (Figure 3.3b). Spores are ejected forcibly except in the order Caliciales in which asci disintegrate at maturity and form a dry, powdery mass called a mazaedium. Apothecia may be borne on stalks of various types and ontogenetic origins. Podetia produced by *Cladonia* species (Figure 3.4) are hollow and derived from tissues peripheral to the ascogonial complex. Solid pseudopodetia produced by *Stereocaulon* species have a thalline origin. Apothecial stipes are produced by *Pilophoron* and *Baeomyces*.

In the Graphidaceae, apothecia are produced in the form of long, narrow furrows called lirellae (Figure 3.2c). They normally have hard, carbonized walls derived from thalline tissue.

Apothecium development (Figure 3.5) proceeds through several stages in various groups, but only after fertilization has taken place in the ascogonium. Henssen (1981) has discussed the main lines of ascocarp development in the Lecanorales and the evolutionary significance of these patterns. They differ a number of ways, but one of the most important is the degree to which somatic and generative tissue participate in ascocarp development. This is because the formation of structures associated with the apothecium (e.g., margin and stipe) are as crucial as the development of the centrum itself. Henssen (1981) and Henssen and Jahns (1974) should be reviewed for discussion of reproductive morphology in lichens.

Perithecia. Immersed, flask-shaped ascocarps with a built-in opening (ostiole) are called perithecia (Figure 3.3a); they characterize the Pyrenomycetous lichens. The hymenium of asci and paraphyses lines the inside of the flask, and spores are expelled through the ostiole, sometimes in a mucilagin-

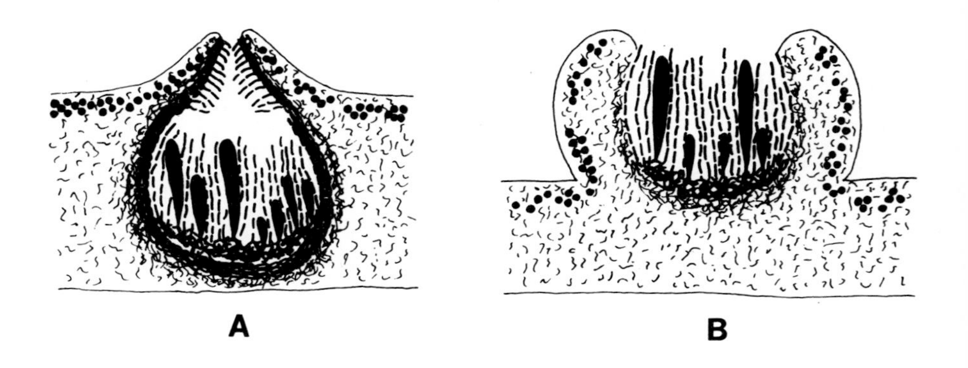

FIGURE 3.3. (a) Lichen perithecium and (b) apothecium.

FIGURE 3.4. Lichen podetia. (a) *Cladina subtenuis,* bar = 1 cm; (b) *Cladonia cenotea,* bar = 1 cm; (c) *Cladonia deformis,* bar = 1 cm.

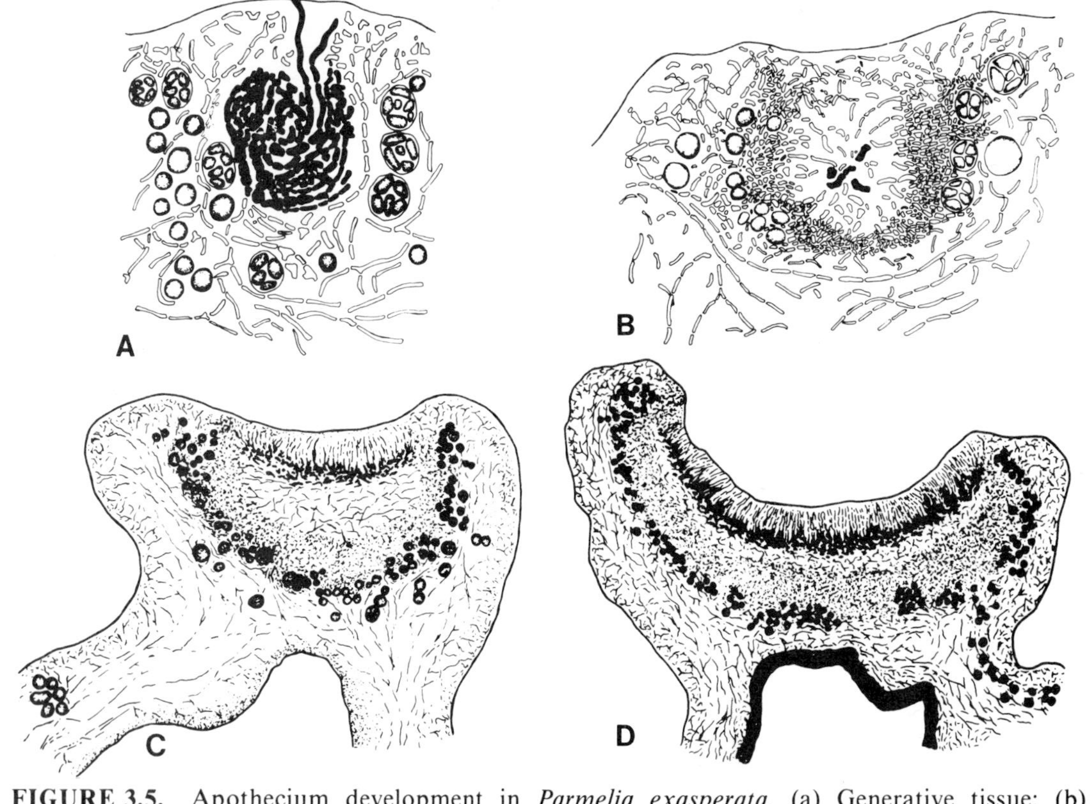

FIGURE 3.5. Apothecium development in *Parmelia exasperata.* (a) Generative tissue; (b) Primordium; (c) Young apothecium; (d) Mature apothecium. Redrawn from A. Henssen & H.M. Jahns. 1974. *Lichenes. Eine Einfuhrung in die Flechtenkunde.* Stuttgart: Georg Thieme Verlag. Reprinted with permission of the author.

ous matrix. Other paraphysislike sterile hyphae (periphyses) are produced near the ostiole. The perithecial wall is formed from sterile cells derived from hyphae surrounding the ascogonial apparatus during development.

Pseudothecia. Organized ascocarps produced by ascolocular lichens are called pseudothecia because they sometimes superficially resemble perithecia, apothecia, or lirellae. An example is the apotheciumlike pseudothecia of *Dendrographa* species (Figure 3.2d). These lichens lack a true hymenium; instead, they bear scattered asci within locules imbedded in a vegetative stroma.

Spores. Virtually all groups of ascolichens produce eight spores per ascus, although some groups (*Acarospora, Candelaria, Sarcogyne,* and *Anzia*) produce more than eight. Simple and colorless throughout their development in most groups, they become darker and variously subdivided in certain cases. Hale's (1974) classification of spores based on septation (Figure 3.6) is a useful one.

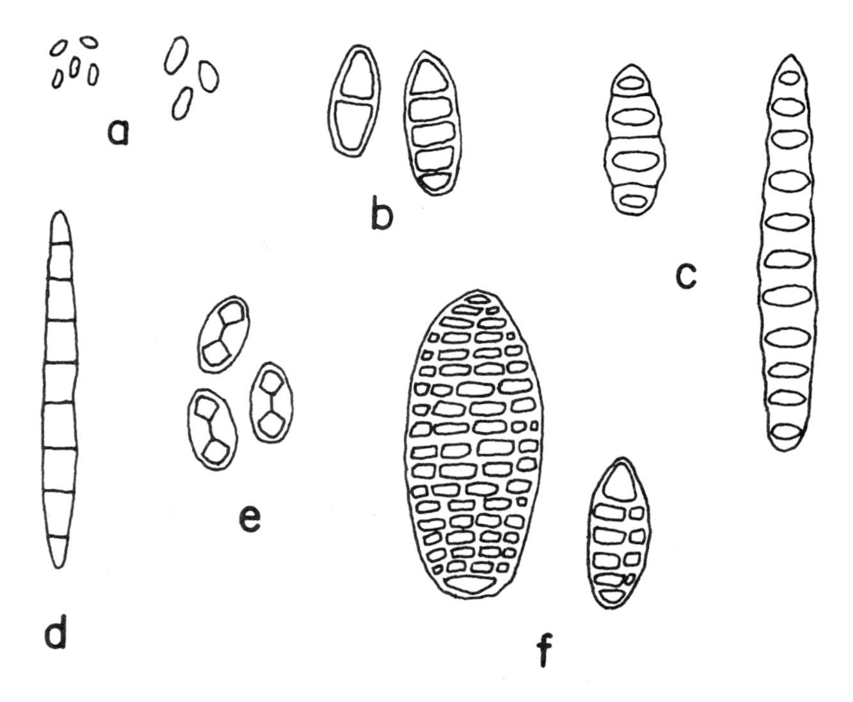

FIGURE 3.6. Lichen ascospore types. (a) Simple; (b,c,d) Transversely septate; (e) Polilocular; (f) Muriform.

1. Simple spores: unicellular and nonseptate, often small and thin-walled (*Lecidea, Lecanora, Parmelia,* and *Usnea*), more rarely very large (to 300 μm) and thick-walled (*Pertusaria*).

2. Transversely septate spores: elongate and multicellular with one to as many as 30–40 cross walls (*Catillaria, Graphis,* and *Pyrenula*).

3. Muriform spores: multicellular with both transverse and longitudinal walls, often large (*Phaeographina, Lopadium, Umbilicaria,* and *Diploschistes*).

4. Polarilocular spores: two-celled spores with a thick median wall and a thin isthmus, or conversely a single-celled spore with a median constriction (Teloschistaceae).

Pycnidia and Conidia. Pycnidia are flask-shaped structures that resemble perithecia (Figure 3.7). However, they do not contain ascospores; rather, they surround simple or branched hyphae (conidiophores) which produce numerous asexual buds or spores, variously called pycnospores, pycnidiospores, conidiospores, spermatia, or microconidia (Hale, 1961)—Vobis (1980) calls them conidia, which is the term most consistent with current mycological usage.

Vobis (1980) and Vobis and Hawksworth (1981) have reviewed the available information on conidiogenesis. There are apparently several types of developmental patterns. The *Umbilicaria* type is widespread in the Lecan-

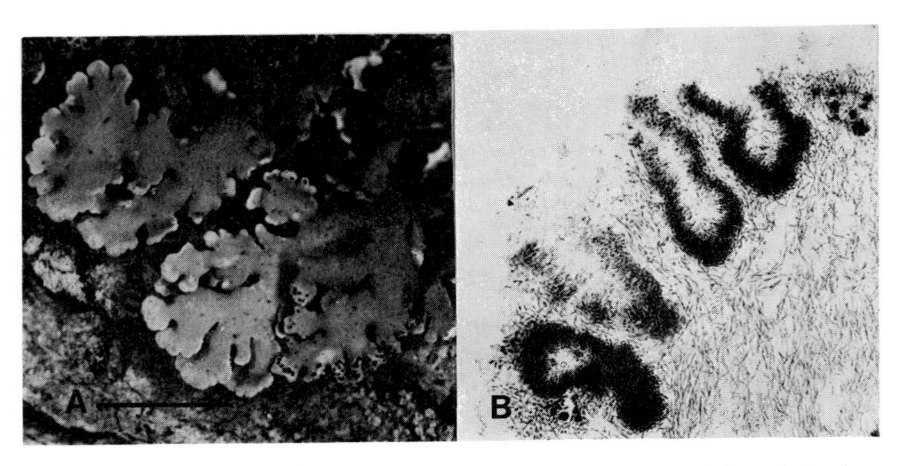

FIGURE 3.7. Lichen pycnidia. (a) *Anzia colpodes* with pycnidia on lobe tips, bar = 4 mm; (b) Pycnidia from *Santessonia namibensis* (×200). Figure 3.7b from M.E. Hale, Jr. & G. Vobis. 1978. "*Santessonia,* a new lichen genus from southwest Africa." Source: *Bot. Notiser* 131:1–5, The Lund Botanical Society.

orales (Vobis & Hawksworth, 1981). In this type, the pycnidium originates from a ball of more-or-less isodiametric cells within the algal layer. While the inner cells show little activity, outer cells divide to expand the ball and create a small, central cavity. As the cavity enlarges, conidiophores develop in the available spaces. Simultaneously, cells nearest the surface grow through the upper cortex creating an ostiole. Both the basal cells of the conidiophores and the cells of the pycnidial wall arise from the same pseudoparenchymatous tissue.

Conidiophores are produced in a range of morphological types; Glück (1899; see also Vobis & Hawksworth, 1981) devised a system of eight types, some of which produce conidia apically and some laterally.

Conidiogenesis is not very well known in lichen-forming fungi. What is known suggests that it is largely phialidic, although holoblastic development has been reported in *Conotrema urceolatum* (Gilenstam, 1969; see also Vobis & Hawksworth, 1981).

The number of conidial lichen-forming species is large—perhaps as many as 8000 have conidial or anamorphic states, but most of these occur on the sexual, teleomorphic ascocarp or on thalli indistinguishable from those that produce only ascocarps (Vobis & Hawksworth, 1981). The biology of the conidial anamorph has therefore been paid little attention.

In addition to those species for which both teleomorphic and anamorphic states are known, Vobis and Hawksworth (1981) identify 41 genera of conidial lichen-forming fungi for which no teleomorphic phase is known. There are also parasymbiotic fungi and entirely sterile lichenized taxa—very little information is available on these groups.

Some conidial lichen-forming fungi are hyphomycetous; that is, the conidia are not produced in an immersed or superficial pycnidium. Hawksworth (1979) has published a comprehensive survey of the lichenicolous Hyphomycetes. These have been called parasymbionts by many lichenologists, the term being used by Smith (1921) to refer to any harmless but not mutually beneficial growth of two organisms. In reality, the parasymbiotic association is a "three-membered symbiosis" (Poelt, 1977) involving the two lichen symbionts and a third fungal associate. Hawksworth (1978) considers the parasymbiont to actually be lichenized since it obtains its carbohydrate needs from the same algal partner as the natural mycobiont.

Other lichen-forming hyphomycetous anamorphs are not as well known, but there have been a few accounts of lichen conidial production not involving pycnidia. Hasenhüttl and Poelt (1978) discovered that some *Umbilicaria* species produce conidia on the undersurfaces of lobes and rhizines. These Brutkörnen (brood grains) are apparently able to adhere to damp surfaces and germinate, possibly generating new lichenized thalli. They occur only in thalli that rarely if ever produce apothecia.

There have also been reports of conidial production in pure culture by

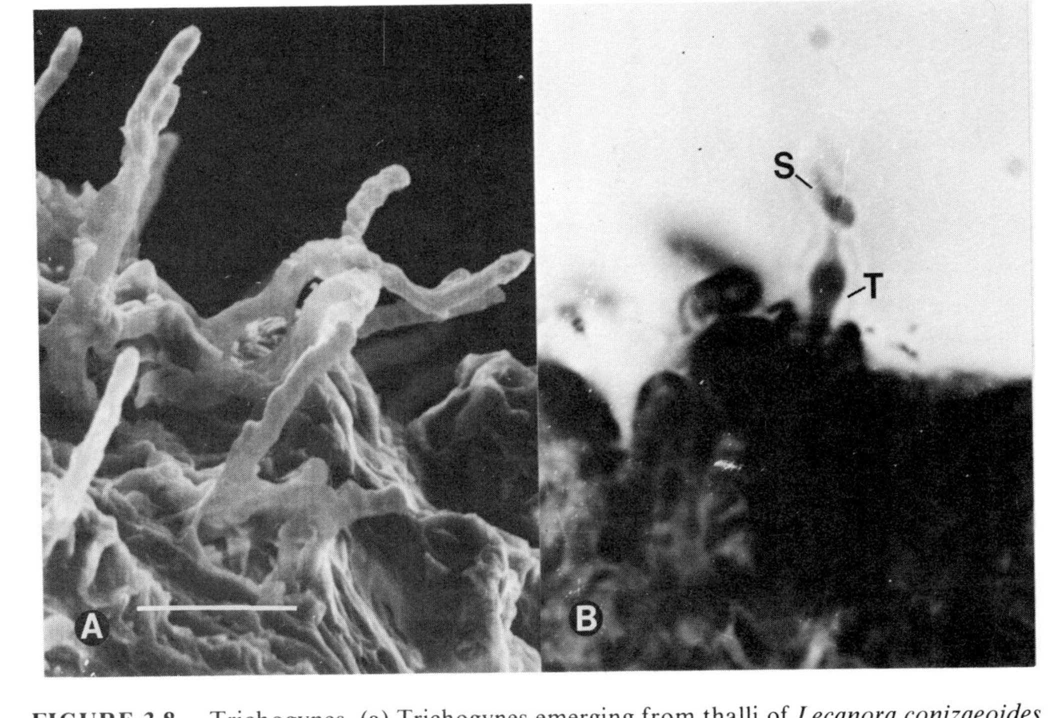

FIGURE 3.8. Trichogynes. (a) Trichogynes emerging from thalli of *Lecanora conizaeoides* growing on tree bark, bar = 25 μm. From H.M. Jahns et al. 1979. "Die Neubesiedlung von Baumrinde durch Flechten. I." *Natur und Museum* 109:40–51. Reprinted with permission from Natur und Museum. (b) Trichogyne (T) of *Pilophorus robustus* with attached spermatium (S) (×4400). From H.M. Jahns. 1970. "Untersuchungen zur Entwicklungsgeschichte der Cladoniaceen." *Nova Hedwigia* 20:1–177. Reprinted with permission of Cramer and the author.

isolated lichen-forming fungi (Hale, 1957; Ahmadjian, 1963) and by germinating ascospores of *Vezdaea aestivalis* (Tschermak-Woess & Poelt, 1976).

Lichen conidia are thought to have two biologically important roles, spermatization and dispersal of the mycobiont. Figure 3.8b gives evidence for their role as spermatia which comes from observations of conidia adhering to trichogynes (Stahl, 1877; Bachmann, 1912; Stevens, 1941; Ahmadjian, 1966a; Jahns, 1970; Vobis, 1977). However, no evidence exists at this time for actual nuclear transfer from conidia to trichogyne.

The role of conidia in dispersal has been documented by Vobis (1977), who showed that conidia of three crustose species germinated in culture and produced a luxuriant mycelium (Figure 3.9). These results substantiated

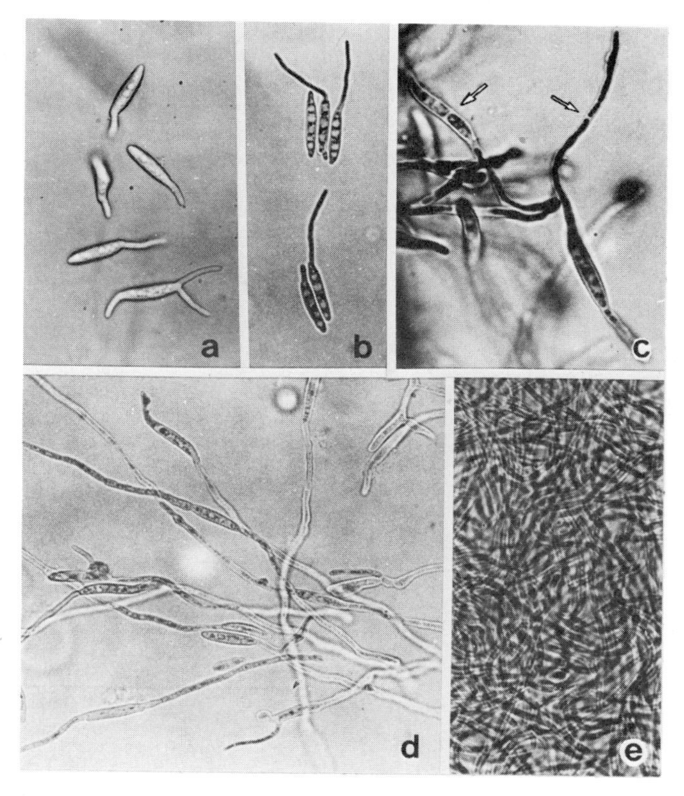

FIGURE 3.9. Germinating conidia of *Lecanactis abietina*. (a) 14 days (unstained) (×700); (b) 15 days (×700); (c) 2 months with septa (arrows) (×100); (d) 10 months, hyphae developed from conidia (×700); (e) 1 year, mycobiont colony. All on bark extract agar. From G. Vobis. 1977. "Studies on the germination of lichen conidia." *Lichenologist* 9:131–136. Reprinted with permission of Academic Press.

earlier laboratory studies (Möller, 1887; Hedlund, 1895). Vobis' work gave preliminary support for the following generalizations:

1. Conidia from crustose lichens germinate more frequently than those from foliose and fruticose lichens. Vobis tested 12 foliose and fruticose species and observed no definite germination.

2. Pycnidial age is important; conidial germination is more likely in newly matured pycnidia.

3. Conidial germination is stimulated on media containing substrate extracts. Also, additions of glucose, saccharose, fructose, or maltose in small concentrations (1 percent or less) and 0.5 percent yeast extract stimulated germination.

Sexual Structures—Basidiomycetes

There are approximately 20 species of basidiolichens belonging to the orders Agaricales or Aphyllophorales (Letrouit-Galinou, 1973). They are primarily tropical, but a few have extratropical ranges.

Very little is known about the reproductive biology of basidiolichens, but it is assumed that the process is the same as in nonlichenized species. Basidiocarps are either of the Agaricales type (*Omphalina* category of

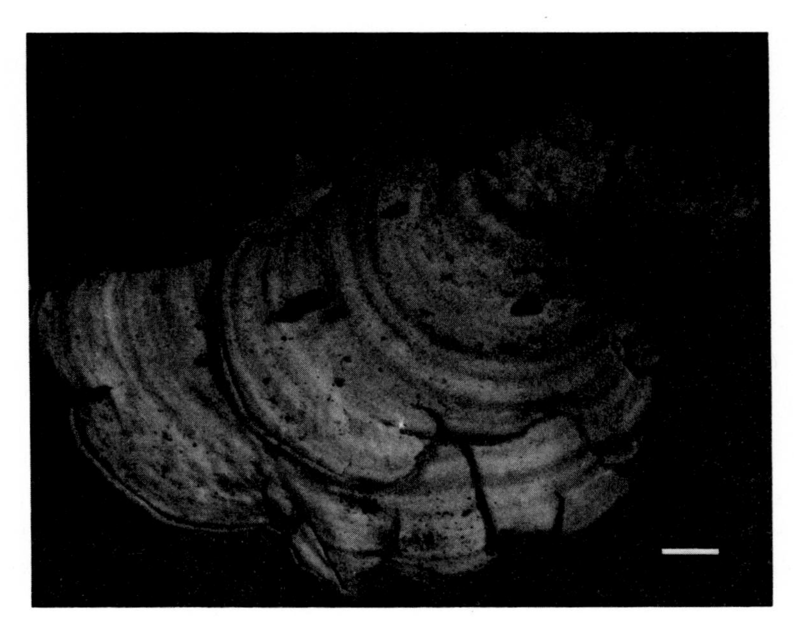

FIGURE 3.10. The basidiolichen, *Cora pavonia,* bar = 5 mm.

Oberwinkler (1970)) or, more frequently, of the Aphyllophorales type (including the corticioid *Athelia-Dictyonema* category and the clavarioid *Multiclavula* category of Oberwinkler (1970)).

In the bracket-forming species *Cora pavonia* (Figure 3.10), the basidiocarp is formed of several layers of tightly appressed fungal hyphae and a distinct *Scytonema* zone, a heteromerous condition quite different from the homiomerous arrangement observed in *Dictyonema* (Slocum & Floyd, 1977; Slocum, 1980). The hymenium of both species develops on the undersurface of the bracket. In *Cora pavonia* the basidia can be seen quite clearly in scanning electron micrographs (Figure 3.11). Here, basidiospores are produced as in other nonlichen groups.

FIGURE 3.11. Ventral surfaces of *Cora pavonia*. (a) Sterile hyphae (×500); (b) Sterile hyphae (×2000); (c) Fertile hyphae (×500); (d) Fertile hyphae (×2000). Figure 3.11d from M.E. Hale, Jr. 1976. "Lichen structure viewed with the scanning electron microscope." In *Lichenology: Progress and Problems* (D.H. Brown, D.L. Hawksworth, & R.H. Bailey, eds.), pp. 1–15. London: Academic Press. Reprinted with permission of Academic Press.

SEXUAL REPRODUCTION IN ASCOLICHENS

Figure 3.12 illustrates the general process of sexual reproduction in ascolichens which has been reviewed by Letrouit-Galinou (1973) (Figure 3.12). It involves alteration of three distinct phases:

1. a gametophyte that produces male structures (pycnidia or parascogonial filaments in the ascogonial apparatus)

2. a prosporophyte (sporophyte I) with hyphae containing numerous nuclei, some male and some female

3. an ascosporophyte or sporophyte II that is dikaryotic and able to undergo meiosis to produce asci with ascospores.

Sporophytes I and II together are considered the sporophytic apparatus (Figure 3.12d) and develop on the gametophyte within the ascocarp.

Since the male elements are either borne in pycnidia or are associated with ascogonial filaments in the ascogonial apparatus, two distinct types of plasmogamy or fusion of male and female elements can occur. If the male and female elements are produced together in the ascogonial apparatus, fusion is likely to be direct and homomictic, involving genetically similar nuclei produced by the same gametophyte. However, if the male element is produced in a pycnidium, the conidiospores are considered spermatia. Then transfer to the female ascogonial filament by wind, water, or animals is less direct and the entire process is potentially heteromictic, since spermatia from widely separated gametophytes may be transported to any given ascogonial filament.

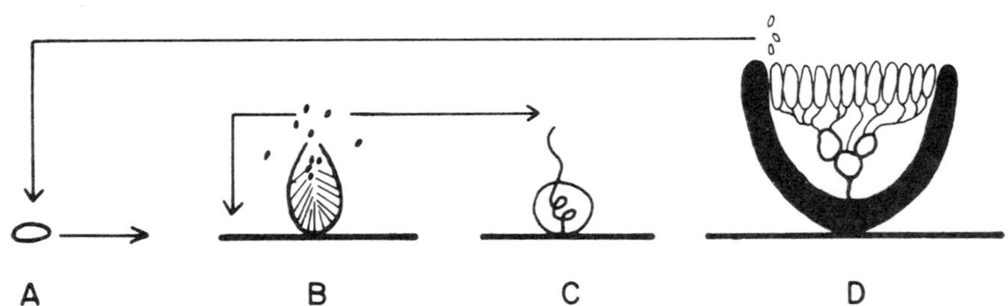

FIGURE 3.12. General process of sexual reproduction in ascolichens. (a) Ascospore germinates to produce a thallus; (b) Mature pycnidium with asexual conidia that can also function as spermatia; (c) Ascogonial apparatus with trichogyne; (d) Mature sporophytic apparatus with asci. Adapted from M.-A. Letrouit-Galinou. 1973. "Sexual reproduction." In *The Lichens* (V. Ahmadjian & M.E. Hale, Jr., eds.), pp. 59–90. New York: Academic Press, with permission of the publisher.

This possibility has led Robinson (1975) to suggest that hybridization between presumably asexual (sorediate or isidiate) morphs that have pycnidia and sexual morphs that produce apothecia may be responsible in part for the complexes of species that exhibit similar morphological, chemical, and ecological characters, but have different reproductive characters.

Plasmogamy has never definitely been demonstrated in lichen fungi, but there have been reports of presumptive spermatial trichogamy. Trichogynes are frequently observed in whole thalli and isolated mycobionts (Figure 3.8a), and spermatia have sometimes been found adhering to the tips of trichogynes (Stahl, 1877; Baur, 1898; Ahmadjian, 1966a; Jahns, 1970). This process is depicted in Figure 3.8b.

Lichenization

The sexual process in ascolichens results in production and release of ascospores. Once these ascospores germinate and begin producing hyphae, a suitable algal host must be obtained fairly quickly, although it is possible that saprophytic or very slow growth may sustain some young independent mycobionts under certain conditions. Given the likelihood that

1. most ascospores fail to germinate because of unfavorable environmental conditions,
2. of those that do germinate, only a few are probably able to obtain a suitable algal host,

one would predict that lichenization is a relatively infrequent event in nature. However, the fact that many lichen species are strictly sexual and have no other means of dispersal demonstrates that it is not only possible, but successful.

The actual mechanisms that result in lichenization in nature are not well known. The question of algal sources for germinating lichen spores is perhaps the most troublesome problem. Some common phycobionts are capable of independent existence and are frequently found in lichen habitats. However, trebouxoid algae, hosts of a wide variety of lichenized fungi, are only occasionally found free-living in nature (Tschermak-Woess, 1978). Hale (1974) and other lichenologists suggest that the primary source of algal hosts may be vegetative diaspores and lichen thallus fragments that contain suitable algae. Slocum et al. (1980) have also recently reported that zoosporogenesis and liberation of zoospores is possible in lichenized *Trebouxia* species, suggesting another source of algal hosts for germinating ascospores.

Lichenization has never been observed in nature; however, studies suggest that our technical ability to observe the process has increased tremendously in recent years. An example of such a study is an SEM

FIGURE 3.13. Clumps of fungal hyphae and algal cells on pine needles. (a) bar = 250 μm; (b) bar = 50 μm. From H.M. Jahns et al. 1979. "Die Neubesiedlung von Baumrinde durch Flechten. I." *Natur und Museum* 109: 40–51. Reprinted with permission of Natur und Museum.

investigation of the early colonization of pine needles and bark by lichen propagules (Jahns et al., 1979). Although the investigators failed to observe germinating lichen ascospores, they were able to find numerous clumps of fungal hyphae, presumably from *Lecanora varia* spores, and algal cells (Figure 3.13). They also observed masses of soredia produced by *Hypogymnia physodes* and *Lecanora conizaeoides,* which gradually differentiated into identifiable thalli. Use of such techniques in future investigations will undoubtedly increase our knowledge of this little-known process.

VEGETATIVE REPRODUCTIVE STRUCTURES

Given the apparent difficulties involved in lichenization, it is not surprising that a number of unique vegetative structures have evolved in the lichenized fungi that disperse the mycobiont and phycobiont together. Production of these structures, commonly referred to as vegetative diaspores, is assumed to be under genetic control; they are constant characters that are used to show important systematic relationships between groups (Poelt, 1973).

This section will only discuss the most frequently observed vegetative diaspores. More comprehensive treatments are available, however (Jahns, 1973; Poelt, 1973; Henssen & Jahns, 1974; Hale, 1983).

Soredia

Soredia are small (25–100 μm) balls of a few phycobiont cells wrapped in mycobiont hyphae. They originate in the medulla and algal layer and are released through pores or cracks in the upper surface of the thallus (Figure 3.14).

In some groups, soredia are produced over the entire thallus surface; in *Lepraria* species, the thallus is essentially a continuous layer of soredia (Figure 3.15c). In other groups, soredia are produced in delimited zones called soralia—the shape and position of which are of taxonomic use (Figure 3.15b).

Isidia

Isidia are peglike protuberances of the upper surface that are composed of algal and medullary tissues surrounded by a cortex (Figures 3.15b & 3.16). Their shapes vary considerably from group to group, ranging from short and globose to long and polypshaped. In most groups, they are easily broken off and obviously serve a dispersal role. In other groups, they may rarely break off. Jahns (1973) has suggested that isidia in these groups may increase thallus surface area thereby increasing assimilative capacity as well, an idea that requires experimental testing.

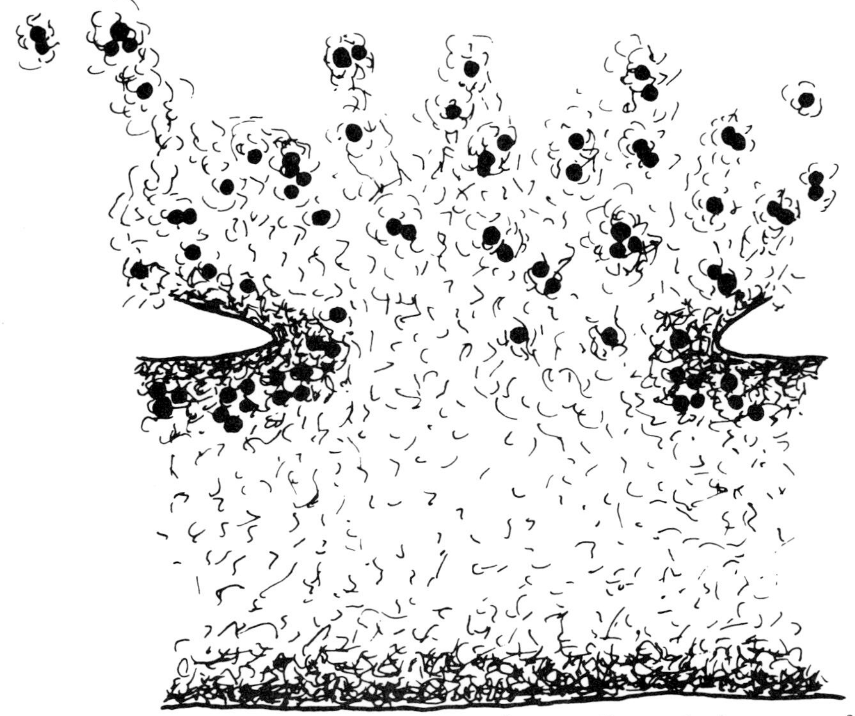

FIGURE 3.14. Schematic illustration of a soredia-producing area of a lichen thallus.

Hormocysts

A unique type of diaspore is produced by some species of *Lempholemma.* These are hormocysts, short *Nostoc* filaments enclosed by a gelantinous sheath sometimes penetrated by fungal hyphae. They are produced in apotheciumlike organs called hormocystangia. Since not all hormocysts are lichenized, this type of diaspore may serve to disperse mainly the phycobiont (Jahns, 1973).

Rhizomorphs

Rhizomorphs are thread- or cord-shaped structures made of numerous closely appressed fungal hyphae. They are commonly produced by various groups of terrestrial fungi and are generally found just below the ground surface. They have not been studied by lichenologists, but a recent study of the soil lichen, *Endocarpon pusillum,* by Malone (1977) showed that they may function as asexual reproductive structures.

FIGURE 3.15. Vegetative reproductive structures. (a) Isidia produced by *Pseudevernia consocians;* (b) Orbicular soralia of *Punctelia subrudecta;* (c) *Lepraria* sp., bar = 1 mm.

FIGURE 3.16. Isidia of *Parmelia awasthii* CS (×200). Provided by M.E. Hale, Jr.

Malone collected thalli of *E. pusillum* from various prairie sites in western Iowa by removing chunks of soil containing numerous thalli and carefully separating the lichen thalli from the soil substrate. He found that what appeared to be separate thalli on the soil surface were actually clusters of thalli connected by a network of subterranean rhizomorphs. Microscopic examination showed that the rhizomorphs arose from thick-walled globose hyphae in the lower cortex. Although they were found to contain no algal cells, rhizomorphs were able to extend the fungus over a large area; fungal extensions can presumably become lichenized through chance contact with phycobiont cells released by the parent thallus or other nearby thalli.

SEXUAL AND ASEXUAL REPRODUCTIVE STRATEGIES

Bowler and Rundel (1975) have considered lichen reproductive strategies in terms of two divergent pathways: sexual and asexual (vegetative). Their analysis of the distribution of these pathways in 24 families in the order Lecanorales revealed that:

1. Asexual groups are derived from sexual ancestors.

2. Species rarely produce both sexual and asexual structures on the same thallus, apparently due to energetic constraints.

3. Species with asexual reproductive structures generally have broader geographic distributions than sexual species, a pattern also observed by Hale (1967) and W.L. Culberson (1972).

Bowler and Rundel suggested a number of adaptive advantages of lichen asexuality, which has arisen in practically all lichen groups. Among these are increased survival of propagules and rapid invasion of new habitats.

Another obvious factor favoring vegetative modes of dispersal in lichens is that single requirement characteristic of all lichens—lichenization. Except in a few rare cases in which sexually produced lichen ascospores are released with adhering phycobiont cells (e.g., *Pertusaria pertusa* and *Lecidea limitata* discussed by Pyatt (1973)), sexual propagules disperse only the mycobiont which must then locate and incorporate cells of the appropriate phycobiont. Vegetative modes of dispersal circumvent this requirement by dispersing the mycobiont and phycobiont together in specialized thallus fragments (soredia and isidia). In lichens, therefore, sexual reproduction and dispersal are not as closely coupled as in other groups of organisms.

Williams (1975) has made a number of predictions about the natural history characteristics of sexual and asexual organisms (Table 3.1). A sexual strategy is expected in harsh, unpredictable selective environments; an

TABLE 3.1. Characteristics Predicted for Sexually and Asexually Produced Propagules

Asexual	*Sexual*
Large	Small
Produced continuously	Seasonally limited
Develop close to parent	Widely dispersed
Develop immediately	Dormant
Develop directly to adult stage	Series of developmental stages
Environment and optimum genotype predictable	Environment and optimum genotype unpredictable
Low mortality rate	High mortality rate
Natural selection mild	Natural selection intense

George C. Williams, *Sex and Evolution*. Copyright © 1975 by Princeton University Press. Table 1, p. 4.

asexual strategy is expected in more predictable environments. Since many organisms reproduce both ways under various conditions, apportionment of resources between the two should reflect the degree to which each is successful in the environment.

What we know about lichen reproductive characteristics generally supports Williams' predictions. Sexually produced spores are small and seasonally produced (Pyatt, 1973), whereas vegetative diaspores are large and continuously produced. Also, on the basis of size alone sexual spores are probably dispersed farther than asexual diaspores, although asexual species have wider distributions and there is some evidence that they colonize new substrates earlier than sexual species (Topham, 1977). However, these observations could simply reflect limitations on the dispersability of a sexual species due to the lichenization requirement.

The fact that lichens seldom produce both sexual and asexual reproductive structures has limited the study of mixed strategies in lichens, but these strategies are apparently important in other plant groups. For example, Harper, Lovell, and Moore (1970) found that many seed plants have mixed (sexual and asexual) reproductive strategies, and these are best suited to variable environmental conditions. For some colonizing moss species, Joenje and During (1977) found that both asexual and sexual structures were formed during the life cycle: asexual structures formed during the early phase of colonization and sexual structures formed after colony establishment. Williams (1975) provides a number of other examples.

Lawrey (1980a) found a number of reproductive patterns in the large foliose genus *Parmotrema* that correlated with latitude. Particularly apparent was the fact that temperate species as a group exhibited mixed strategies much more frequently than tropical species. Lawrey suggested that this may reflect strong selection in temperate species for mechanisms that maintain genetic diversity through the sexual process in relatively unpredictable environments, while also facilitating lichenization through the asexual process in habitats where availability of suitable phycobionts is low.

If this explanation is correct, then why are there so few lichens with mixed reproductive strategies? As Bowler and Rundel (1975) have suggested, the likely answer involves the high energetic cost of producting both sexual and vegetative structures; however, nothing is presently known about the relative costs of producing sexual and asexual structures.

Other reproductive patterns in lichens apparently relate in a general way to changes in environmental predictability during succession. Topham (1977) reported an interesting pattern in *Cladonia* species suggesting that species producing larger diaspores invade as woodland succession results in progressively less open habitats. This observation would lend support to Williams' (1975) predictions. Also, *Peltigera* species from disturbed areas apparently devote a larger proportion of thallus area to reproduction than species from more stable woodland habitats (Topham, 1977). This would follow from *r*- and *K*-selection theory (discussed in Chapter 10), one prediction of which is that species with relatively high reproductive efforts occur early in succession and species with relatively high competitive abilities (and consequently low reproductive efforts) occur late in succession. Unfortunately, it is not possible at this time to adequately document the relationship between lichen reproductive strategies and environmental changes during succession, although Topham (1977) is of the opinion that lichens do display *r*- and *K*-strategies during succession.

DISPERSAL

The wide global distribution of many lichens attests to the efficiency of their dispersal mechanisms. However, lichen dispersal in nature is a subject about which we know very little. The main problem in conducting field studies on lichen propagule dispersal is their small size, which also limits the ability of investigators to identify material in the field.

What is known about lichen dispersal comes primarily from laboratory study of sexual and vegetative lichen propagules, including production, discharge, germination, and establishment of lichen spores and diaspores (Pyatt, 1973). Despite the relatively large number of published investigations, however, the work as a whole is inconclusive with numerous inconsistencies, gaps, and contradictions.

This section will briefly review dispersal mechanisms and discuss recent analyses of lichen propagules in terms of dispersal ecology. A relatively recent review by Bailey (1976) should also be consulted.

Dispersal by Wind

Lichen spores, soredia, and thallus fragments have been collected in Hirst traps by several investigators (Pettersson, 1940; Bailey, 1966; 1976; Rudolph, 1970). Du Rietz (1931) observed soredia and thallus fragments in snowfalls and commented that the abundance of the various propagules was more or less proportional to the abundance of species producing them in the region, suggesting that most wind-dispersed propagules traveled relatively short distances.

Wind dispersal has also been studied in the laboratory, although to a far lesser extent for lichens than for nonlichen fungi. Brodie and Gregory (1953), Gregory (1961), and Bailey (1966) have demonstrated that relatively low wind speeds are capable of liberating spores and soredia from various species. Bailey's (1966) experiments also showed that thallus wetness influences the pattern of soredia release. Dry thalli released large numbers of soredia in a single initial burst, whereas wet thalli released soredia slowly over a longer period of time; wet thalli also released more soredia overall than dry thalli.

Dispersal by Water

Bailey (1966) and Brodie (1951) demonstrated that soredia can be dispersed by means of reflected splash droplets. This mechanism appears to be particularly efficient in lichens with cup-shaped fruiting structures (e.g., *Cladonia pyxidata* complex). No studies of splash dispersal of spores have been done.

Water flowing over rock and tree trunk surfaces is probably also responsible for carrying propagules from the parent thallus to suitable colonization sites, although considerable field testing remains to be done. Bailey (1966) demonstrated in the laboratory that soredia can be liberated from thalli of *Lecanora conizaeoides* and *Lepraria incana* by trickling water; he also observed soredia in stem flow collected from *Fraxinus excelsior.*

Armstrong (1978) observed a colonization pattern by *Melanelia glabratula* on a vertical slate rock surface in Wales that suggested a downward transport of propagules by rainwash. Thalli of *M. glabratula* were significantly larger at the top of the rock face than at the bottom, suggesting that they initially colonized the top and gradually spread downward. More recently, Armstrong (1981) found lichen fragments in run-off collected from the rock and conducted a number of experiments to determine what factors influenced their establishment. He found that:

1. Presence of cracks and other irregularities increased the likelihood of establishment.

2. Survivorship of fragments was greatest in cracks and lowest on smooth slate.

3. Isidia from *Xanthoparmelia conspersa* placed on smooth slate survived longer than other fragments.

These results demonstrated that propagules of various types are abundant in rainwash, but that establishment depends upon the availability of safe sites on the substrate and the ability of each propagule to tolerate the specific conditions found at these sites.

Dispersal by Animals

Numerous accounts of lichen propagule dispersal by animals are available in the literature (Gerson, 1973; Bailey, 1976; Gerson & Seaward, 1977; Richardson & Young, 1977). Most are anecdotal, with little experimental evidence to show:

1. Propagules are viable.

2. The probability of propagule establishment is increased as a result of the interaction.

3. As a result of animal dispersal, a lichen's range is extended beyond its normal range or an ecological niche not normally available is opened up.

Given the small size of lichen propagules, they are probably not dispersed any more effectively by animals than by other means. Lichens are opportunistic colonists and are apparently able to achieve long-distance dispersal without need of passive transport. As Armstrong's (1981) study showed, colonization success probably depends most on the suitability of the substrate; there seems to be no lack of propagules reaching most substrates.

It may be, however, that animal dispersal is important to lichens in subtle ways. For example, animals may increase the likelihood of lichenization by dispersing phycobionts over a broader range than would be achieved without them. Support for this idea comes from a report by McCarthy and Healy (1978) that lichen-eating slugs frequently defecate viable algal cells. Phycobionts may also be transported long distances by larger animals. For example, Scharf (1978) found that many birds and mammals acted as passive transporters of lichen phycobionts, including *Trebouxia*.

The importance of participation by animals in lichen dispersal has never been evaluated experimentally, but the problems involved in designing such experiments are considerable. The most rigorous approach requires long-term laboratory experiments in which animal dispersal can be tightly controlled. It is not likely such approaches will soon be taken.

Active Spore Discharge

Despite the number of studies of lichen ascospore discharge,

> ...work so far published is fragmentary, inconclusive and greatly in need of critical reevaluation based on further experimental work. No comparison of discharge from apothecia and perithecia has yet been undertaken in the lichenized fungi and the liberation of basidiospores and conidia from lichenized fungi has not been studied (Bailey, 1976, pp. 224–225).

Several relatively comprehensive reviews are available (Bailey & Garrett, 1968; Garrett, 1971; Pyatt, 1973; Bailey, 1976).

Spore discharge is apparently rhythmic in many lichens, although numerous environmental factors are probably involved and no one has yet been able to show how they interrelate. Pyatt (1968; 1973) observed distinct circadian rhythms of discharge in several species; however, some species discharged spores at night, others during the day. Many investigators have also reported distinct seasonal patterns of spore discharge.

The most important factor appears to be moisture, although there is still no agreement about how this factor specifically influences spore discharge. Pyatt (1973) suggested that any change in the water balance of asci will initiate spore discharge.

Other factors may also be important. Pyatt (1973) found that light stimulated sporulation in some species and suggested that temperature extremes may also have a generally repressive effect. Spore discharge is also reduced in the presence of various air pollutants (Pyatt, 1973), suggesting another way in which lichens can be used as biological monitors of air quality.

Once discharged, lichen ascospores presumably germinate under favorable conditions, although this has seldom been observed in the field. There have been numerous laboratory studies of ascospore germination, but few generalizations can yet be made. Pyatt (1968; 1973) noted that ascospores from crustose species generally had higher percentage viabilities than spores from foliose and fruticose species. He suggested that the higher elevation of the foliose/fruticose species into the turbulent boundary layer of the atmosphere favored dispersal of larger vegetative diaspores. Also, the fact that crustose species seldom produce vegetative diaspores makes them more dependent upon ascospores for dispersal, presumably favoring a greater spore viability or longevity following release. This idea deserves further experimental testing.

4 Isolation and Culture of Lichen Symbionts

The fact that whole lichen thalli are notoriously difficult to maintain under laboratory conditions has led lichenologists to devote much research effort to physiological investigations of cultured lichen algae and fungi. Unfortunately, it is often difficult to relate results obtained with isolated symbionts to the unique and totally different situation that results from lichenization. As in many tightly regulated symbiotic systems, the lichen thallus is a physiologically integrated entity with characteristics quite unlike those of the isolated symbionts. Nevertheless, a complete understanding of lichenization will not be possible without a better understanding of symbiont behavior in isolation. This chapter reviews the physiology of isolated lichen phycobionts and mycobionts, emphasizing isolation and culture techniques, nutritional requirements, and mycobiont secondary compound production.

THE PHYCOBIONT

Isolation

The most common method used to isolate lichen phycobionts is the micromanipulator or micropipette method first described by Jaag (1929). Washed thalli are crushed between two glass slides in sterile water, and algal cells are chosen for culture under a compound microscope. Healthy cells with fungal hyphae still attached are chosen to be sure that the actual phycobiont, not an algal epiphyte, is being removed (Quispel, 1959; Ahmadjian, 1967a; 1967b; 1973b). After transfer by micropipette through a series of four or five successive sterile water washes, cells are inoculated in an appropriate culture medium (Table 4.1).

Isolation of filamentous phycobionts, especially blue-green algae, is accomplished by placing thallus fragments in an inorganic medium in light and allowing the alga to grow away from the thallus. Several transfers of filaments may be required to reduce contamination problems. Care must be taken to ensure that the isolated filaments are not thallus epiphytes.

Richardson (1971) described a centrifugation technique used to isolate larger quantities of algal cells from lichen thalli. It involves grinding up whole thalli in distilled water and centrifuging the suspension at various speeds depending on the size and type of alga. For blue-green algae, the suspension is centrifuged at 375 g for three minutes; the algal portion of the pellet is then resuspended and centrifuged at 125 g for 30 seconds. This procedure is repeated several times, increasing the speed and time up to 90 seconds until the algal suspension is relatively clean. Individual cells are then inoculated onto nutrient agar slants. For green algae, thallus suspensions are centrifuged at 60 g for ten seconds, and the supernatant is recentrifuged at 375 g for five minutes; the algal pellet is resuspended and centrifuged for another five minutes. This technique works best for lichens with large differences in mass between the alga and fungus.

A less reliable isolation method must be employed if single-cell isolates fail to produce colonies in culture. Pieces of whole thallus are surface sterilized with alcohol or a plasmolytica (Bogusch, 1944) and placed in an inorganic medium in light. Obviously there are numerous problems with this technique and it should only be used when others fail.

It is often very difficult to isolate blue-green algae in axenic culture because many have gelantinous sheaths with associated bacterial contaminants. One way to decontaminate these algae is to use ultraviolet light at a high enough intensity to kill the bacteria but not the algae. Henriksson (1951) was able to obtain axenic cultures of *Nostoc muscorum* with UV exposures of 3 and 3.5 minutes. Other workers have used a combination of UV light and antibiotics (Watanabe & Kiyohara, 1963).

The use of a spray-plate method allows the axenic isolation of unicellular green algae. The algae are suspended in sterile water, picked up in a capillary pipette, and atomized by a fine stream of air onto a nutrient agar plate. The plate is observed periodically until bacteria-free colonies are seen. These are transferred to a new nutrient plate or slant. Presence or absence of bacteria in cultures can be assayed by inoculating suspended cultures on nutrient broth agar slants.

Many lichen phycobionts have been isolated and maintained in culture

TABLE 4.1. Media Commonly Used for the Cultivation of Lichen Phycobionts

1. Bold's Basal Medium (BBM)

Six stock solutions (each in 400 ml water):

$NaNO_3$	10 g
$CaCl_2 \cdot 2\ H_2O$	1 g
K_2HPO_4	3 g
KH_2PO_4	7 g
$MgSO_4 \cdot H_2O$	3 g
NaCl	1 g

Addition of 10 ml each of the above six stock solutions to 940 ml water. To this add 1 ml of each of the four following stock trace element solutions.

1. EDTA: Dissolve 50 g EDTA and 31 g KOH in 1 liter water.
2. Stock iron: Dissolve 4.98 g $FeSO_4 \cdot 7H_2O$ in 1 liter acidified water (1 ml conc H_2SO_4 in 999 ml water).
3. Stock boron: Dissolve 11.42 g H_3BO_4 in 1 liter water.
4. Miscellaneous stock: Dissolve, in 1 liter acidified water:

a. $ZnSO_4 \cdot 7\ H_2O$	8.82 g
b. $MnCl_2 \cdot 4\ H_2O$	1.44 g
c. MoO_3	0.71 g
d. $CuSO_4 \cdot 5\ H_2O$	1.57 g
e. $Co(NO_3)_2 \cdot 6\ H_2O$	0.49 g

Addition of 15 g agar to the liter of this inorganic solution yields a satisfactory solid medium.

2. Ahmadjian's *Trebouxia* medium I

BBM (see above)	970 ml
Proteose peptone	10 g
Glucose	20 g

For solid medium, add 15 g agar to 1 liter of medium.

3. Ahmadjian's *Treboxia* medium II

BBM (see above)	980 ml
Casamino acids (vitamin free)	10 g
Glucose	10 g

For solid medium, add 15 g agar to 1 liter of medium.

V. Ahmadjian, 1967a. In *The Lichen Symbiosis.* Waltham, MA: Blaisdell. Reprinted with permission of the author.

for use by investigators. Ahmadjian (1980b) listed these sources of lichen phycobionts:

1. Culture Collection of Algae at the University of Texas, Austin, Texas, U.S.A.

2. Culture Centre of Algae and Protozoa, Institute of Terrestrial Ecology, Cambridge, England.

3. Culture Collection of Lichen Symbionts at Clark University, Worcester, Massachusetts, U.S.A.

Interested investigators may obtain pure cultures from these sources at minimal cost.

Culture

The cultural requirements of lichen phycobionts are not very different from those of free-living algae (Table 4.1). Blue-green and green phycobionts can be maintained in standard defined media like Bold's Basal Medium. Some lichen algae, especially *Trebouxia,* grow better if a carbon source is supplied. The reason for this is not clear, but may involve a general sensitivity to fluctuations in the physical environment. Lange (1953), for example, found that isolated lichen phycobionts exhibited elevated sensitivities to heat and drying.

Trebouxia and *Pseudotrebouxia* species are also known to be very sensitive to light. Fox (1966) observed bleaching of phycobiont cells of *Ramalina ecklonii* at light intensities of 2000 lux. Intensities below 1000 lux eliminated this problem. Ahmadjian (1967a) recommended even lower light intensities (< 200 ft.-c.) or the use of organic media in total darkness. Fox (1966) also found that high temperatures ($> 26°$ C), unaerated inorganic liquid media or old, glucose-peptone media caused cell bleaching. Optimal pH varies for different phycobionts; however, *Trebouxia* species apparently tolerate wide ranges of pH (Quispel, 1943–45; Mish, 1953). Temperature optima vary from 4° to 24° C depending on the species, its position in the thallus (Thomas, 1939), and the climatic conditions under which it naturally grows (Ahmadjian, 1962b). There is generally a great deal of heterogeneity in nutrition, growth rate, and temperature optima exhibited by clonal cultures of *Trebouxia* and *Pseudotrebouxia* species isolated from a single lichen thallus (Ahmadjian, 1960b). This heterogeneity suggests a broad physiological plasticity, but it needs to be better documented.

Lichen phycobionts, particularly *Trebouxia* species, grow very slowly in culture and can be transferred less frequently than other algae. Phycobiont growth is inhibited on old media (Fox, 1966), suggesting that autoinhibitors are produced. Archibald (1977) found that both *Trebouxia* and *Pseudotre-*

bouxia species generally grew better on agarized media than in liquid media. Nutritional requirements will be considered in the following sections.

Distribution of Phycobionts

Of the 29 genera of lichen phycobionts known, most are green algae (Table 4.2). The overwhelming majority of green phycobionts are *Trebouxia* or *Pseudotrebouxia* species (Figures 4.1c and 4.1d). These two genera have been discussed by Archibald (1975; 1977) and Hildreth and Ahmadjian (1981). In tropical lichen associations, *Trentepohlia* and *Phycopeltis* species are common phycobionts (Ahmadjian, 1967a). *Nostoc* and *Scytonema* species make up most of the known blue-green phycobionts. Only a few algal genera besides blue-greens and greens are known to participate in lichen associations. One is *Heterococcus* (= *Monocilia*), a yellow-green alga associated with *Verrucaria;* another is *Petroderma,* a marine brown alga also found in *Verrucaria.*

Nutritional Requirements

Carbon Utilization. The nutritional requirements of lichen algae have been discussed by Quispel (1959), Ahmadjian (1977; 1978), and Archibald (1977), among others.

Early culture work with *Trebouxia* species showed that little or no growth occurred on inorganic media in light. Addition of sugars and/or peptone increased growth, both in darkness and light (Quispel, 1959). Archibald (1977) compared the carbon utilization patterns of *Trebouxia* and *Pseudotrebouxia* species in light and darkness and found that most species could use glucose, although a number of *Pseudotrebouxia* species were found to grow better on fructose (Table 4.3). Growth of most species was the same in light and darkness, though some species (*T. erici, T. crenulata* from *Parmelia*) preferred dark conditions and some species (*T. arboricola* and *T. crenulata* from *Xanthoria*) preferred light conditions. The difference in behavior between *T. crenulata* from *Parmelia* and the same species from *Xanthoria* suggests a certain degree of physiological plasticity inherent in many of these algae, possibly mycobiont induced; other behavioral differences were observed in identical species isolated from different lichens.

Blue-green lichen phycobionts are best cultured in inorganic media in light. Their behavior is similar to that of free-living species (Quispel, 1959). Attempts to cultivate blue-green phycobionts on organic media in darkness have been unsuccessful, suggesting that these algae are obligate autotrophs. Far less nutritional work has been done with these algae, however.

Nitrogen Utilization. Quispel (1959) found that most lichen algae grew better on peptone than on inorganic sources of nitrogen, although there does

TABLE 4.2. Algal Genera Known to Participate as Primary Symbionts in Lichen Associations

Algal Genera	*Fungal Associations*
I. Cyanophyta (blue-green algae)	
A. Chroococcales	
1. *Chroococcus*	*Phylliscum, Pyrenopsidium*
2. *Gloeocapsa*	*Gonohymenia, Peccania, Psorotichia, Pyrenopsis, Synalissa, Thallinocarpon, Thyrea*
B. Chamaesiphonales	
1. *Hyella*	*Arthopyrenia*
C. Hormogonales	
1. *Calothrix*	*Calotrichopsis, Lichina, Porocyphus*
2. *Dichothrix*	*Lichina, Placynthium*
3. *Nostoc*	*Hydrothyria, Lempholemma,* Arctomiaceae, Collemataceae, Pannariaceae, Peltigeraceae
4. *Scytonema*	*Coccocarpia, Cora, Dictyonema, Erioderma, Heppia, Koerberia, Lichenothrix, Lichinodium, Parmeliella, Petractis, Polychidium, Thermutis, Vestergrenopsis, Zahlbrucknerella*
5. *Stigonema*	*Ephebe, Spilonema*
II. Chlorophyta (green algae)	
A. Chlorococcales	
1. *Chlorococcum*	*Conotrema, Bacidia*
2. *Chlorella*	*Calicium, Lecidea, Lepraria*
3. *Coccomyxa*	*Baeomyces, Coriscium, Epigloea, Icmadophila, Multiclavula, Omphalina, Peltigera, Solorina*
4. *Gloeocystis*	*Gloeolecta, Gyalecta*
5. *Myrmecia*	*Bacidia, Catillaria, Dermatocarpon, Lecidea, Phlyctis, Psoroma, Sarcogyne, Verrucaria, Lobaria*
6. *Pseudochlorella*	*Lecidea*

Alga	Lichen genera/families
7. *Trebouxia* and *Pseudotrebouxia*	*Alectoria, Buellia, Caloplaca, Cetraria, Cladonia, Conotrema, Lecanora, Lecidea, Parmelia, Physcia, Ramalina, Stereocaulon, Umbilicaria, Usnea, Xanthoria*
8. *Trochiscia*	*Polyblastia*
B. Ulotrichales	
1. *Cephaleuros*	*Raciborskiella, Strigula*
2. *Chlorosarcina*	*Lecidea*
3. *Coccobotrys*	*Lecidea, Verrucaria*
4. *Leptosira*	*Thrombium*
5. *Phycopeltis*	*Arthonia, Mazosia, Opegrapha, Porina, Trichothelium*
6. *Physolinum*	*Coenogonium*
7. *Pleurococcus*	*Dermatocarpon, Endocarpon, Lecidea, Maronella, Polyblastia, Staurothele, Thelidium, Verrucaria*
8. *Pseudopleurococcus*	*Verrucaria*
9. *Stichococcus*	*Calicium, Chaenotheca, Coniocybe, Lepraria*
10. *Trentepohlia*	*Chaenotheca, Cystocoleus,* Arthoniaceae, Dirinaceae, Graphidaceae, Gyalectaceae, Lecanactidaceae, Opegraphaceae, Pyrenulaceae s.l., Roccellaceae
III. Xanthophyta (yellow-green algae)	
A. Heterotrichales	
1. *Heterococcus*	*Verrucaria*
IV. Phaeophyta (brown algae)	
A. Ectocarpales	
1. *Petroderma*	*Verrucaria*

A. Henssen & H.M. Jahns. 1974. *Lichenes. Eine Einführung in die Flechtenkunde.* Stuttgart: Georg Thieme Verlag. Reprinted with permission of the authors.

TABLE 4.3. Growth of Species of *Trebouxia* and *Pseudotrebouxia* after Four Weeks in 1 ×N Bold's Basal Medium with Selected Sources of Carbon in the Light (L) and Dark (D)

		Glucose		Fructose	
Origin	*Species*	*L*	*D*	*L*	*D*
Unknown	*T. arboricola*	G*	F	G	F
Cladonia	*T. erici*	F	G	F	F
	T. pyriformis	G	G	F	—**
Parmelia	*T. crenulata*	F	E	F	—
	T. anticipata	F	F	F	F
	T. gelatinosa	F	F	T	T
Physcia	*T. flava*	0	0	0	0
Pilophorus	*T. magna*	G	G	F	—
Stereocaulon	*T. pyriformis*	G	G	F	—
	T. excentrica	G	G	G	G
	T. glomerata	G	G	F	T
Xanthoria	*T. crenulata*	G	T	G	T
	T. italiana	F	F	F	—
Buellia	*P. decolorans*	F	F	G	G
Lecanora	*P. incrustata*	F	F	F	T
	P. potteri	G	G	G	G
Physcia	*P. impressa*	G	G	G	G
Xanthoria	*P. decolorans*	F	F	F	F
Free-living	*P. corticola*	F	T	F	T

*E = excellent, G = good, F = fair, T = trace, 0 = no growth.

**Information not available.

P.A. Archibald. 1977. "Physiological characteristics of *Trebouxia* (Chlorophyceae, Chlorococcales) and *Pseudotrebouxia* (Chlorophyceae, Chlorosarcinales)." *Phycologia* 16: 295–300. Reprinted with permission of Blackwell Scientific Publications.

not seem to be a specific requirement for organic nitrogen. Ahmadjian (1977) summarized the inorganic nitrogen requirements of lichen phycobionts (Table 4.4) and found that ammonium was generally preferred to nitrate, although some algae had poor utilizations of both. Farrar (1976d) has suggested that fungal preference for ammonium may allow nitrate use by the alga, since mycobionts do not appear to use nitrate as efficiently. If this is true, it may follow that phycobionts unable to use nitrate efficiently (Table 4.4) are somehow able to obtain organic or environmental sources of nitrogen. This idea needs experimental testing.

Archibald (1977) found that *Pseudotrebouxia* species tended to be more

TABLE 4.4. Utilization of Ammonium and Nitrate by Algae Isolated from Various Lichen Thalli. Utilization Determined by Comparing Growth in Presence of Nitrogen Source to Growth in a Nitrogen-Free Medium.

Origin of Alga	*Ammonium*	*Nitrate*
Trebouxia Symbiont		
Alectoria implexa	+++*	+
Anaptychia ciliaris (= *Physcia ciliaris)*	+++	+
Buellia pernigra	+	+
Caloplaca cerina	—	+++
C. murorum	—	+++
Chaenotheca chrysocephala	+++	+++
Cladonia bacilliformis v. *irregularis* (= *C. irregularis)*	+++	—
C. cristatella	+++	+++
C. digitata	—	+++
C. endiviaefolia	+++	—
C. fimbriata	—	+
C. pyxidata	+++	+++
C. squamosa	—	+++
Parmelia acetabulum	+++	+++
P. caperata	+++	+++
P. saxatilis	+++	+++
P. scortea	+	+
P. sulcata	+	—
Physconia pulverulenta (= *Physcia pulverulenta)*	+++	+
Polycauliona citrina	+	+
Pseudevernia furfuracea (= *Parmelia furfuracea)*	+	+
Ramalina fraxinea	+++	+
Stereocaulon dactylophyllum	+++	+

TABLE 4.4. *continued*

Origin of Alga	*Ammonium*	*Nitrate*
Umbilicaria deusta	+	+
Xanthoria parietina	+++	+++
Chlorella Symbiont		
Calicium chlorinum	+++	+++
Coccomyxa Symbiont		
Peltigera aphthosa	+++	+++
Solorina sp.	+++	+++
Stichococcus Symbiont		
Chaenotheca stemonea	+++	+++
Coniocybe furfuracea	+++	+++
Dermatocarpon miniatum	+++	+++

*+++ = strongly utilized; + = poorly utilized.

Modified from Ahmadjian, 1977. Reprinted with permission from *CRC Handbook Series in Nutrition and Food,* M. Rechcigl, Jr., ed. Copyright 1977, CRC Press, Boca Raton, FL. and the author.

selective in their nitrogen utilization than *Trebouxia* species. None of the species she studied was able to use urea, although other organic sources of nitrogen were used. Ahmadjian's (1977) survey of phycobiont nutritional requirements showed that only one species, *Polycauliona citrina,* was able to use urea (Schofield & Ahmadjian, 1972); utilization was described as poor. For some as yet unknown reason, *Pseudotrebouxia* species use casamino acids better than other organic sources.

Archibald (1977) also found that the same phycobiont species isolated from different lichens sometimes had different nitrogen preferences. For example, *Trebouxia magna* isolated from *Parmelia* was more tolerant of various nitrogen sources than the same species from *Pilophorus.* Also, whereas *T. crenulata* from *Xanthoria* was more sensitive to carbon sources, *T. crenulata* from *Parmelia* was apparently more sensitive to nitrogen sources. Archibald (1977) frequently found greater physiological similarity between different phycobionts isolated from the same lichen than between the same phycobiont from different lichens. This apparent physiological plasticity should be investigated to see if lichenization processes somehow change phycobiont behavior.

Growth Factors

Quispel (1943–45) studied the influence of various growth substances on isolates of *Trebouxia* from *Xanthoria parietina, Physçia pulverulenta,* and *Parmelia acetabulum* on organic medium in darkness and in light. Under heterotrophic conditions, no requirement for growth substances was indicated; however, growth was increased in the presence of either yeast autolysate or mycobiont homogenates. In some cases, high concentrations of nicotinic acid stimulated cell growth; even better growth was obtained when nicotinic acid was combined with other compounds (m-inositol, thiamine, pyridoxine, β-alanine, pantothenic acid, and biotin). In light, algae hardly grew at all. Growth was found to be stimulated by high concentrations of ascorbic acid and a closely related compound, dioxymaleic acid. Isolated lichen mycobionts are not known to produce ascorbic acid, but there has been a report of ascorbic acid production by two arctic lichens (Gustafson, 1954).

The entire field of lichen phycobiont growth factors needs more experimental investigation. There are a number of interesting observations, but far too little information exists to even speculate on the role these substances may play in regulating symbiont interactions in whole thalli. As an example, Ahmadjian (1967a) cited a number of instances of phycobiont growth stimulation by yeast extract, indolacetic acid (IAA), and streptomycin. Indolacetic acid has also been shown to increase growth of *Trebouxia albulescens* from *Xanthoria parietina* and *T. humicola,* a free-living strain (Di Benedetto & Furnari, 1962). On the other hand, other *Trebouxia* strains are known to be inhibited by IAA (Leonian & Lilly, 1937; Zehnder, 1949). With such seemingly contradictory evidence, it is difficult to generalize about the role of IAA; the same is true for other presumed algal growth factors.

THE MYCOBIONT

Isolation

The most reliable mycobiont isolation method involves the use of ascospores (Ahmadjian, 1961; 1973b). The ascocarp is washed, removed from the thallus, and attached with petroleum jelly to a petri plate cover. The cover is then placed over the bottom that contains a plain agar gel. Spores are discharged onto the surface of the agar and can be observed without removing the cover by placing the entire plate under a dissecting or compound microscope (Figure 4.1a and b). The plate is sometimes inverted allowing spore discharge up onto the agar surface to minimize contamination. However, care must be taken to allow a suitable distance between the ascocarp and the agar surface.

Freshly collected thalli are generally best for isolation, but spores can be

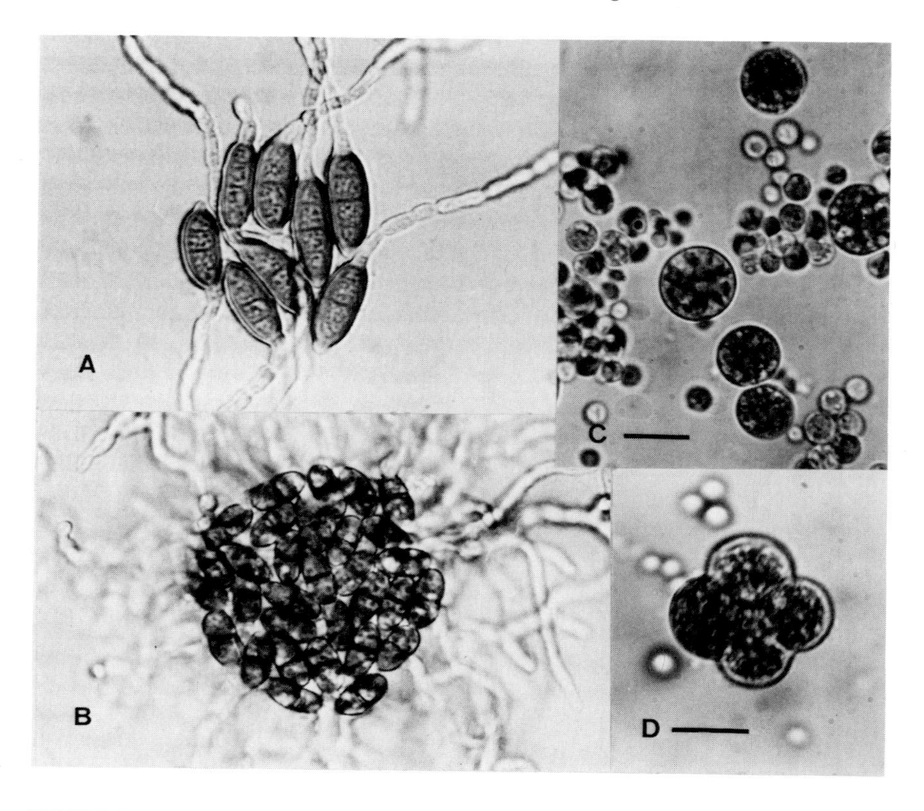

FIGURE 4.1. Isolated lichen symbionts. (a) Polyspore isolate of *Pyxine eschweileri* (×400); (b) Polyspore isolate of *Buellia schaereri* (×400); (c) *Trebouxia excentrica* from *Cladonia bacillaris,* bar = 10 μm; (d) *Pseudotrebouxia aggregata* from *Lecanora fuscata,* bar = 10 μm. Figures 4.1c and 4.1d provided by V. Ahmadjian.

collected from specimens up to eight months (Ahmadjian, 1967a); viable spores have even been discharged from air-dried material stored up to two years (Lawrey, unpublished). Discharge is much less reliable in older material, however.

Vobis (1977) has observed conidiospore germination from *Calicium adspersum, Lecanactis abietina,* and *Opegrapha vermicellifera,* demonstrating that pycnidia function in asexual mycobiont reproduction as well as in gamete transfer during sexual reproduction (see discussion in Chapter 3). The importance of this method as an isolation technique is probably limited, however, since conidia from 12 of the 15 species Vobis worked with did not germinate. It would be interesting to compare cultural characteristics of

TABLE 4.5. Media Commonly Used in Cultivation of Lichen Mycobionts

1.	Malt yeast agar:	
	Malt extract	20 g
	Yeast extract	2 g
	Agar	20 g
	Distilled water	1000 ml
2.	Ahmadjian's modified Lilly and Barnett medium:	
	Dextrose	10 g
	K_2HPO_3	1 g
	$MnSO_4 \cdot 4\ H_2O$	1.1 g
	$MgSO_4 \cdot 7\ H_2O$	0.5 g
	$FeCl_4 \cdot 6\ H_2O$	0.2 g
	$ZnSO_4 \cdot 7\ H_2O$	0.2 g
	Biotin	5 μg
	Thiamine	100 μg
	Distilled water	1000 ml

V. Ahmadjian, 1967a. In *The Lichen Symbiosis*. Waltham, MA: Blaisdell. Reprinted with permission of the author.

mycobiont cultures obtained via sexual and asexual spores to see if greater variability is observed in sexually derived cultures. Ahmadjian (1964) has documented a great deal of variability in single ascospore cultures of *Cladonia cristatella*. One would expect less variability in conidiospore cultures.

When ascospore germination is observed on plain agar, spores can be transferred to an appropriate nutrient medium (Table 4.5) by cutting and tranferring small blocks of agar containing obviously viable spores. If single spore cultures are desired, single spores must be transferred with a dissecting needle.

Problems with this standard mycobiont isolation procedure are:

1. The lichen does not produce spores.
2. Spores are not discharged.
3. Discharged spores do not germinate, or if they germinate, mycelial growth stops shortly after germination.

Many lichens produce and discharge ascospores that never germinate under culture conditions. Fewer than half the species Ahmadjian (1961; 1964) worked with produced mycobiont colonies in culture. Even if viable spores are produced, however, many species exhibit delayed spore germination. For example, Fox (1966) found that at least five weeks were required for germination of *Ramalina ecklonii* spores. *Cetraria ciliaris* spores did not germinate until they had been left in culture six weeks (Lawrey, unpublished).

Attempts to break spore dormancy by altering pH, temperature, and culture medium have been largely unsuccessful (Ahmadjian, 1967a). Extracts of phycobionts often stimulate spore germination; also, bark extracts and compounds such as erythritol, pectin and glycerol hasten spore germination in *Xanthoria parietina* (Am Ende, 1950).

If spores are produced but not discharged, it is possible to squash apothecia and micropipette washed spores to an agar surface. However, there are usually contamination problems with this procedure. If spores are not produced, or if discharged spores never germinate, then the only way to isolate the mycobiont is by transfer of thallus fragments to an agar medium (Ahmadjian, 1967a; 1973b). Pieces of thallus pulled from the medulla of fresh specimens are surface sterilized in sodium hypochlorite or alcohol and placed onto an agar surface. Contamination by bacterial and fungal epiphytes is a common problem with this technique; it is also impossible to say with certainty that the fungus isolated with this procedure is the lichen mycobiont.

As is the case with lichen phycobionts, mycobionts can be obtained from culture collections. Ahmadjian (1980b) lists the following sources:

1. American Type Culture Collection, Rockville, Maryland, U.S.A.

2. Culture Collection of Lichen Symbionts at Clark University, Worcester, Massachusetts, U.S.A.

3. Culture Collection of Mycobionts at George Mason University, Fairfax, Virginia, U.S.A.

Isolated lichen mycobionts may be obtained from these sources at minimal cost.

Culture

Standard defined and complex culture media used for lichen mycobionts are shown in Table 4.5. Not all lichen mycobionts do well on nutrient agar. For example, Ahmadjian (1973b) found that *Endocarpon pusillum* grows best on media with limited organic nutrients; a malt-yeast agar medium inhibits the growth of this mycobiont.

Mycobionts tolerate wide variations in temperature, but do best at around 20° C. They also exhibit various pH optima. Ahmadjian (1961) and Thomas (1939) found that most species exhibited pH optima between 4.5 and 6.5.

Because mycobiont growth rates are very slow compared to other fungi in culture, transfer of isolates is usually not required more than once or twice a year—even when cultures are maintained under optimal growth conditions. This slow growth limits the use of mycobionts in nutritional and physiological studies in the laboratory.

Nutrition

Carbon Sources. Ahmadjian (1977) has summarized the information presently available on carbon utilization by lichen mycobionts. Uses of the five sugars most commonly studied are shown in Table 4.6. Maltose is used most by almost all mycobionts studied; only *Stereocaulon saxatile* exhibits poor utilization of this sugar. Other sugars are used less than maltose for the most part, but can be used nonetheless. The only sugar not used at all was lactose, but only for two mycobionts, *Buellia stillingiana* and *Xanthoria parietina.* It appears from the information available that lichen mycobionts are capable of growth on a wide variety of carbon sources (Quispel, 1959).

In addition to simple sugars, a number of complex carbohydrates can be utilized by isolated mycobionts. Ahmadjian (1964) observed visible growth of *Anthracothecium* sp., *Arthonia* sp., *Graphina bipartita,* and *G. virginalis* on filter paper, but was unable to show that the fungi exhibited cellulolytic activity. He also observed growth of *Arthonia* sp. and *G. virginalis* on a soluble starch substrate. The ability of lichen mycobionts to digest cellulose would suggest that at least short-term saprophytic nutrition is possible.

TABLE 4.6. Utilization of Various Carbon Sources by Isolated Lichen Mycobionts

Species of Origin	*Glucose*	*Lactose*	*Maltose*	*Sorbose*	*Sucrose*
Acarospora fuscata	+++*	+	+++	+	+++
Anthracothecium sp.	+++	+++	+++	+	+
Arthonia sp.	+	+	+++	+++	+
Buellia stillingiana	+++	0	+++	—**	+++
Caloplaca decipiens	+	+	+++	+	+
Cladonia cristatella	+++	+	+++	+	+++
Collema tenax	+++	+++	+++	—	+
Glyphis lepida	+	+++	+++	+++	+++
Graphina bipartita	+	+++	+++	+++	+
G. virginalis	+++	+++	+++	+	+++
Ramalina ecklonii	+++	—	—	—	+++
Sarcographa rechingeri	+	+++	+++	+++	+++
Sarcographina sandwicensis	+++	+++	+++	+++	+++
Sarcogyne similis	+++	+++	+++	+	+++
Stereocaulon saxatile	+	+	+	+++	+++
S. vulcani	+	+++	+++	+	+++
Xanthoria parietina	+++	0	+++	—	—

*+++ = utilized; + = poorly utilized; 0 = not utilized.

**Information not available.

V. Ahmadjian, 1977. Reprinted with permission from *CRC Handbook Series in Nutrition and Food,* M. Rechcigl, Jr., ed. Copyright 1977, CRC Press, Boca Raton, FL. and the author.

Nitrogen Nutrition. Farrar (1976d) observed that fungi generally exhibit a more efficient utilization of ammonium than nitrate. If this is true of lichens, then trapped algal cells in a lichen thallus should be able to obtain inorganic nitrogen as nitrate. Table 4.7 demonstrates that lichen mycobionts do utilize ammonium more efficiently than nitrate, but relatively few studies of this sort have been done to date.

Although synthesis of urease has been observed in both phycobionts and mycobionts (Xavier Filho & Vicente, 1978; Vicente & Xavier Filho, 1979), urea was not found to be used by cells of *Trebouxia* and *Pseudotrebouxia* species studied by Archibald (1977). Some mycobionts use this nitrogen source (Ahmadjian, 1977), and urease activity is known for free-living lichens (Moissejeva, 1961; Shapiro, 1977). Ahmadjian (1966b) suggested that hydrolysis of urea to ammonium and carbon dioxide could influence the metabolism of phycobiont cells, providing them with nitrogen and CO_2, thereby increasing photosynthetic rates. Urease activity has also been found to be regulated to a certain extent by caperatic acid produced by the mycobiont of *Parmelia roystonea* (Xavier Filho & Vicente, 1978), which suggests that the fungus controls certain aspects of algal metabolism.

Amino acids are used by lichen mycobionts as nitrogen sources (Gross & Ahmadjian, 1966) and can influence growth rates and pigment production. The amino acids used most frequently are alanine, arginine, asparagine, glutamic acid, and proline; those not frequently used are methionine, phenylalanine, and tryptophane. It is not known if these compounds are broken down by the fungi before being incorporated into fungal proteins;

TABLE 4.7. Utilization of Nitrogenous Substrates by Isolated Lichen Mycobionts.

Species of Origin	*Urea*	*Ammonium*	*Peptone*	*Nitrate*
Acarospora fuscata	+++*	+++	+	+++
Cladonia cristatella	+++	+++	+++	+++
Lecanora tephroeceta	+++	+++	0	+
Lecidea macrocarpa (= *L. steriza*)	—**	—	—	+
Ramalina ecklonii	—	+++	—	+
Sarcogyne similis	—	—	—	+++
Stereocaulon saxatile	0	+++	+++	+++
Xanthoria parietina	0	+++	+++	+++

*+++ = utilized; + = poorly utilized; 0 = not utilized.

**Information not available.

V. Ahmadjian, 1977. Reprinted with permission from *CRC Handbook Series in Nutrition and Food.* M. Rechcigl, Jr., ed. Copyright 1977, CRC Press, Boca Raton, FL. and the author.

however, there appears to be no requirement for exogenously supplied amino acids.

Vitamins and Growth Factors. Additions of vitamins are known to influence the growth of mycobionts in culture (Tobler, 1944; Zehnder, 1949; Quispel, 1943–45; Hale, 1958; Ahmadjian, 1961; 1964). Ahmadjian (1961) plotted mycobiont growth curves of *Acarospora fuscata* (Auburn 1) in glucose-asparagine media with various vitamin supplements (Figure 4.2). Biotin and thiamine were found to be particularly effective in increasing mycobiont growth. Ahmadjian (1977) summarized the information presently available on biotin and thiamine requirements of isolated lichen fungi (Table 4.8). Only one fungus tested, *Stereocaulon tomentosum,* was able to grow without exogenously supplied biotin and thiamine. Most mycobionts required both vitamins for growth. Where do they obtain these vitamins? Is the phycobiont somehow involved?

Quispel (1943–45) observed that the mycobiont of *Xanthoria parietina* grew much better on media without added growth substances if algae had

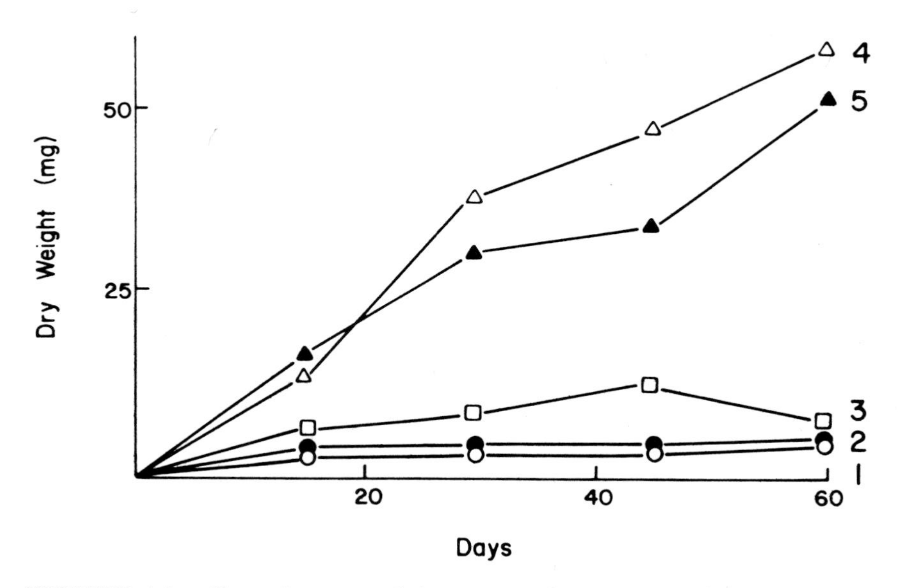

FIGURE 4.2. Growth curves of *Acarospora fuscata* mycobionts in glucose-asparagine vitamin supplement media at 18° C. Additions: 1—control (no additions); 2—thiamine; 3—biotin; 4—thiamine + biotin; 5—thiamine, biotin, inositol + pyridoxine. From V. Ahmadjian. 1961. "Studies on lichenized fungi." *Bryologist* 64:168–179. Redrawn with permission of the American Bryological and Lichenological Society.

TABLE 4.8. Thiamine and Biotin Requirements of Isolated Lichen Mycobionts

Species of Origin	*Thiamine*	*Biotin*
Acarospora fuscata	R*	R
Buellia stillingiana	R	R
Caloplaca aurantiaca	N	—**
C. decipiens	R	R
Collema tenax	R	N
Graphina virginalis	R	R
Graphis scripta	N	—
Hypogymnia physodes	R	R
Lecanora muralis (= *Placodium saxicola*)	N	—
Lecidea elaeochroma (= *L. parasema*)	N	—
Physconia pulverulenta	N	—
Ramalina ecklonii	R	R
Sarcographina sanwicensis	R	R
Sarcogyne similis	N	R
Stereocaulon tomentosum	N	N
S. vulcani	R	R

*R = required; N = not required.

**Information not available.

V. Ahmadjian. 1977. Reprinted with permission from *CRC Handbook Series in Nutrition and Food,* M. Recheigl, Jr., ed. Copyright 1977. CRC Press, Boca Raton, FL. and the author.

been cultivated on media *before* fungal inoculation, leading him to conclude that the algae were capable of producing whatever growth substances were required. Algae are known to produce many growth factors; for example, Taylor and Wilkinson (1977) have documented production of gibberellins and gibberellinlike substances by numerous free-living algae. Apparently, gibberellins do not influence mycobiont growth, however (Hess, 1959a; 1959b; Furnari & Luciani, 1962).

Even if they do not produce growth factors, phycobionts may produce vitamins required by the mycobiont. The isolated phycobiont of *Peltigera aphthosa,* a *Coccomyxa* species, was found to produce and excrete 16 times more biotin into the culture medium than the free-living *Chlorella pyrenoidosa* (Bednar, 1963; Bednar & Holm-Hansen, 1964), and Zehnder (1949) found that *Trebouxia* and *Coccomyxa* species excrete thiamine (Ahmadjian, 1966b). Henriksson (1961) has also observed vitamin excretion into the culture medium by the blue-green phycobiont of *Collema.*

Lichen fungi are also known to produce auxins or auxinlike substances. Fortin and Thibault (1972) used purified extracts of lichenized *Cladonia alpestris* and isolated *Acarospora fuscata, Lecanora cinerea,* and *Stereocaulon vulcani* in coleoptile elongation tests of auxinic activity. The isolated mycobionts were cultured on tryptophane-rich media. All extracts induced significant coleoptile elongation, suggesting that mycobionts are capable of auxinic activity. It is unclear what the significance of these results is in terms of the lichen symbiosis; however, control of algal growth by the fungus is clearly possible if auxins are synthesized and released in the lichen thallus.

At the risk of generalizing beyond the available data, it appears that the growth of each symbiont can be stimulated by extracellular growth factors produced by the associated symbiont. This is obviously an area worthy of future concentrated effort.

Secondary Compound Production by Mycobionts

Whole thallus chemistry is one of the most interesting and rapidly developing areas of experimental lichenology. Despite the wealth of information on the synthesis and phylogenetic significance of secondary compounds in lichenized thalli, however, almost nothing is known about the chemistry of isolated mycobionts. There are a number of reasons for this. Isolated mycobionts seldom produce compounds in culture like those found in lichenized thalli. In ten years of analyzing for typical lichen substances, Ahmadjian (1980a) has never observed chemical production by mycobionts alone. There have been reports of chemical production by mycobionts (Ahmadjian, 1980a in Table 4.9), including an observation that low temperatures seem to facilitate compound production in mycobionts (Komiya & Shibata, 1969). However, these appear to be isolated cases; it is presently not understood why observed chemical production by mycobionts can almost never be duplicated. The fact that it does occur in isolated mycobionts demonstrates that the phycobiont is not necessary for production, but it is clear that typical compound production is somehow facilitated by lichenization.

Culberson and Ahmadjian (1980) have hypothesized the existence of important biochemical interactions between symbionts that influence compound production (Figure 4.3). For example, there is some indirect evidence to suggest that the lichen alga secretes an inhibitor to the production of certain fungal phenolic compounds. Inhibition of this pathway results in the accumulation of phenolic acid precursors that are converted by the fungus to characteristic lichen products. If the alga is removed or disturbed, the inhibitor is no longer secreted and phenolic precursors are used to synthesize fungal phenolic compounds. This may be adaptive, since fungal phenolics are more water soluble and may be able to function as antimicrobial or antiherbivore compounds. Characteristic lichen compounds are relatively

TABLE 4.9. Reports of Lichen Substances Synthesized by Mycobionts in Culture

Lichen Mycobiont	*Substances Produced*	*Source*
Candelariella vitellina	Pulvinic acid, pulvinic dilactone, calycin, vulpinic acid	Thomas (1939); Mosbach (1969)
Xanthoria and *Caloplaca* spp.	Parietin, erythritol, emodin, teloschistin, fragilin, fallacinal, 2-chloroemodin	Thomas (1939); Tomaselli (1957); Piatelli & de Nicola (1968); Nakano et al. (1972)
Cladonia macilenta, C. pyxidata, C. chlorophaea	Orsellinic acid, haematomic acid	Hess (1959)
Lecanora rupicola	Roccellic acid, eugenitol, eugenitin, sordidone	Fox & Huneck (1969)
Lecanora chlarotera	Gangaleoidin	Santesson (1969)
Cladonia cristatella	Usnic acid, didymic acid, rhodocladonic acid	Castle & Kubsch (1949)
Ramalina crassa, R. yasudae	Usnic acid, salazinic cid	Komiya & Shibata (1969)
Bombyliospora japonica, Anaptychia hypoleuca	Zeorin	Ejiri & Shibata (1974); Shibata (1974)
Cladonia crispa	Squamatic acid	Ejiri & Shibata (1975)
Acroscyphus sphaerophoroides	Calycin, skyrin	Shibata (1974)
Cladonia bellidiflora	Bellidiflorin, skyrin	Shibata (1974)

Reprinted from "Separation and Artificial Synthesis of Lichens," by Vernon Ahmadjian, in *Cellular Interactions in Symbiosis and Parasitism,* edited by Clayton B. Cook, Peter W. Pappas, and Emanuel D. Rudolph (Ohio State University Biosciences Colloquia, no. 5).

water-insoluble and probably have to be hydrolyzed to become biologically active. There is more speculation on this subject in Chapter 9.

Umezawa et al. (1974) suggested that lichen compounds are not observed in isolated mycobiont cultures because they are readily hydrolyzed. Lecanoric acid produced by a species of the free-living fungus *Pyricularia* in culture is hydrolyzed to orsellinic acid and orcinol. An accumulation of simple phenolic acid precursors in the medium, however, could be explained by Culberson and Ahmadjian's (1980) hypothesis as well.

Many isolated lichen mycobionts produce pigments and other compounds that are not found in the lichenized thallus (Ahmadjian, 1964). Some

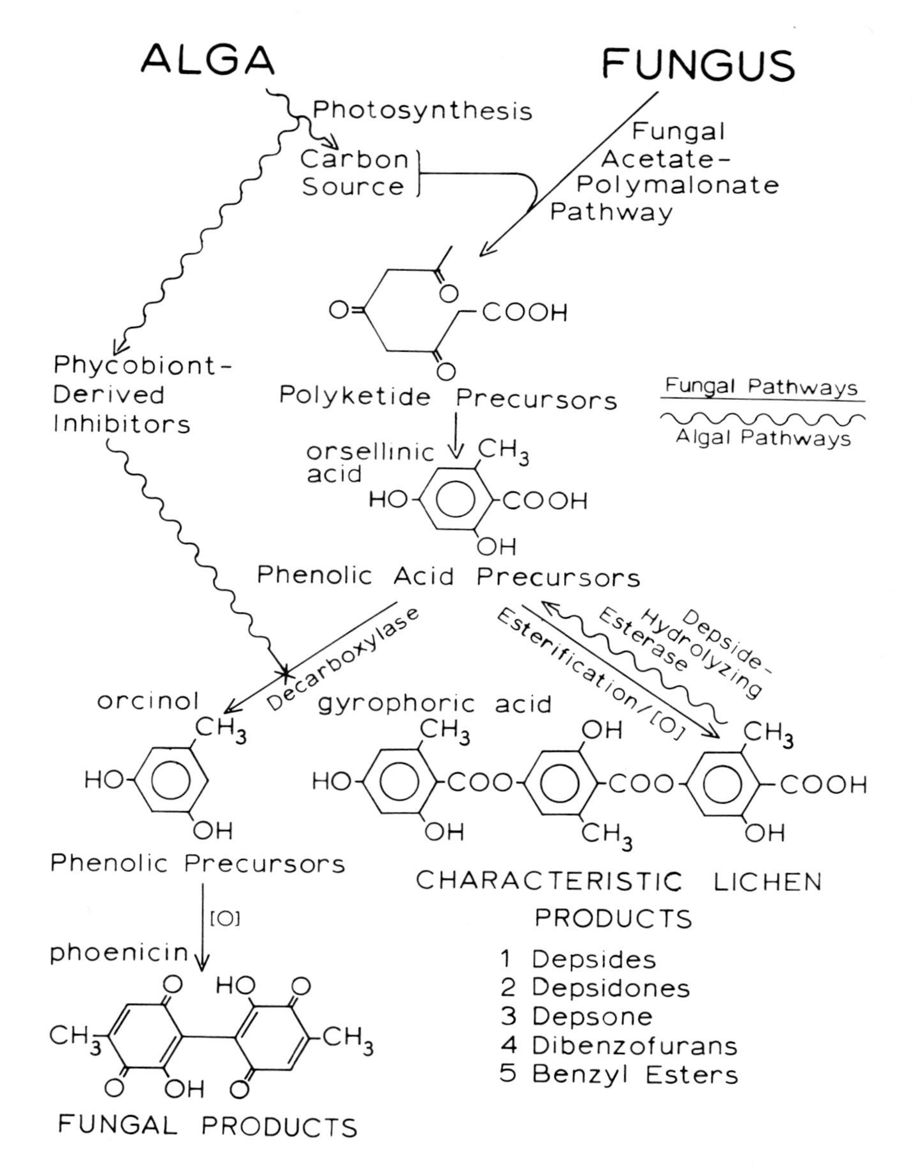

FIGURE 4.3. Hypothesized role of the alga in production of characteristic lichen products by inhibition of biosynthetic steps leading to products of nonlichenized fungi. Reprinted by permission from *Mycologia* 72:90–109. Copyright 1980 by C.F. Culberson & V. Ahmadjian and The New York Botanical Garden.

of these compounds exhibit marked antibiotic activity against gram-positive bacteria and free-living molds; others are stimulatory (Ahmadjian & Reynolds, 1961). These compounds are sometimes similar to typical lichen secondary compounds. The mycobiont of *Acarospora smaragdula* produces a compound similar to usnic acid (Ahmadjian & Reynolds, 1961; Ahmadjian, 1961); the mycobiont of *Lecanora rupicola* produces a fatty acid similar to protolichesterinic acid (Ahmadjian, 1966b).

Pigment production by mycobionts is variable and depends upon culture conditions. Ahmadjian (1964) found that the mycobiont of *Graphina bipartita* produced a yellow pigment in lactose and a darker colored pigment in maltose; in sucrose, there was no pigment. Biotin was found to be involved in the production of a red pigment by *Acarospora fuscata* in culture (Ahmadjian, 1961). This fungus was later found to produce a number of pigments (Diner, Ahmadjian, & Rosenkrantz, 1964), and the great variability observed in pigment production was found to be related to a number of cultural conditions (temperature, vitamins, and nitrogen source). This is an area, like so many in lichen symbiont physiology, that requires further experimentation.

5 Synthesis

Attempts to resynthesize lichens from their isolated symbionts have been made since 1867, when Schwendener first suggested that lichen thalli were composed of both fungi and algae. Ahmadjian (1973a), Richardson (1975), and Ahmadjian and Jacobs (1983) have reviewed some of these early resynthesis experiments, emphasizing the slow progress and often contradictory results. Although some degree of association was frequently observed between recombined symbionts, thalli typical of those seen in the field were rarely formed.

Recent evidence reviewed in this chapter suggests that these difficulties may in some way involve the nature of the lichen association itself. Ahmadjian (1973a) suggested that lichen symbionts are so completely adjusted to a symbiotic condition that their isolation and subsequent recombination cause severe, possibly irreversible, changes that inhibit normal thallus development. He noted that successful lichen resynthesis generally has resulted only when investigators attempted to simulate conditions under which the symbionts coexist in nature.

We presently know more about lichen symbiosis than we ever have, thanks in large part to the increasing attention being paid to lichen synthesis. The difficulties with resynthesis investigations are formidable, however. As a consequence, "the numerous hypotheses... proposed to elucidate... interactions between the symbionts have tended to be general, speculative, sometimes highly imaginative, and always long-lived" (Ahmadjian, 1966b, p. 36).

There are many problems in lichenology that can only be solved by the long-term laboratory culture of lichen thalli. However, only one species, *Endocarpon pusillum,* has shown spore-to-spore development in axenic culture. Very few others have been investigated at all. Until cultured thalli can be maintained routinely in the laboratory, many of the most interesting and challenging problems of lichenization will remain unsolved.

This chapter will focus on the development of lichen thalli under controlled laboratory conditions and the environmental factors that may be necessary for thallus development. Speculation on the mechanisms of lichenization under natural conditions in the field—despite the dearth of information on this subject—will also be included.

ARTIFICIAL RESYNTHESIS

Environmental Requirements for Resynthesis

The stages in the development of lichenized thalli are similar for all lichens that have been resynthesized in the laboratory, but not all lichens examined to date exhibit all stages of development. These stages include:

1. an initial contact of fungal hyphae with algal cells that is probably a general response,

2. an as yet unknown recognition process that allows the fungus to associate only with suitable algae,

3. early synthesis that includes a number of important physiological changes but relatively little tissue differentiation,

4. vegetative tissue development that results in a typical thallus composed of cortex, algal layer and medulla,

5. reproductive tissue development, beginning with pycnidia and ultimately leading to ascocarps with fertile ascospores.

The lichen species most commonly used in resynthesis experiments are *Acarospora fuscata* (Ahmadjian, 1962a), *Cladonia cristatella* (Ahmadjian, 1966a) and *Endocarpon pusillum* (Stahl, 1877; Bertsch & Butin, 1967; Ahmadjian & Heikkilä, 1970). *Acarospora fuscata* exhibits development of a vegetative thallus only, *C. cristatella* shows further development of pycnidia and sterile apothecia, and only *E. pusillum* is known to exhibit complete spore-to-spore development in culture.

A number of different methods have been used successfully in resynthesis experiments. Ahmadjian commonly places mixed suspensions of fungal hyphae and algal cells onto purified agar gels or sterilized soil in clay flower pots. Recently (Ahmadjian & Jacobs, 1981), mica strips soaked in Bold's Basal Medium were used as a substrate. Marton and Galun (1976) have used a nutrient solution added to silica gel as a medium for the synthesis and culture of *Heppia echinulata.*

It has often been suggested that conditions of slow drying and low nutrient supply are necessary for complete development of artificially resynthesized lichen thalli (Ahmadjian, 1962b; 1966b; 1967a; 1967b; 1973a; Pearson, 1970). For *Cladonia cristatella,* Ahmadjian (1966a; 1967b) found that alternating wet and dry conditions were necessary for thallus development, and slow drying stimulated the formation of reproductive structures. In a study of an artificial synthesis using a *C. cristatella* mycobiont and a *Trebouxia* phycobiont from *Huilia albocaerulescens,* Ahmadjian (1980a) found it necessary to transfer cultures from agar gels to soil in clay pots to obtain further thallus development. Similar responses were observed when the *C. cristatella* mycobiont was combined with other *Trebouxia* phycobionts.

Symbiont Specificity

Until recently, the only information available about the degree of algal specificity exhibited by lichen fungi was obtained by sectioning thalli and identifying the phycobiont. This technique has been used frequently and has demonstrated that a single lichen species may have more than one phycobiont. Recent resynthesis studies have attempted to document the level of specificity and the mechanisms involved.

Ahmadjian (1980a) found that the mycobiont of *Cladonia cristatella* formed associations with its own phycobiont, *Trebouxia erici,* and 12 other species of *Trebouxia,* but not with ten species of *Pseudotrebouxia* or the free-living terrestrial *Pleurastrum terrestre* (Table 5.1). This appparent genus-level specificity is not absolute, however. Other resynthesis studies have shown that some mycobionts are able to successfully associate with phycobionts from more than one genus. For example, *Rhizoplaca chrysoleuca* associated with five isolates of *Pseudotrebouxia* and one isolate of *Trebouxia* (Ahmadjian, Russell, & Hildreth, 1980). In all cases, development proceeded only to an early soredial stage.

In a more extensive later investigation, Ahmadjian and Jacobs (1981) attempted to synthesize lichen thalli using *Cladonia cristatella* mycobionts and numerous algal species, some lichen phycobionts, and others free-living (Table 5.2). They observed some degree of lichenization with six *Trebouxia* species and the free-living *Friedmannia israeliensis*; no lichenization occurred with 17 phycobionts, including some *Trebouxia* species, all the *Pseudotrebouxia* species, all the free-living green algae except *F. israeliensis,* and the blue-green phycobiont *Nostoc punctiforme.* Furthermore, they found that the

TABLE 5.1. Artificial Resyntheses between a *Cladonia cristatella* Mycobiont and Different Species of *Trebouxia* and *Pseudotrebouxia* Phycobionts[a]

Mycobiont	*Phycobiont*	*Lichen from which Phycobiont Was Isolated*	*Degree of Synthesis*[b]
C. cristatella	*Trebouxia erici*	*Cladonia cristatella*	+++
	T. sp.	*C. bacillaris*	+++
	T. sp.	*C. boryi*	+++
	T. sp.	*C. leporina*	+++
	T. sp.	*C. tenuis*	+++
	T. sp.	*Lecidea albocaerulescens*	+++
	T. sp.	*L. tumida*	+++
	T. sp.	*Lepraria zonata*	+++
	T. sp.	*Protoblastenia metzleri*	+++
	T. sp.	*Stereocaulon* sp.	+++
	T. sp.	*S. pileatum*	+++
	T. sp.	*Lepraria membranacea*	++
	T. sp.	*Stereocaulon evolutoides*	++

C. cristatella	*Pseudotrebouxia* sp.	*Buellia straminaea*	–
	P. sp.	*Caloplaca cerina*	–
	P. sp.	*Lecania* sp.	–
	P. sp.	*Lecanora hageni*	–
	P. sp.	*Lecidea fuscoatra*	–
	P. sp.	*L. tenebrosa*	–
	P. sp.	*Pertusaria* sp.	–
	P. sp.	*Physcia stellaris*	–
	P. sp.	*Ramalina complanata*	–
	P. sp.	*Usnea* sp.	–
	Pleurastrum terrestre	Free-living	–

[a]Observations after seven months culture.

[b]+++ Squamules present; barbatic and didymic acids detected; pycnidia and young podetia produced.

++ Squamules present but no pycnidia or podetia.

– Initial contacts between symbionts; no soredial formation.

Reprinted from "Separation and Artificial Synthesis of Lichens," by Vernon Ahmadjian, in *Cellular Interactions in Symbiosis and Parasitism,* edited by Clayton B. Cook, Peter W. Pappas, and Emanuel D. Rudolph (Ohio State University Biosciences Colloquia, no. 5).

TABLE 5.2. Interactions between the Mycobiont *Cladonia cristatella* and Various Species of Phycobionts and Free-living Algae

Species	*Source*
Algae lichenized by mycobiont	
Friedmannia israeliensis	Free-living
Trebouxia erici	*Cladonia cristatella*
T. excentrica	*C. bacillaris*
	C. leporina
	C. subtenuis
	Lecidea metzleri
	Lepraria sp.
	Stereocaulon dactylophyllum
T. glomerata	*Cladonia boryi*
	Huilia albocaerulescens
	Stereocaulon saxatilis
T. italiana	*Xanthoria parietina*
T. magna	*Pilophorus acicularis*
T. pyriformis	*Stereocaulon pileatum*
Algae parasitized by mycobiont	
Bracteacoccus minor	Free-living
Chlorella vulgaris	Free-living
Chlorococcum ellipsoideum	Free-living
C. gelatinosum	Free-living
Myrmecia biatorellae	*Dermatocarpon tuckermanii*
Nautococcus pyriformis	Free-living
Neochloris terrestris	Free-living
Nostoc punctiforme	*Peltigera canina*
Pleurastrum terrestre	Free-living
Pseudochlorella sp.	*Lecidea granulosa*
	L. scalaris
	Stereocaulon strictum
Pseudotrebouxia aggregata	*Lecidea fuscoatra*
	Xanthoria sp.
P. corticola	Free-living
P. decolorans	*Buellia punctata*
P. galapagensis	*Ramalina* sp.
P. gigantea	*Caloplaca cerina*
P. higginsiae	*Buellia straminea*
P. impressa	*Physcia stellaris*
P. jamesii	*Schaereria tenebrosa*
P. potteri	*Pertusaria* sp.
	Rhizoplaca chrysoleuca

TABLE 5.2. *(Continued)*

Species	Source
P. showmanii	*Lecanora hageni*
P. usneae	*Usnea filipendula*
P. sp.	*Aspicilia calcarea*
	Astroplaca sp.
	Caloplaca holocarpa
	Diploschistes scruposus
	Lecania sp.
	Lecanora tenera
	Lecidea cryptallifera
	L. sarcogynoides
	Omphalodium arizonicum
	Parmelia tinctorum
	Rhizocarpon geographicum
	Toninia caerulonigricans
	Xanthoria parietina
Radiosphaera dissecta	Free-living
Spongiochloris excentrica	Free-living
S. llanoensis	Free-living
Trebouxia anticipata	*Parmelia rudecta*
T. crenulata	*P. acetabulum*
	Xanthoria aureola
T. gelatinosa	*Parmelia caperata*

V. Ahmadjian & J.B. Jacobs. 1981. Reprinted by permission from *Nature,* Vol. 289, pp. 169–172.

fungus parasitized and killed algae unsuitable for lichenization. Compatibilty was apparently determined by the degree of algal resistance to lichenization rather than fungal choice, since many of the suitable algae in successful resyntheses were penetrated and killed by fungal haustoria. Indeed, Ahmadjian and Jacobs (1981) found that resynthesized *C. cristatella* thalli had a lower frequency of algal cells penetrated by haustoria than natural lichen thalli (Table 5.3). These observations have suggested to Ahmadjian (1982b) that lichen associations are "controlled parasitisms" rather than true mutualisms.

Symbiont Recognition

Resynthesis experiments suggest that algal suitability is recognized by both partners early in the lichenization process. Unsuitable algal partners are

TABLE 5.3. Frequency of Haustoria in Natural and Resynthesized Squamules of *Cladonia cristatella*

	Total Cells Observed	*Cells Penetrated by Haustoria*	*Percent Cells Penetrated*
Natural lichen			
Squamule 1	200	113	57
Squamule 2	100	65	65
Squamule 3	184	107	58
Resynthesized lichens			
C. cristatella s.s. (single-spore culture) 8 X *Trebouxia excentrica* *(Cladonia leporina)*	200	54	27
C. cristatella s.s. 8 X *T. erici*	100	24	24

V. Ahmadjian & J.B. Jacobs. 1981. Reprinted by permission from *Nature,* Vol. 289, pp. 169–172.

rapidly destroyed by the fungus and suitable algae are somehow recognized as suitable, thereby allowing or stimulating normal thallus development. Ahmadjian and Jacobs' (1981) studies on resynthesis between *Cladonia cristatella* mycobionts and numerous algae indicated that the recognition may be a very general response of an alga to the mycobiont. Mycobionts do not appear to be attracted to appropriate algae. Indeed, they are known to encircle any round object of suitable size (Ahmadjian, 1980a; Ahmadjian & Jacobs, 1981). Thus, the initial exploitation of algae is nonspecific. The process that determines specificity between symbionts occurs in the next developmental stage when suitable algae are incorporated into the developing thallus and unsuitable algae are destroyed. This suggests that specificity in lichenization may be the by-product of a differential sensitivity to fungal exploitation by potential phycobionts. The question is: What causes this difference in algal sensitivity?

Recent ultrastructural and biochemical studies indicate that fungal discrimination between suitable and unsuitable algae takes place at symbiont cell surfaces, a possibility that has been mentioned previously for other symbiotic associations (Broughton, 1978; Muscatine, Pool, & Trench, 1975; Smith, 1975; Trench, 1979). A scanning electron microscopy study of early synthesis in *Huilia albocaerulescens* (Ahmadjian et al., 1978) revealed the presence of an extracellular sheath around the phycobiont that may be involved in symbiont recognition. Jacobs and Ahmadjian (1971a) also

TABLE 5.4. Changes that *Trebouxia* Phycobionts Undergo after Isolation from Lichen Thalli

Morphological Changes
1. Cells develop gelatinous sheaths.
2. Cells form a fibrillar sheath.
3. Cells become larger.
4. Cell walls become thinner.
5. Pyrenoglobuli are fewer in number and smaller in size.
6. Polyphosphate bodies and other storage bodies appear.
7. Starch content increases.
8. Pyrenoid less evident.

Physiological Changes
1. Proportion of ribitol to sucrose decreases.
2. More fixed carbon incorporated in ethanol-insoluble compounds.
3. Less fixed carbon incorporated in ribitol.
4. Less photosynthate released from cells.
5. Types of excretory compounds are different.

V. Ahmadjian. 1973. "Resynthesis of lichens." In *The Lichens* (V. Ahmadjian and M.E. Hale, Jr., eds.), pp. 565–579. New York: Academic Press. Reprinted by permission of Academic Press, Inc. and the author.

reported a fibrillar sheath around cells of cultured *Trebouxia erici* that was not observed in lichenized cells. There are apparently a number of modifications to algal cells, morphological and physiological, resulting from lichenization (Table 5.4). What these modifications have to do with symbiont recognition is unclear at the present time.

Recent biochemical investigations suggest that symbiont recognition factors are involved in lichenization. Lockhart, Rowell, and Stewart (1978) isolated phytohaemagglutinins from several N_2-fixing lichens and found that they bound to the phycobionts isolated from their respective source lichens. Bubrick and Galun (1980a) noted differences in the cell walls of freshly isolated and cultured phycobionts of seven lichens using the lectin Concanavalin A (Con A) as a probe. In all cases, freshly isolated phycobionts did not bind Con A, whereas cultured phycobionts did (Table 5.5). Freshly isolated phycobiont cells began binding Con A after 7 days; complete binding was observed after 21 days. The plastic and easily reversible nature of these modifications suggested that they were symbiosis induced. They may have been caused by a change in cell wall biosynthesis or the result of enzymatic alteration of existing cell walls.

In a series of related studies (Bubrick & Galun, 1980b; Bubrick, Galun, & Frensdorff, 1981), a mycobiont protein was isolated from *Xanthoria parietina*. By indirect immunoperoxidase assay this protein was found to be associated

TABLE 5.5 Staining of Freshly Isolated and Cultured Phycobionts of Various Lichen Species with FITC-Con A

	FITC-Con A Staining	
Lichen of origin	*Fresh*	*Cultured*
Xanthoria parietina	−*	+
Caloplaca citrina	−	+
C. aurantia	−	+
Ramalina duriaei	−	+
R. pollinaria	−	+
Lecanora dispersa	−	+
Cladonia convoluta	−	+

*+ = visibly stained; − = not visibly stained.

P. Bubrick & M. Galun. 1980. "Symbiosis in lichens: Differences in cell wall properties of freshly isolated and cultured phycobionts." *FEMS Microbiology Lett.* 7: 311–313. Reprinted with permission of FEMS Microbiology Letters.

with the mycobiont cell wall and visualized in both isolated (Figure 5.1) and and lichenized *X. parietina* mycobionts. It was capable of binding to the cultured phycobiont of *X. parietina* but not to a freshly isolated phycobiont, and it also discriminated to some extent between cultured phycobionts of five other lichens. It therefore meets the requirements for a recognition-type protein suggested by Bubrick et al. (1981).

1. It is produced by the fungus.
2. It is located at or near the fungal cell surface.
3. There are receptor sites on algal cell surfaces.
4. It can discriminate between compatible and incompatible algae.

Only those phycobionts isolated from lichens in the Teloschistaceae (the family to which *X. parietina* belongs) bound the protein; phycobionts from *Cladonia convoluta, Ramalina duriaei,* and *R. pollinaria* did not. It appears from these results the factors that make phycobionts appropriate and inappropriate may be heritable to a certain extent and not necessarily induced by lichenization. However, cell wall binding patterns were not found to be associated with taxonomic status of the phycobionts tested. The three algae that bound the *X. parietina* protein were a *Trebouxia* species from *X. parietina,* and *Pseudotrebouxia* species from *Caloplaca aurantia,* and *C. citrina.* Further cell wall cytochemical experiments indicated that protein binding was highly correlated with high levels of acidic polysaccharide

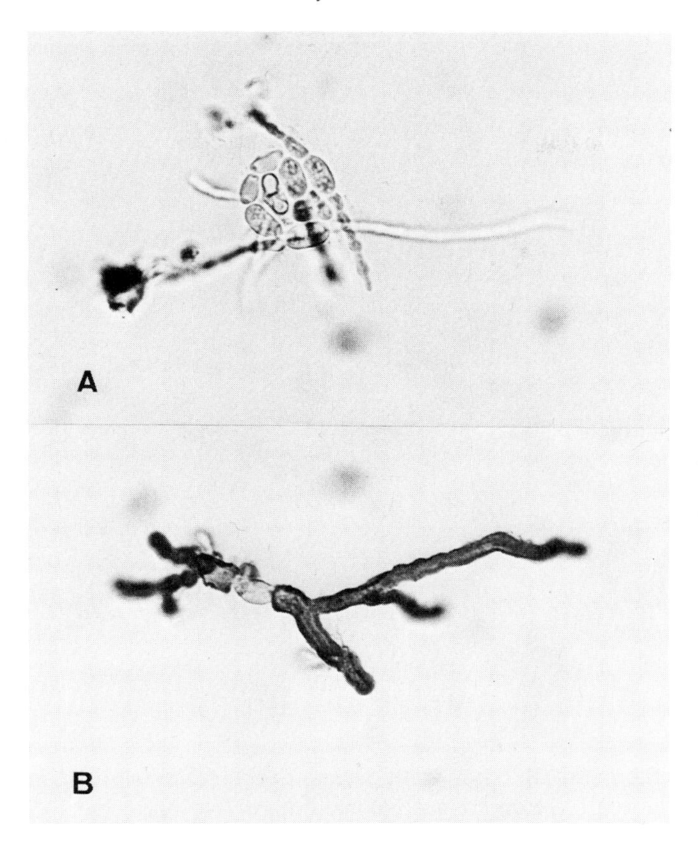

FIGURE 5.1. Visualization of *Xanthoria* protein in the *X. parietina* mycobiont cultured *in vitro*. (a) Untreated mycobiont (×500); (b) Mycobiont treated with anti-Xp and Pap-Gar (horseradish peroxidase labelled goat antirabbit serum). Note darkened hyphal cell walls (×500). From P. Bubrick, M. Galun, & A. Frensdorff. 1981. "Proteins from the lichen *Xanthoria parietina* which bind to phycobiont cell walls: Localization in the intact lichen and cultured mycobiont." *Protoplasma* 105:207–211. Reprinted with permission of Springer-Verlag and the authors.

(Ruthenium red stain) and the presence of a protein coat (Coomassie blue stain) on the cell wall of the phycobiont (Figure 5.2).

Petit (1982) has also isolated from the *Nostoc*-containing *Peltigera horizontalis* a lichen protein fraction that is produced by the mycobiont and binds cultured phycobionts from three *Peltigera* species (Table 5.6). She found that this *P. horizontalis* protein fraction contained phytolectins that

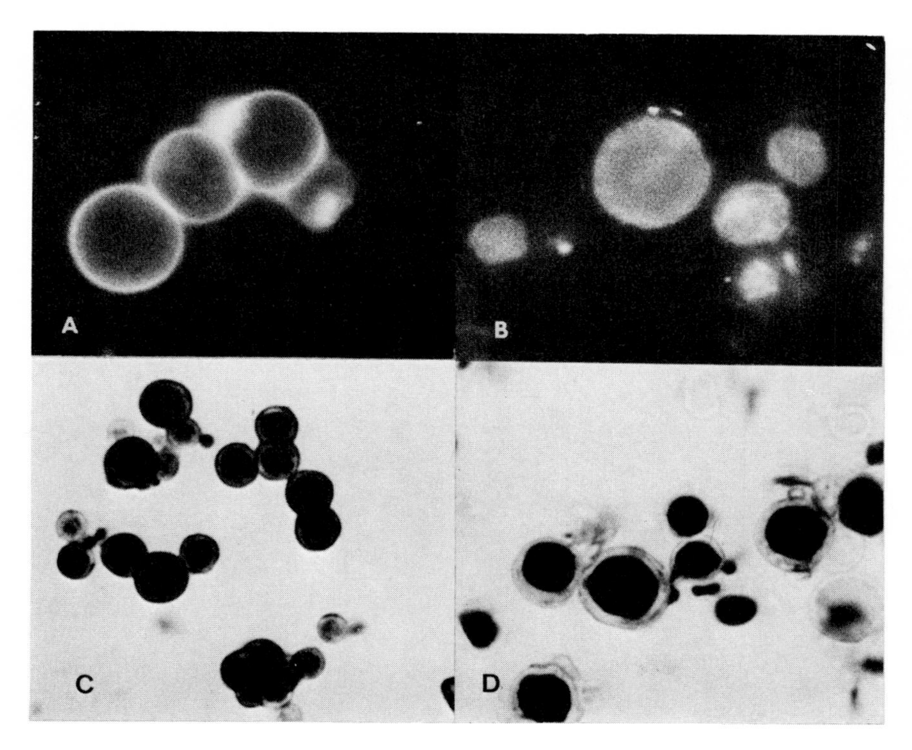

FIGURE 5.2. Possible symbiont recognition factors. (a) Binding of fluorescamine-labelled *Xanthoria*-protein to cultured *Trebouxia* from *X. parietina.* Note intense cell wall fluorescence (×1200); (b) Binding of fluorescamine-labelled *Xanthoria*-protein to freshly isolated *Trebouxia* from *X. parietina.* Note cell walls not fluorescent (×1200); (c) Cultured *Trebouxia* from *X. parietina* stained with Coomassie blue. Note staining on outer cell wall surface (×1000); (d) Freshly isolated *Trebouxia* from *X. parietina* stained with Coomassie blue. Cell wall unstained (×1000). From P. Bubrick & M. Galun. 1980b. "Proteins from the lichen *Xanthoria parietina* which bind to phycobiont cell walls. Correlation between binding patterns and cell wall cytochemistry." *Protoplasma* 104:167–173. Reprinted with permission of Springer-Verlag and the authors.

caused haemagglutination of human erythrocytes in A, B, and O groups. The ability of phytolectins produced by the fungus to bind suitable algae suggested to Petit that they are involved in symbiont recognition.

The importance of these cell surface characteristics in the process of lichenization is not presently known. Bubrick and Galun (1980b) suggested that symbiont recognition during lichenization may involve a number of

TABLE 5.6 Binding of FITC Phytohaemagglutinins Produced by *Peltigera horizontalis* to Various *Nostoc* Spp., *Scytonema*, and One *Trebouxia* Sp.

Algae	*FITC Released (ng μg^{-1}Chl a)*
Nostoc spp.	
(Peltigera horizontalis)	60
(P. canina var. *canina)*	58
(P. canina var. *praetextata)*	40
(Free-living)	3
Scytonema sp.	
(Heppia echinulata)	5
Trebouxia sp.	
(Evernia prunastri)	2

P. Petit. 1982. "Phytolectins from the nitrogen-fixing lichen *Peltigera horizontalis:* The binding pattern of primary protein extract." *New Phytol.* 91: 705–710. Reprinted with permission of Academic Press and the author.

recognition stages during which the phycobiont is slowly modified and induced to conform to the physiological requirements of the fungus. Phycobiont recognition may thus be a stepwise process of fungus-induced modification that can only occur in suitable phycobionts with either the necessary structural and/or physiological characteristics *or* the physiological plasticity to survive these modifications.

Ahmadjian and Jacobs (1983) discussed these studies and others in a review of algal-fungal relationships in lichens and concluded that a lectin-mediated recognition system probably does not exist in lichens. They reasoned that such a system would result in levels of specificity between symbionts that have not been seen in lichen synthesis studies. What little specificity that is observed in these studies is very broad, with one species of phycobiont able to associate with widely different mycobionts (Hildreth & Ahmadjian, 1981). Ahmadjian and Jacobs (1983) argue that if recognition exists between symbionts, it occurs after initial contact is made, perhaps in a way similar to that of host-parasite associations in higher plants (Albersheim & Anderson-Prouty, 1975; Albersheim & Valent, 1978). In these associations, host cells produce toxic compounds called phytoalexins that slow the growth of pathogens in the cell. Formation of these phytoalexins is induced by elicitors located on the cell wall of the pathogen. If elicitors are not recognized by the host cell, defensive compounds are not produced and the pathogen can destroy the cell.

Symbiont recognition may also be physiological. Green and Smith (1974;

FIGURE 5.3. Envelopment of *Trebouxia erici* cell by *Cladonia cristatella* mycobiont. (a) Alga encircled by single hypha, bar = 1 μm; (b) Cell almost fully enveloped by hyphae, bar = 1 μm. From V. Ahmadjian & J.B. Jacobs. 1981. Reprinted by permission from *Nature*. Vol. 289, No. 5794, pp. 169–172. Copyright 1981 Macmillan Journals Limited.

also discussed in Ahmadjian, 1982a), found that the phycobiont of *Xanthoria parietina (Pseudotrebouxia decolorans)* incorporated 52.7 percent of labelled ^{14}C into ethanol-insoluble compounds, whereas the phycobiont of *X. aureola (Trebouxia crenulata)* incorporated only 18.2 percent. This difference between *Trebouxia* and *Pseudotrebouxia* may help to explain why mycobionts sometimes appear to be able to incorporate species of one genus or the other, but not both, since a limited excretion of carbohydrates by a phycobiont could stimulate an increased penetration by the mycobiont (Ahmadjian, 1982a). However, it would not explain why a mycobiont is unable to associate with an alga that released more carbohydrate than normal.

Early Synthesis

The processes involved in early lichen synthesis begin immediately following the location of a suitable phycobiont by the mycobiont. The alga is encircled by fungal hyphae and rapidly enveloped, sometimes being penetrated by haustoria (Figure 5.3). Hyphae apparently branch considerably and begin to clump groups of algal cells together. Ahmadjian, Jacobs, and Russell (1978) investigated the stages of early synthesis of *Huilia albocaerulescens* using scanning electron microscopy (Figure 5.4). They found that fungal hyphae secured clumps of algal cells by forming appressorial contacts (Figure 5.4 d and e) and enveloping the cells with hyphae. Very rarely were algal cells enveloped by several hyphae (Figure 5.4 f); rather, incorporation of as many algal cells as possible into the developing thallus appears to be the predominant characteristic of this early stage of lichenization (Figure 5.4a–c).

Lichenization occurred only in portions of the fungus where there were hyphal outgrowths from the main fungal colony. These outgrowths were sometimes seen in mycobionts cultured on organic media but were most frequently associated with nutrient-poor media. Thus, a change in the algal cell receptivity and growth behavior of fungal hyphae appeared to be stimulated by nutrient-deficient conditions. During this stage, fungal hyphae formed appressoria on other hyphae as well as algal cells. Ahmadjian (1980a) has suggested that fungal self-parasitism is common in lichen mycobionts when grown on organic media and may be a response triggered when organic compounds are initially received by the fungus. Intrahyphal haustorial penetration has been observed in lichens using transmission electron microscopy (Ahmadjian, 1980a).

Early thallus development appears to be a stage during which changes occur in a number of important physiological interactions between alga and fungus. In culture, isolated algal cells exhibit little carbohydrate release; however, in the intact lichen thallus and immediately after isolation, there is a massive release of carbohydrates from lichen phycobiont cells. To determine when this physiological change occurred, Hill and Ahmadjian (1972) studied

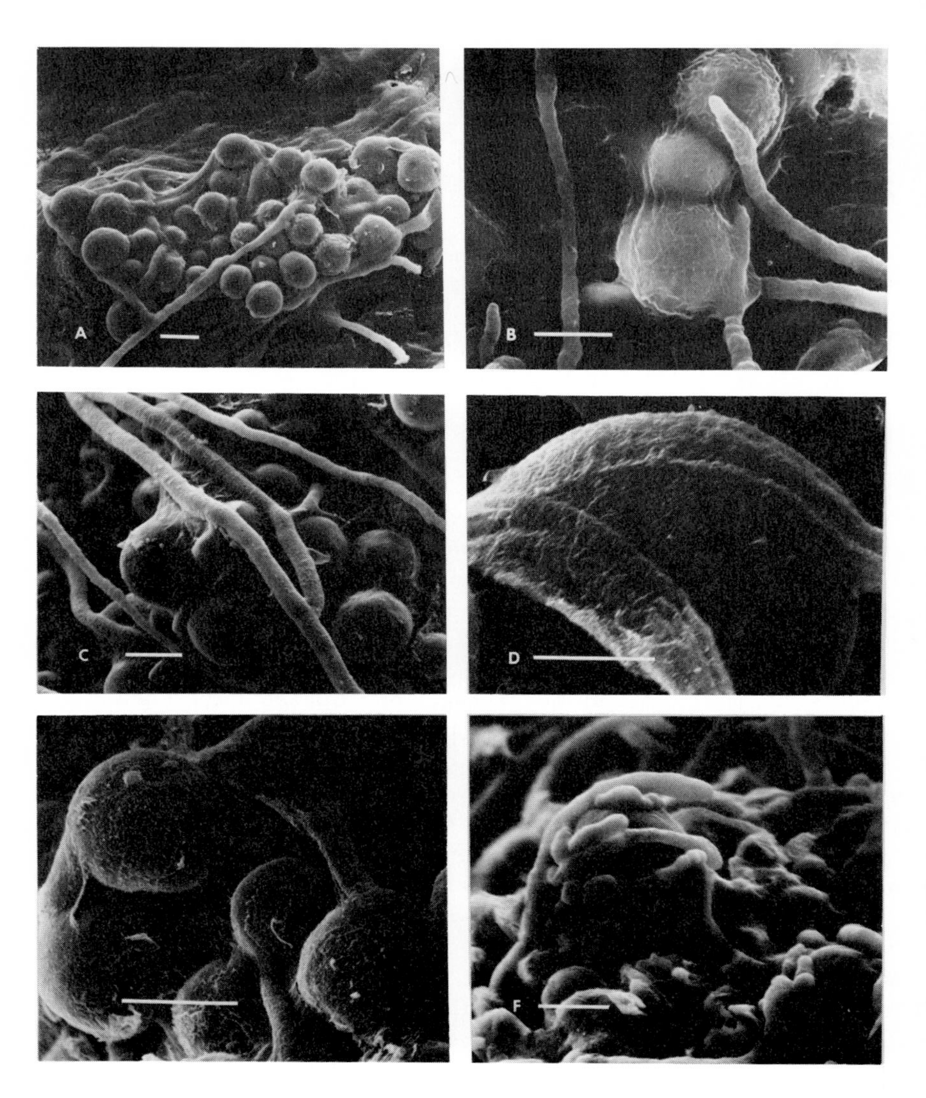

FIGURE 5.4. Early synthesis of *Huilia albocaerulescens.* (a) Cluster of algal cells with entwined fungal hyphae, bar = 10 μm; (b) Several algal cells enveloped by a thin veil of extracellular substance, bar = 10 μm; (c) Fungal hyphae contacting several algal cells, bar = 10 μm; (d) One algal cell with two appressoria, bar = 5 μm; (e) One fungal hypha that has formed appressoria on two algal cells, bar = 10 μm; (f) An algal cell enveloped by several hyphae, bar = 10 μm. From Ahmadjian et al. 1978. Reprinted by permission of the authors, V. Ahmadjian, J.B. Jacobs, & A. Russell. Reprinted from *Science,* Vol. 200, pp. 1062–1064. Copyright 1978 by the American Association for the Advancement of Science.

carbohydrate movement in resynthesized thalli of *Cladonia cristatella*. In inorganic liquid media, the isolated phycobiont *Trebouxia erici* incorporated little of its photosynthate into the transfer carbohydrate ribitol. When algal cells were placed in the presence of the *C. cristatella* mycobiont on agar, the proportion of algal photosynthate incorporated into ribitol was observed to increase significantly. However, even after 12 weeks of thallus development, ribitol release by algal cells was never as high as it was in the fully developed lichen thallus. Whether the presence of the fungus stimulated increased ribitol synthesis or a change from inorganic liquid medium to agar stimulated the change was not clear. No visible contact between fungal and algal cells appeared to be required for the response, but the authors did not believe a fungal factor was involved.

Another physiological change that occurs sometimes during early thallus development is the stimulation of the mycobiont to synthesize secondary compounds. Although these substances are fungal and can sometimes be produced by the mycobiont in isolation (Ahmadjian, 1980a; Culberson & Ahmadjian, 1980), the presence of the phycobiont facilitates compound production. Using thin layer chromatography (TLC) and high-performance liquid chromatography (HPLC), Culberson and Ahmadjian (1980; Ahmadjian, 1980a) investigated secondary compound production in isolated lichen symbionts and newly reestablished thalli involving various combinations of phycobionts and the mycobionts of *Cladonia cristatella* and *Rhizoplaca chrysoleuca*. Successful resyntheses between the *C. cristatella* mycobiont and its normal phycobiont as well as 12 other *Trebouxia* phycobionts all resulted in production of barbatic acid (Table 5.7). Obtusatic and 4-*O*-demethylbarbatic acids were also observed using HPLC. No secondary compounds were produced by isolated symbionts or by resyntheses of *C. cristatella* and *Pseudotrebouxia* phycobionts. However, resyntheses involving *Rhizoplaca chrysoleuca*—which has a *Trebouxia* phycobiont—and several *Pseudotrebouxia* isolates exhibited traces of usnic acid.

The production of secondary compounds by developing lichen thalli appears to be dependent upon successful resynthesis but does not appear to require a specific phycobiont. Furthermore, there is no apparent inhibition of atypical phycobionts by secondary compounds. Therefore, although the process of lichenization includes secondary compound production by the fungus, fungal-algal recognition is probably not influenced in any way by secondary compound production.

Secondary compounds may be important in protecting developing lichen thalli from microorganism attack during early synthesis. Ahmadjian, Jacobs, and Russell (1980) observed numerous free-living fungal contaminants growing on the clay pot surfaces used in the resynthesis of *Cladonia cristatella* thalli. The newly synthesized lichen squamules on these pots seemed to be resistant to fungal contamination, a possible consequence of secondary compound protection, although this was not determined experimentally.

TABLE 5.7. Secondary Products of the Mycobiont *Cladonia cristatella* in Nature and as Resynthesized with *Trebouxia* Phycobionts from Various Lichens

	Usnic	*Barbatic*	*Obtusatic*	*4-O-De-methyl-barbatic*	*Didymic* $-C_2H_4$	*Didymic*	*Didymic* $+C_2H_4$
In nature[a]	+[b]	+++	+	+	+	++	+
With phycobiont from:							
Cladina subtenuis	—	+++	(+)[c]	(+)	—	(+)	(+)
Cladonia bacillaris	—	+++	(+)	(+)	—	(+)	—
C. boryi	—	+++	(+)	(+)	—	+	(+)
C. cristatella	—	+++	(+)	(+)	(+)	++/(+)	(+)
C. leporina	—	+++	(+)	(+)	—	—	—
Huilia albocaerulescens	—	+++	(+)	(+)	(+)	++	(+)
Lecidea tumida	—	+++	(+)	(+)	(+)	(+)	(+)
Lepraria zonata	—	+++	(+)	(+)	(+)	+	(+)
L. sp.	—	+++	(+)	(+)	—	—	—
Protoblastenia metzleri	—	+++	(+)	(+)	—	—	—
Stereocaulon pileatum	—	+++	(+)	(+)	—	(+)	—
S. saxatile	—	+++	(+)	(+)	—	—	—
S. sp.	—	+++	(+)	(+)	—	—	—

[a]The trace products baeomycesic acid and 3-α-hydroxybarbatic acid are not included.

[b]Symbols indicate approximate relative proportions; +++ and ++, major products; +, minor product; —, not detected.

[c]Symbols in parentheses indicate compounds detected by HPLC but not by TLC.

An important structural difference between isolated mycobionts and lichenized fungi is the presence of concentric bodies in lichenized hyphae. Concentric bodies are fungal organelles frequently seen in ultrastructural investigations. They are small (300 nm in diameter), layered, presumably proteinaceous bodies (Galun, Behr, & Ben-Shaul, 1974), and have been reported from numerous lichen species, some free-living ascomycetes and plant pathogens (Ahmadjian, 1980a). Cultured mycobionts do not have concentric bodies; however, Ahmadjian, Russell, and Hildreth (1980) observed them in resynthesized lichen thalli. Marton and Galun (1976) also observed them in resynthesized thalli of *Heppia echinulata*. The biological role of these structures is not known, but they may be involved in thallus development or secondary compound production since all of these processes seem to be related developmentally (Ahmadjian, Russell, & Hildreth, 1980). There is also the suggestion that they are associated with the production of sexual structures (Jacobs & Ahmadjian, 1971a; Tu & Colotelo, 1973; Granett, 1974).

Vegetative Tissue Development

After suitable supplies of phycobiont cells are secured by the mycobiont, vegetative tissues rapidly develop. The presquamule stage of development is usually observed after about four weeks in culture and involves production of soredialike bodies of algal cells encircled to varying degrees by fungal hyphae (Figure 5.5 a). These bodies are then linked together by a network of hyphae that grows over large numbers of presquamules (Figure 5.5 b), eventually forming an upper cortical layer. Further differentiation results after about eight weeks in a well-defined algal layer and medulla. In *Usnea filipendula*, fibril development also occurs about this time (Figure 5.6). The requirement for drying to initiate vegetative tissue development frequently assumed in earlier resynthesis studies (Ahmadjian, 1973a) now appears to be unnecessary (Ahmadjian & Jacobs, 1981).

When various *Trebouxia* phycobionts were used in the resynthesis of *Cladonia cristatella* thalli (Ahmadjian, Russell, & Hildreth, 1980), the squamules that developed were identical to naturally occurring squamules. This suggests there is no phycobiont control over morphogenesis in *C. cristatella* that has been suggested for other lichens (James & Henssen, 1976). However, the type of alga encountered by the mycobiont may in some way influence the rate of thallus development. Scott (1964) studied the natural resynthesis of *Solorina saccata*, which contains a primary green alga (a *Coccomyxa* species) and a secondary blue-green alga (a *Nostoc* species). If germinating ascospores contacted cells of *Coccomyxa*, thallus development proceeded to a more advanced stage than was observed if *Nostoc* cells were initially contacted. This kind of response is most likely due to differences in

FIGURE 5.5. Development of *Cladonia cristatella* squamules in culture. (a) Presquamules formed by mycobiont envelopment of phycobiont cells, bar = 10 μm; (b) Presquamules linked together by a network of superficial hyphae, bar = 10 μm. From V. Ahmadjian & J.B. Jacobs. 1981. Reprinted by permission from *Nature,* Vol. 289, No. 5794, pp. 169–172. Copyright © 1981 Macmillan Journals Limited.

FIGURE 5.6. Early development of thallus fibril by *Usnea florida* synthesized with *Pseudotrebouxia usnea* isolated from *Usnea filipendula*, bar = 10 μm. From Ahmadjian & Jacobs. 1982. "Artificial reestablishment of lichens. III. Synthetic development of *Usnea strigosa*." J. Hattori Bot. Lab. 52: 393–399. Reprinted with permission from Hattori Botanical Laboratory and the author.

the process of photosynthate transfer between symbionts, although more complex feedback mechanisms may also be involved. No thallus resynthesis studies done to date have suggested a morphogenetic role of the alga in lichen thallus development.

A hypothalluslike zone of fungal hyphae without algal cells has been observed on the edges of the squamules in some resynthesized lichen thalli that attain the squamule stage (Ahmadjian, Russell, & Hildreth, 1980). This was particularly apparent in resyntheses between *Cladonia cristatella* mycobionts and *Trebouxia* phycobionts from *Stereocaulon* species. It has been suggested that *Cladonia* species have extensive underground mycelial mats (Thøgersen, 1977 in Adhmadjian, Russell, & Hildreth, 1980); the physiological and/or ecological role of this mycelium is not known. The fact that it develops most strongly in resynthesized thalli containing *Stereocaulon* phycobionts rather than the typical *Cladonia* phycobiont may be due to an insufficient nutrient flow between symbionts that stimulates hyphal behavior similar to that observed during the earliest stages of phycobiont incorporation.

Reproductive Tissue Development

Development of reproductive tissue has been observed in resynthesized *Cladonia cristatella* (Ahmadjian, 1966a) and *Endocarpon pusillum* (Bertsch & Butin, 1967; Ahmadjian & Heikkilä, 1970). Reproductive structures have also been observed in artificially resynthesized thalli involving the mycobiont *C. cristatella* and numerous phycobionts (Ahmadjian, Russell, and Hildreth, 1980). Development of apothecial initials and pycnidia has been observed 10 to 11 weeks after the development of squamules in resynthesized *C. cristatella* (Ahmadjian, Russell, & Hildreth, 1980). In this species, ascocarps are frequently borne at the tips of short podetia (Figure 5.7). Pycnidia have also been observed on soredia in *C. cristatella* thalli resynthesized with the phycobiont of *Huilia albocaerulescens* (Ahmadjian et al., 1980), demonstrating that squamule development is not required for the initiation of reproductive structures.

Substrate drying has been assumed to stimulate formation of apothecia and pycnidia. Ahmadjian, Russell, and Hildreth (1980) found that transfer of newly synthesized *Cladonia cristatella* thalli from agar to a soil surface was required for formation of reproductive structures. Low pH has also been found to be stimulatory (Ahmadjian, 1973b).

Development of pycnidia appears to precede that of apothecia. This may be important in the sexual process of lichens, inasmuch as microconidia produced in pycnidia are assumed to function as spermatia. Genetic transfer within or between colonies of lichen fungi has never been documented; however, spermatia have been observed attached to tips of trichogynes produced by juvenile apothecia on lichen squamules in culture (Ahmadjian, 1973a).

LICHENLIKE ASSOCIATIONS

Associations between green algae and various microorganisms have been produced in the laboratory by Lazo (1961; 1964; 1966). He found that three

FIGURE 5.7. Development of mature squamules and podetia by synthesized lichen thalli. (a) Mature squamules. *Cladonia cristatella* mycobiont X *Trebouxia glomerata* (from *Huilia albocaerulescens*), soil synthesis; (b) Podetia. *C. cristatella* mycobiont X its own phycobiont *(Trebouxia erici),* soil synthesis. From V. Ahmadjian, L.A. Russell & K.C. Hildreth. 1980. Reprinted by permission from *Mycologia,* 72: pp. 73–87. Copyright 1980; V. Ahmadjian, L.A. Russell, & K.C. Hildreth and the New York Botanical Garden.

species by *Chlorella* were able to enter into apparent symbiotic associations with the bacteria-free plasmodia of the myxomycetes, *Physarum didermoides* and *Fuligo cinerea,* forming green plasmodia that could be maintained in light. It was suggested that the algal cells may have substituted for the bacteria in stimulating normal plasmodial growth. In the case of *F. cinerea,* the association between algae facilitated the fruiting process. Alexopolous (1962) later called these associations myxolichens. Using labelled phosphorus Zabka and Lazo (1962) demonstrated that transfer of materials occurred between the associates, a significant observation, since no transfer of radiophosphorus occurred between different species of myxomycetes grown together on the same medium.

In 1964, Lazo reported the formation of a lichenlike crust from an experimentally obtained association between *Chlorella xanthella* and a *Streptomyces* species. The formation of this actinolichen stimulated the search for other types of interactions in other *Streptomyces* species. In 1966, Lazo reported three types of interactions between several *Streptomyces* species and *Chlorella xanthella* on various media. One type of interaction resulted in the reduced growth or complete inhibition of the alga by the actinomycete. In another set of interactions, both associates developed independently. Only one *Streptomyces* strain, a gray-spored species in the section *Spira,* formed a stable, crustlike thallus that could be maintained indefinitely in culture. It only formed on potato sections, sand, or sawdust soaked with potato broth; when transferred to a nutrient or potato agar, the association broke down and the symbionts grew independently. Sections through the thallus revealed an organized thallus structure with a dark-colored surface layer of spores, a green algal layer, and a mycelial layer on the bottom.

These observations suggested that many microorganisms are capable of entering into lichenlike associations with algae in nature. The difficulties in initiating and maintaining these associations in the laboratory are very similar to those encountered when working with lichens. Further study of these associations and comparisons with lichen associations may reveal similarities that will contribute greatly to our understanding of the nature and origin of the lichen symbiosis.

These laboratory studies are not generally considered ecologically important because they probably do not represent natural conditions. However, naturally occurring associations between microorganisms and algae have been investigated on occasion, and there have been some interesting findings. Parker and Bold (1961) observed a number of interactions between soil microorganisms and algae isolated from Texas soil samples. Of the 143 two-organism combinations between soil algae and various heterotrophs they studied under laboratory conditions, half resulted in algal stimulation, 10 percent exhibited clear-cut algal inhibition, and the rest were neutral. No cases of thallus formation or symbiont specificity were documented. Further studies of this sort will probably contribute significantly to our understanding of the earliest stages in the evolution of lichenization on land.

Kohlmeyer and Kohlmeyer (1979) described a number of facultative and obligate associations between marine algae and fungi. The obligate associations (mycophycobioses) are quite lichenlike, despite the absence of unique morphological structures or chemical compounds that characterize many lichens. In mycophycobiosis, the alga is dominant and the fungus grows intercellularly in the alga tissues. Algal cells are not penetrated by the fungus and no transfer of metabolites has been reported. An example of obligate mycophycobiosis is the alga, *Ascophyllum nodosum,* which is associated with hyphae of *Mycosphaerella ascophylli.* The alga apparently requires the fungus for normal development, but the reasons for this are not clear. Kohlmeyer and Kohlmeyer (1979) have suggested that the fungus enables the alga to better withstand periods of dessication. Since neither the alga nor the fungus is harmed in any way by the association, mycophycobioses may be better examples of mutualisms than lichen associations.

LICHEN SYNTHESIS IN NATURE

Although synthesis of lichen thalli has never been observed in nature, it is logical to assume that it occurs in much the same way as in the laboratory. The absence of asexual methods of reproduction in many lichens, especially crustose species, suggests that lichenization is a common and frequently successful natural process. Although *Pseudotrebouxia* and *Trebouxia* species are not assumed to exist commonly outside of lichen associations, they are frequently found as components of asexual progagules. Also, both isolated and lichenized *Trebouxia* cells (Slocum et al., 1980) can produce motile zoospores that may provide an algal source for germinating ascospores.

Although they are available in nature, however, phycobionts necessary for most lichen syntheses are not thought to be particularly abundant, and there is a good deal of speculation about why this is so. The best explanation is that the evolutionary development of the lichen association has involved a physiological modification of the phycobiont to meet the needs of the fungus. In many groups of lichenized fungi, this process has progressed to the point where the independent existence of the alga is far less likely. As a consequence of this, lichenization in nature has become more difficult, favoring complex modes of reproduction and dispersal that involve both symbionts. It would seem, therefore, that the study of lichenization in nature will be most illuminating when lichen groups exhibiting both primitive and advanced dispersal characteristics are compared.

CEPHALODIUM SYNTHESIS AND THE PHYCOSYMBIODEME CONCEPT

Cephalodia are associations of lichen fungi with foreign algae, usually blue-green species, different from the host phycobiont. They are found in a number of lichen groups, but are most numerous in the Peltigeraceae and Stictaceae. A synopsis of the occurrences and types of cephalodia is provided by James and Henssen (1976). Cephalodia appear to develop through fungal entrapment and incorporation of free-living algae, usually from the lower cortex. They may be internal and relatively inconspicuous (Figure 5.8) or external, assuming a variety of growth forms (Figure 5.9). Since cephalodial blue-green algae can fix nitrogen, they may supply needed nitrogen to fungi capable of forming cephalodia.

Cephalodial synthesis is not very different from normal lichen synthesis. For example, the presence of cephalodial algae can induce tissue differentiation in the fungus. This is most pronounced in external cephalodia, where upper and lower cortical tissues may form. Internal cephalodia develop by blue-green algal proliferation in the medulla, with little fungal involvement. However, there is some fungal differentiation in the medulla around the periphery of the developing cephalodium, and continued growth may eventually interrupt the green algal layer (Jordan, 1970).

The initiation of cephalodial formation is likely chemically induced (Jordan, 1970; Ahmadjian, 1982a). There is some indirect evidence that substances excreted by free-living, blue-green algae stimulate lichen fungi to overgrow and incorporate these cells (Jahns, 1972; Jordan, 1970; Jordan & Rickson, 1971). This is one area of symbiont recognition that warrants further study.

There has been much recent interest in certain lichens that produce well-differentiated cephalodia capable of independent growth. Species of *Peltigera, Lobaria, Sticta,* and numerous other genera (James & Henssen, 1976) have occasionally been reported to have shrublike growths on the dorsal surface. These were frequently assumed to be epiphytes and were assigned separate names. However, a few workers suggested that these epiphytes were actually cephalodia (Forssell, 1883; Galløe, 1939; Moreau, 1956).

James (1975) has called such thalli chimerae because they appear to be distinctly different morphological manifestations of the same fungus in association with different algae. Other authors have used other terminology, reviewed by Renner and Galloway (1982), who proposed the term "phycosymbiodeme," which will be used throughout this discussion.

If there are dimorphic lichens that can take on at least two distinctly different forms, this suggests that the phycobiont has a much greater influence on lichen morphological development than has been assumed. Most of the recent evidence for the existence of phycosymbiodemes is anatomical and morphological (James & Henssen, 1976; Brodo & Richardson, 1978; Renner

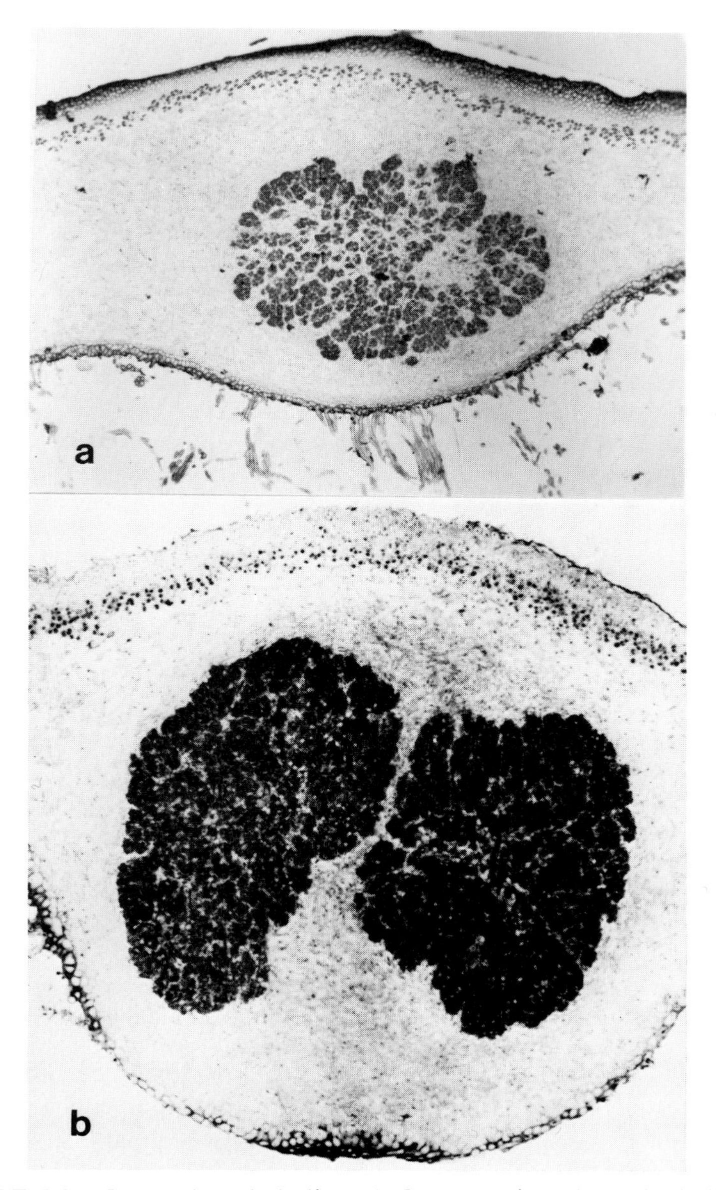

FIGURE 5.8. Internal cephalodia. (a) Cross section through thallus of *Lobaria linita* showing intrusive cephalodia (×140). From W.P. Jordan. 1970. "The internal cephalodia of the genus *Lobaria*." *Bryologist* 73:669–681. Reprinted with permission of the American Bryological and Lichenological Society and the author. (b) Cross section through mature cephalodium of *Nephroma expallidum* (×240). From W.P. Jordan & F.R. Rickson. 1971. *Am. J. Bot.* 58:562–568. Reprinted with permission of the American Journal of Botany and the authors.

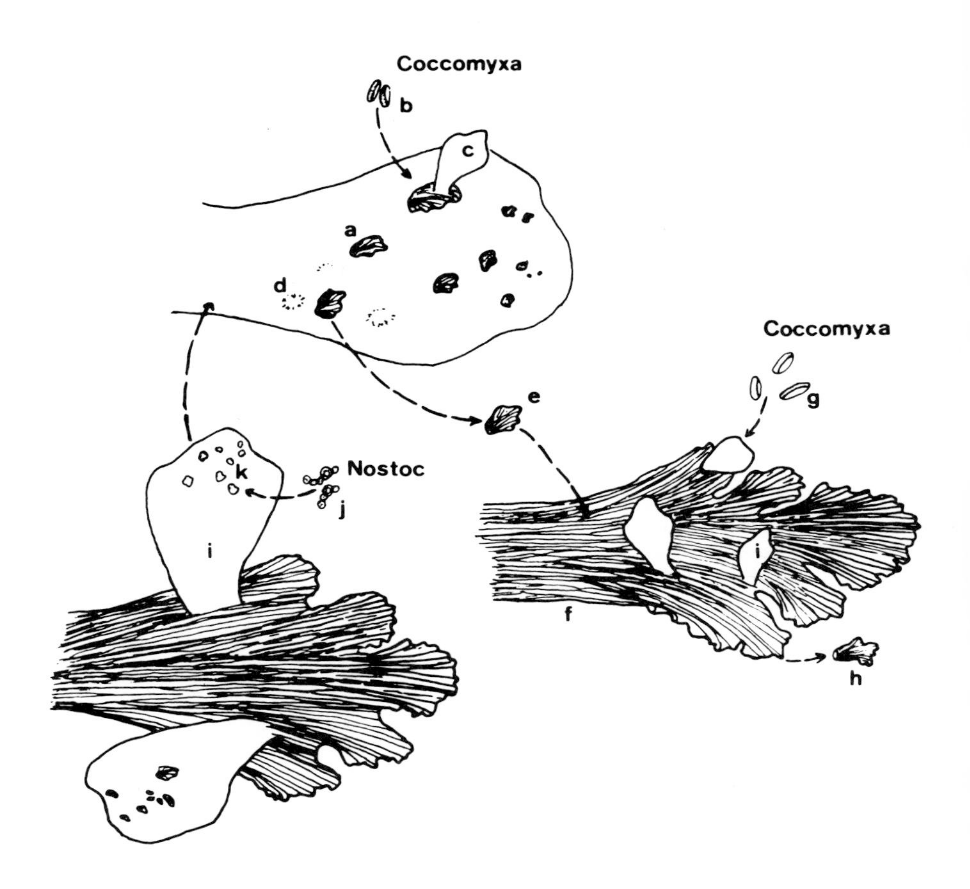

FIGURE 5.9. External cephalodia in *Peltigera*. Small cephalodia of a *P. aphthosa* thallus (a) enlarge and become subfoliose, finally breaking loose, leaving a white spot (d), and survive independently (e) and (f). These *Nostoc*-containing morphotypes can accept *Coccomyxa* cells from the environment (b) and (g) and grow out into green alga-containing lobes (c) and (i). The *Nostoc*-containing thallus (f) can also reproduce itself by marginal squamules (h). The green algal lobes continue to develop, finally accepting *Nostoc* cells (j) which form new cephalodia on the thallus surface (k). The green morphotype outgrows the blue-green parent thallus and completes its development as an independent thallus of *P. aphthosa*. From I.M. Brodo & D.H.S. Richardson. 1978. "Chimeroid associations in the genus *Peltigera*." *Lichenologist* 10: 157–170. Redrawn with permission of Academic Press.

& Galloway, 1982). Thin sections of the attached phycosymbiodemes with their associated algae seem to show continuity of fungal hyphae; chemical analyses also suggest that the same mycobiont is involved in each of the associations.

Resynthesis studies involving one mycobiont and two phycobionts are obviously required to test the phycosymbiodeme hypothesis experimentally. As Ahmadjian (1982a) has pointed out, however, it is impossible to do this at present because of difficulties in isolating mycobionts of species that characteristically produce phycosymbiodemes. The resynthesis studies that have been done suggest that a fungus will kill a phycobiont that is drastically different from its normal phycobiont (Ahmadjian, 1982b). The only direct study of the role of phycobiont on lichen development was done with *Usnea strigosa,* a species that does not produce phycosymbiodemes (Ahmadjian & Jacobs, 1982). In this study, the investigators found no evidence for algal control over lichen development.

Until this entire question can be resolved experimentally, there are two equally plausible explanations for presumptive phycosymbiodemes. First, a single fungus is involved and two lichen morphs result from the influence of different algae (the phycosymbiodeme theory). Second, two different fungi are involved, one capable of lichenization with a green phycobiont and the other with a blue-green phycobiont; the appearance of two morphs is due to a close epiphytic association of separate species (the epiphyte theory). Whichever of these two theories is correct, the presence of apparent phycosymbiodemes needs to be explored further. If they are really one fungus, how does the alga elicit such drastically different developmental patterns? If they are really two fungi, why are they so closely associated in nature? Are they related phylogenetically? If so, under what selective pressures did they differentiate? How is the similarity in chemistries best explained? Unresolved questions about the phycosymbiodeme problem are among the more interesting in lichen biology.

6 Whole Thallus Physiology

Despite their phylogenetic diversity, all lichens have one characteristic in common—their growth and metabolic activities are directly dependent upon photosynthates produced by algae. An understanding of algal photosynthesis is therefore necessary in any discussion of lichenized fungal physiology. The physiological interactions between fungus and alga that result in the maintenance of the lichen thallus seen in nature are not entirely understood for *any* lichen species. However, for *most* species studied, the association appears to be a tightly coevolved one that allows both the alga and fungus to coexist in a relatively balanced association.

As Smith (1976) and Lewis (1973) have pointed out, lichen associations are *biotrophic*. That is, unlike parasitic or necrophytic associations, they are characterized by nutrient flow between living cells of the associates. Furthermore, they are coevolutionary systems driven by selective forces that tend to make the associates more dependent upon each other. Other familiar examples of biotrophic associations include mycorrhizae, root nodules, animal gut microflora symbioses, and alga-invertebrate associations. Smith (1976) constructed a framework for comparing biotrophic associations that included the following characteristics:

1. primary exploitation,
2. ancillary modification to facilitate nutrient transfer,
3. balanced growth between the symbionts,
4. structural changes,
5. dependence upon complex metabolites,
6. specificity.

In terms of these characteristics, lichens are functionally similar to other associations and may therefore be particularly well suited to the general study of symbiotic systems.

There is a constant tendency, even among lichenologists, to regard lichenized thalli as single organisms. This is because lichens exhibit few of their normal characteristics when the associates are isolated and cultivated in the laboratory. A whole lichen thallus is a functionally integrated unit with properties quite different from the simple sum of its component parts. This is especially true of the physiological characteristics of lichens. This chapter examines some of these physiological attributes, including: the source of nutrients for lichenized fungi, environmental factors that modify algal photosynthesis and respiration, and the communication and control between fungus and alga that maintains the lichen association in a balanced equilibrium.

PHOTOSYNTHESIS

Photosynthetic Rates

Photosynthetic rates measured for lichen thalli are due entirely to the activity of the captured algal cells. Since the volume of algal cells is usually small in relation to the fungal biomass in a typical lichen thallus, photosynthetic rates measured on a whole thallus dry-weight basis are small compared to vascular leaves (Table 6.1). However, when photosynthetic rates are measured in terms

TABLE 6.1. Maximal Rates of Net Photosynthesis (on an Area and a Dry Weight Basis) Exhibited by Various Groups of Plants

	CO_2 Assimilation	
	$mg\ dm^{-2}\ h^{-1}$	$mg\ g^{-1}\ (dw)\ h^{-1}$
Herbaceous plants		
C_4 plants	50–80	60–140
Crop plants	20–40	30–60
Sun plants	20–50	30–80
Shade plants	4–20	15–25
Deciduous trees		
Sun leaves	10–20 (25)*	15–25 (30)
Shade leaves	5–10	—
Evergreen trees	4–15	3–18
Mosses	~3.0	2–4
Lichens	0.5–2.0	0.3–2.0 (3.0)
Algae	—	~3.0

*Values in parentheses are unusually high values reported in the literature. P. Bannister. 1976. *Introduction to Physiological Plant Ecology*. Oxford: Blackwell Scientific Publications. Reprinted with permission of Blackwell Scientific Publications, Ltd.

of the chlorophyll content of the whole lichen thallus (Table 6.2), it is evident that lichen algae are very efficient producers.

Although few reports are available, it appears that algal cells typically constitute only around 5–10 percent of the total volume of a lichen thallus (Bednar, 1963; Drew & Smith, 1967a). Apparently, algal cells are continuously turned over in lichen thalli, but little direct evidence is presently available. Harris (1971) found that the number of algal cells per unit area of *Parmelia caperata* thalli changed seasonally. Electron microscopy has demonstrated that algal cells frequently divide in the lichen thallus (Ahmadjian & Jacobs, 1970), and there appear to be cells of various ages in the thallus (Jacobs & Ahmadjian, 1973). Algal cell stratification by age has also been observed in electron micrographs (Galun, Paran, & Ben-Shaul, 1970a). If it is eventually determined that marked changes in algal volume occur in nature, it will be necessary to document experimentally, not only the effects such changes have on lichen production capabilities, but also the internal and external mechanisms that control algal biomass fluctuations.

Aside from production abilities inherent in the captured algal cells, rates of whole thallus photosynthesis are regulated by a number of factors. One set of factors has to do with morphological characteristics of the thallus. The absorption of CO_2 by algae is controlled by the diffusion resistance through the upper cortical tissue of the thallus (Collins & Farrar, 1978; Green & Snelgar, 1981), and this may change as the thallus dries (Ertl, 1951; Hill, 1976). Ried (1960b) found that crustose lichens exhibited great variability in photosynthetic rates; he attributed this to differences in thallus thickness. It has only been recently established that morphological features of some lichens (cyphellae, pseudocyphellae, pores, etc.) may influence gas-exchange rates (e.g., Green, Snelgar, & Brown, 1981).

Other important factors that control lichen photosynthetic rates are environmental, and the combined effect of environmental factors on lichen physiological processes has frequently been studied along complex environmental gradients. A recent example of such a study is that of Lechowicz (1982a), who evaluated photosynthetic data for 42 fruticose and foliose lichen species from different latitudes and climatic conditions. Four aspects of the photosynthetic response were compared.

1. the maximal net photosynthetic rate, (P_{max}),
2. the optimum temperature for net photosynthesis, (T_{opt}),
3. the photon flux density for P_{max}, ($PhAR_{sat}$),
4. the relative water content, (RWC_{opt}), or percentage dry weight, (DW_{opt}), for P_{max}.

Only the optimal temperature for net photosynthesis (T_{opt}) exhibited a

TABLE 6.2. **Photosynthetic Rates of Selected Lichen Species Based on Chlorophyll Content**

Species	*Environmental Conditions*	*Rate of CO_2-fixation (μmole CO_2/mg chl/hr)*	*Source*
Hypogymnia physodes	20° C, 20,000 lux	160.0	Farrar (1973)
H. physodes	3.5° C, 60 lux	132.0	Schulze & Lange (1968)
Cladonia subtenuis	23° C, 420 W/m^2	41.0	Rundel (1972)
Sticta filix	20° C, 150 μE m^{-2}sec^{-1}	41.1	Rundel et al. (1979)
S. filix	10° C, 150 μE m^{-2}sec^{-1}	34.5	"
Pseudocyphellaria delisei	10° C, 400 μE m^{-2}sec^{-1}	20.4	"
P. delisei	20° C, 400 μE m^{-2}sec^{-1}	15.4	"

statistically significant trend with latitude; T_{opt} decreased with increasing latitude (Figure 6.1). None of the other variables measured showed a significant latitudinal pattern. Lechowicz suggested that the lack of clear latitudinal trends in lichen photosynthesis is due to a close coupling of lichen activity to temporal sequences of environmental conditions rather than mean conditions approximated by macroclimatic data. Since lichens are known to exhibit quite different photosynthetic reponses in the same habitat, due to differences in physiological responses to particular microhabitat conditions and to morphology (Lechowicz & Adams, 1974a; 1974b; Larson, 1979a; 1979b; Lechowicz & Adams, 1979), it is not too surprising that clear-cut photosynthetic trends along broad macroclimatic gradients are not seen.

Numerous environmental variables have been considered in the study of whole thallus physiology. The level of thallus water saturation has long been known to influence whole thallus metabolic rates. Temperature, light intensity, and availability of nutrients have also been frequently studied. These factors will be considered separately in the sections that follow.

Measuring Lichen Photosynthetic Rates

A number of different methods have been used to measure whole thallus photosynthetic and respiration rates, and there are disadvantages associated with all of them (Green & Snelgar, 1981). Recent methods have made use of an infrared gas analyzer system (IRGA). There are also problems with this approach (Larson & Kershaw, 1975a; Green & Snelgar, 1981), but IRGA represents the most practical and reliable method for use in ecophysiological studies of numerous environmental factors.

There has been some recent discussion about the best IRGA system to use in lichen studies. Larson and Kershaw (1975a) introduced a simple method that involved placing lichen thallus fragments in small, unventilated cuvettes and equilibrating them under constant environmental conditions. Samples are then taken from the cuvette with a syringe and injected directly into the analysis tube of the IRGA. The maximum deflection of the IRGA is used to determine the concentration of CO_2 in the injected sample. Rates of photosynthesis can be determined by measuring the rate of CO_2 concentration decrease within the cuvette.

In their preliminary description of the technique, Larson and Kershaw reported finding no significant differences in gas exchange measurements made with ventilated and unventilated cuvettes. They suggested that changes in CO_2 concentration in unventilated cuvettes have little effect on lichens because lichens have relatively low CO_2 photosynthetic saturation points. Their measurements supported this hypothesis.

Other investigators have criticized the use of the cuvette technique, however. Green and Snelgar (1981) reported results that called into question

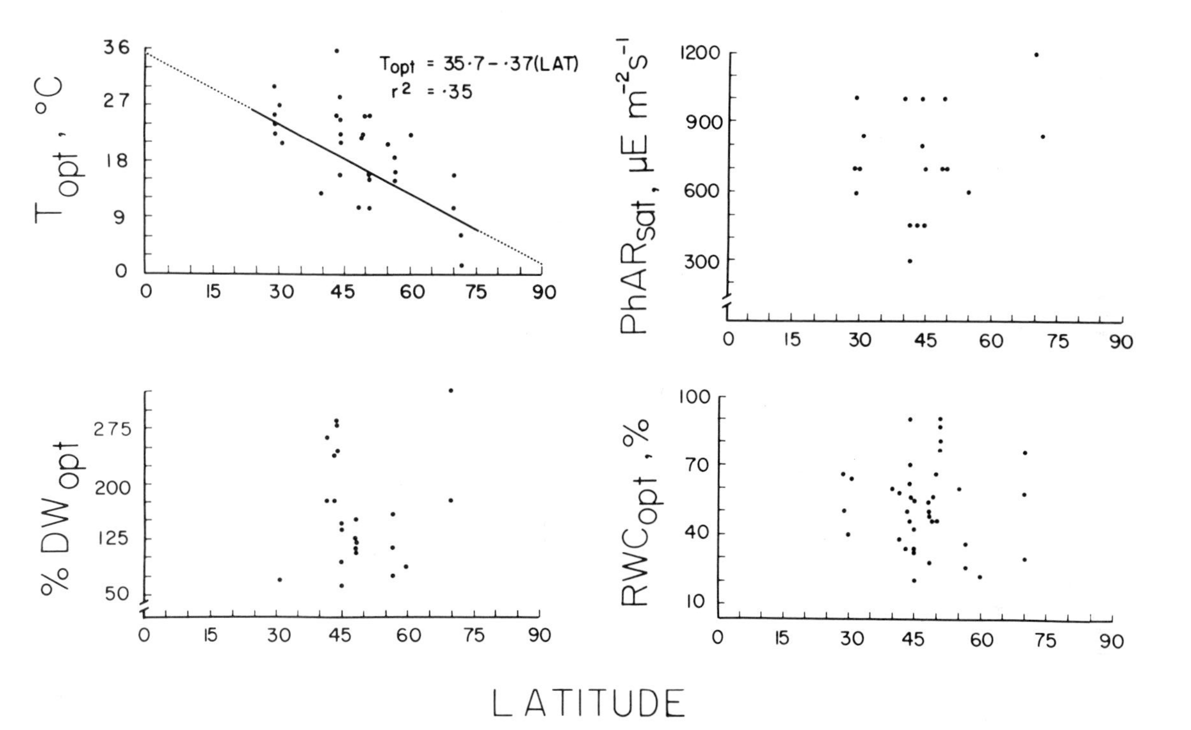

FIGURE 6.1. Scatter diagrams of photosynthetic response parameters on latitude from which the lichens were collected. The only significant relationship is that of T_{opt} on latitude. From M.J. Lechowicz. 1982a. "Ecological trends in lichen photosynthesis." *Oecologia* (Berlin) 53:330–336. Reprinted with permission of Springer-Verlag and the author.

the CO_2 independence of lichen photosynthesis at high CO_2 levels. Their gas-exchange investigations of various members of the Stictaceae showed that relatively high CO_2 concentrations were required to saturate photosynthesis. They also found that lichen photosynthetic rates were closely coupled with CO_2 concentration at or near ambient levels, calling into question the use of unventilated cuvettes, at least in certain situations.

Using a ^{14}C method, Nash et al. (1983) also found that CO_2 dependencies were basically the same as those reported for higher plants. However, they suggested that this does not necessarily proscribe the use of the cuvette method. Further support for this idea comes from Link et al. (1983), who obtained essentially the same CO_2 dependency curves using $^{14}CO_2$ techniques and both flow-through and cuvette IRGA techniques.

At the present time, it appears that the advantages of the cuvette system, including speed and low cost, outweigh the major disadvantage—reduced precision.

Dark Carbon Fixation

Estimates of lichen production seldom take into account rates of dark CO_2 fixation, although it has been suggested (Drew & Smith, 1967a; Farrar, 1973; 1976d) that dark fixation may constitute 5–10 percent of light reaction rates. Recently, Kershaw et al. (1979) examined various environmental controls over dark CO_2 fixation in *Pseudoparmelia caperata* and *Peltigera polydactyla* in an attempt to assess the role of dark CO_2 fixation in the ecology of these lichens. They measured dark rates that were 10–12 percent of equivalent light rates for *P. caperata* and 15–27 percent for *P. polydactyla*, suggesting that dark CO_2 fixation is an important metabolic process in lichens. They also found lower temperature optima for dark rates than for light rates, suggesting a certain degree of physiological adaptation, since night temperatures are lower than day temperatures.

Age Dependence of Photosynthesis

Since lichens are relatively long-lived organisms, there is the obvious question of age dependence over whole thallus photosynthetic production. Do younger tissues contribute more to whole thallus production than older ones? Does thallus senescence result when the proportion of dependent old tissue to productive young tissue reaches some critical level? These questions have stimulated some interesting research into the subject of age-dependent physiological responses of lichens and the associated subject of lichen senescence.

Only a few studies can be cited here. Nash, Moser, and Link (1980) observed nonrandom gas exchange patterns in arctic lichens from Anaktuvuk Pass, Alaska. The youngest, actively growing portions of the thalli of *Cladina rangiferina, Cladina stellaris*, and *Parmelia separata*, had the highest rates of photosynthesis and respiration, whereas the older, basal portions were relatively inactive. Moser and Nash (1978) had previously found similar patterns of activity in *Cetraria cucullata* populations from the same area. For the *Cladina* species, Nash, Moser, and Link (1980) also measured a decrease in total chlorophyll concentration from growing tips to basal portions of the thalli, a result also obtained by Kärenlampi (1970) for *Cladina alpestris* from northern Finland.

Lechowicz (1983) has made the most comprehensive study of age-dependent whole thallus photosynthesis in lichens. He subdivided thalli of *Cladina stellaris* into individual, branched whorls of known age and measured photosynthetic rate—on a thallus dry weight basis—and biomass supported by each whorl. He found that photosynthesis declined steadily with age, whereas whorl biomass increased (Figure 6.2). He suggested that as thalli

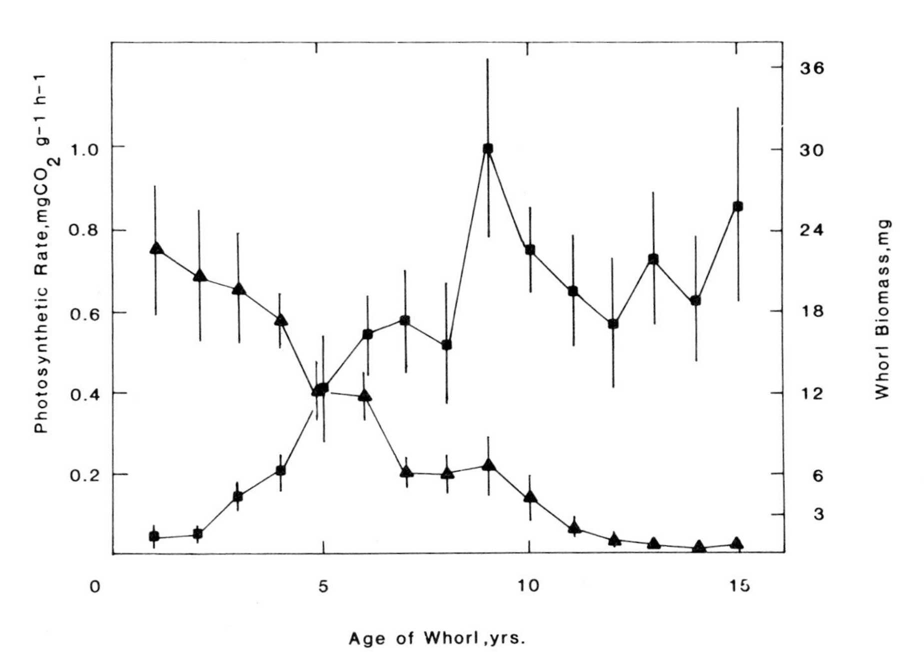

FIGURE 6.2. The age dependence of dry biomass (■) and photosynthetic rate (▲) in individual branch whorls within a thallus of *Cladina stellaris*. The values are the mean ± standard error of 10 replicates. From M.J. Lechowicz. 1983. "Age dependence of photosynthesis in the caribou lichen *Cladina stellaris*." *Plant Physiol.* 71:893–895. Reprinted with permission of the American Society of Plant Physiologists and the author.

grow in height, basal portions became more shaded, reducing photosynthetic capacity—an interpretation offered before by other investigators (Kershaw & Harris, 1971). Lichen thalli with growth forms similar to *C. stellaris* are apparently supported mainly by photosynthetic production of the youngest tissues. There must be considerable translocation of photosynthates from growing tips to older basal portions of these lichen thalli, and there is probably an upper limit to the amount of unproductive basal tissue that can be maintained. However, as Lechowicz (1983) has pointed out, this is an area where little empirical evidence is available.

RESPIRATION

Respiration Rates

Unlike photosynthetic rates, rates of respiration measured for whole lichen thalli are due to the activity of both the fungal and algal components. Since it is difficult to measure metabolic rates of the components separately without disturbing the association, most respiration rates for lichens are reported on a whole thallus dry weight basis. Values in the literature are variable, but generally range under optimal conditions from 0.2 to 2.0 mg CO_2 g^{-1} h^{-1} (Quispel, 1960; Hale, 1974). Quispel (1960) mentions that patterns of lichen respiration rates are not readily apparent. Species belonging to the families Stictaceae and Peltigeraceae generally have high rates, while those from Ramalinaceae generally have the lowest rates. However, no essential differences in optimal rates are seen when thalli collected at different latitudes or seasons are compared. Ahmadjian (1966b) also commented that respiration rates of lichens and their isolated symbionts collected from Antarctic, temperate, and tropical habitats revealed no important differences or indications of climatic or ecological adaptations.

Respiratory rates have not frequently been measured in thalli of different ages, nor for lichen tissues of different ages in the same thallus. Ried (1960b) found no difference in respiratory rates of juvenile and older thalli of *Umbilicaria pustulata*, although photosynthetic rates were different. In a gas-exchange study of the carbon balance of three arctic lichens, Nash, Moser, and Link (1980) found that the youngest, actively growing tissue exhibited the highest rates of photosynthesis and respiration, and that rates declined in older tissue. This study demonstrated that lichen thalli are not physiologically homogeneous entities and that senescence can occur in lichens. This needs to be documented in other species as well.

Respiration rates are strongly coupled with environmental temperature and water conditions. Peet and Adams (1972) measured the photosynthetic and respiratory responses of *Cladina subtenuis* from southern Missouri at various temperatures and hydration levels. They observed the highest net photosynthesis at 18° C and 80 percent relative saturation; however, at the same temperature, respiration declined rapidly from 100 percent to 80 percent relative water saturation. As drying continued, there was a gradual decrease in respiration rate until 15 percent relative water content, at which point gas exchange declined sharply. These results indicated that at 18° C, the optimal temperature for photosynthesis, dark respiration was less sensitive to water content than was net photosynthesis. A similar response was found for *Cladina rangiferina* (Adams, 1971).

The fact that lichens exhibit such low productivities suggests that the fixed carbon acquired during photosynthesis is almost all lost to dark respiration. As Farrar (1976b) has remarked, this carbon loss can be accounted for by three distinct processes: basal respiration, resaturation respiration that occurs when dry thalli are wetted (Smith & Molesworth, 1973), and leakage of carbon due to reversible membrane damage to fungus and alga during resaturation. Therefore, low lichen productivities appear to be a consequence of respiratory responses to an unpredictable, fluctuating environment.

Net carbon balances for lichens are not constant throughout the year. Negative balances can be measured quite frequently on days when net photosynthesis during the day cannot compensate for respiratory losses at night. Kappen et al. (1979) measured the diurnal carbon balance for *Ramalina maciformis* in the Negev desert from March 1971 to February 1972. They found high productivities during April and the winter months and high respiratory losses during the summer and fall. On 218 days the balance was positive, on 49 days it was zero, and on 88 days it was negative. The positive CO_2 balance that occurred over any 24-hour period for this species was found by Lange, Schultze, and Koch (1970) to be most dependent upon the early morning rate of thallus desiccation. Almost all production was found to occur within three hours after dawn. Thereafter, low thallus water content resulted in reduced gas exchange until evening thallus hydration allowed dark respiration to begin. If the thallus desiccated too rapidly in the early morning hours, photosynthetic production could not balance the amount of carbon lost from respiration.

Respiration in Various Lichen Tissues

Ultimately, more research effort will need to be focused on the relative contributions to total thallus dark respiration made by the fungal and algal components of lichens in various environments. Harley and Smith (1956) and Smith (1960a; 1960b; 1960c) found that the algal region of *Peltigera polydactyla* has a greater respiration rate than the medulla and that it absorbs more glucose, asparagine, phosphate, and ammonia than the medulla. The

TABLE 6.3. Mean Dark Respiration Rates of Lichen Tissues from *Peltigera canina* Measured Using a Cartesian Diver Technique

	Dark Respiration Rate ($\mu l\ 0_2$ *uptake* mm^{-3} *tissue* hr^{-1})
Algal Layer	1273
Medulla	157
Cortex	681
Rhizines	260
Apothecia	14

L.C. Pearson & E. Brammer. 1978. "Rate of photosynthesis and respiration in different lichen tissues by the Cartesian Diver technique." *Am. J. Bot.* 65:276–281. Reprinted with permission of American Journal of Botany.

medulla had a greater water-holding capacity and may therefore function as a water storage region. Pearson and Brammer (1978) measured dark respiration in various tissues of the lichens *Peltigera canina* and *Evernia prunastri* using a Cartesian diver technique. For both lichens, the algal-containing tissues were metabolically more active than medullary tissue. In *Peltigera canina* (Table 6.3), cortical tissue as well as algal-layer tissue had fairly high dark respiration rates. Medulla and rhizines exhibited low rates, and reproductive tissue had the lowest rates. An apparent light-induced decrease in respiration was also observed in the medullary tissue of both lichens, a finding that was difficult to explain. Apparently, lichen tissues make different contributions to whole thallus respiratory rates, and the lichen symbionts respond differently to environmental fluctuations.

Seasonal Acclimation

Acclimation occurs in lichens when rates of photosynthesis, respiration, or other metabolic processes are maintained throughout the year despite seasonal fluctuations in optimum temperature for maximum carbon gain (Larson, 1980a). There have been a number of studies of seasonal changes in lichen metabolic activity including a few that have focused on seasonal ultrastructural changes (Jacobs & Ahmadjian, 1969; Holopainen, 1982), but very few have demonstrated that seasonal acclimation occurs in lichens. Larson (1980a) summarized the information from these studies and found only three cases in which seasonal acclimation unequivocally can be said to occur. All three species (*Alectoria ochroleuca, Cetraria nivalis, and Bryoria nitidula*) are subarctic and all have green phycobionts. Other species show distinct seasonal changes in photosynthetic, respiratory, or nitrogen-fixation rates in response to temperature, moisture, or light intensity. Although it had earlier been suggested (Larson and Kershaw, 1975b) that acclimation should occur in habitats where there are pronounced seasonal variations in microclimatic conditions and that lichens with green phycobionts may have greater potential to acclimate (Kershaw, 1975b), neither of these suggestions appears to be supported by the evidence collected so far. There are no strong relationships between the pattern of seasonal metabolic activity and latitude or type of phycobiont. As Larson (1980a) has remarked, we are far from understanding these patterns despite the relatively large amounts of data collected.

WATER RELATIONS

As poikilohydric plants, lichens have no specific organs for water conservation and are not able to control their moisture content to the same degree that vascular plants can (Blum, 1973). Thallus moisture content thus appears to be strongly coupled to environmentally available moisture. Since lichen physiological activity, particularly algal photosynthesis, is controlled by available water, lichens exhibit a number of morphological and distributional patterns that reflect water availability. For most lichens, liquid water absorption is a very rapid process, similar to that exhibited by a hydrophilic gel. Some desert lichens are relatively nonwettable due to thallus surface characteristics that inhibit water absorption. This may be an adaptation to prevent water loss rather than to repel water.

Physiological responses to changing water conditions are also rapid in many lichens. An example of this is provided by Martin Lechowicz and Michel Groulx, who studied desiccation recovery of photosynthetic activity after wetting in a number of lichen species (Figure 6.3). In all species, photosynthesis reaches a saturation level rapidly, usually well within two hours after rewetting, suggesting that many lichens are physiologically opportunistic. Such recovery behavior is of obvious adaptive value in stressful habitats that fluctuate widely in water content.

Snelgar, Green, and Wilkins (1981) have found that thallus water content profoundly influences the resistance to CO_2 uptake in numerous lichen species. They measured the net assimilation rates of four New Zealand lichens at varying thallus water contents. In all cases, the relationship between CO_2 resistance and water content was found to be triphasic, with high resistances observed at both low and high water contents and low resistances (with associated high assimilation rates), observed at intermediate water contents. The investigators suggested that the increased resistance at high water contents was due in part to the presence of water on the lower cortex, particularly in tomentose species capable of holding large amounts of water.

Absorption of water vapor by lichens is perhaps more important than absorption of liquid water. Most studies of lichen water content in the field

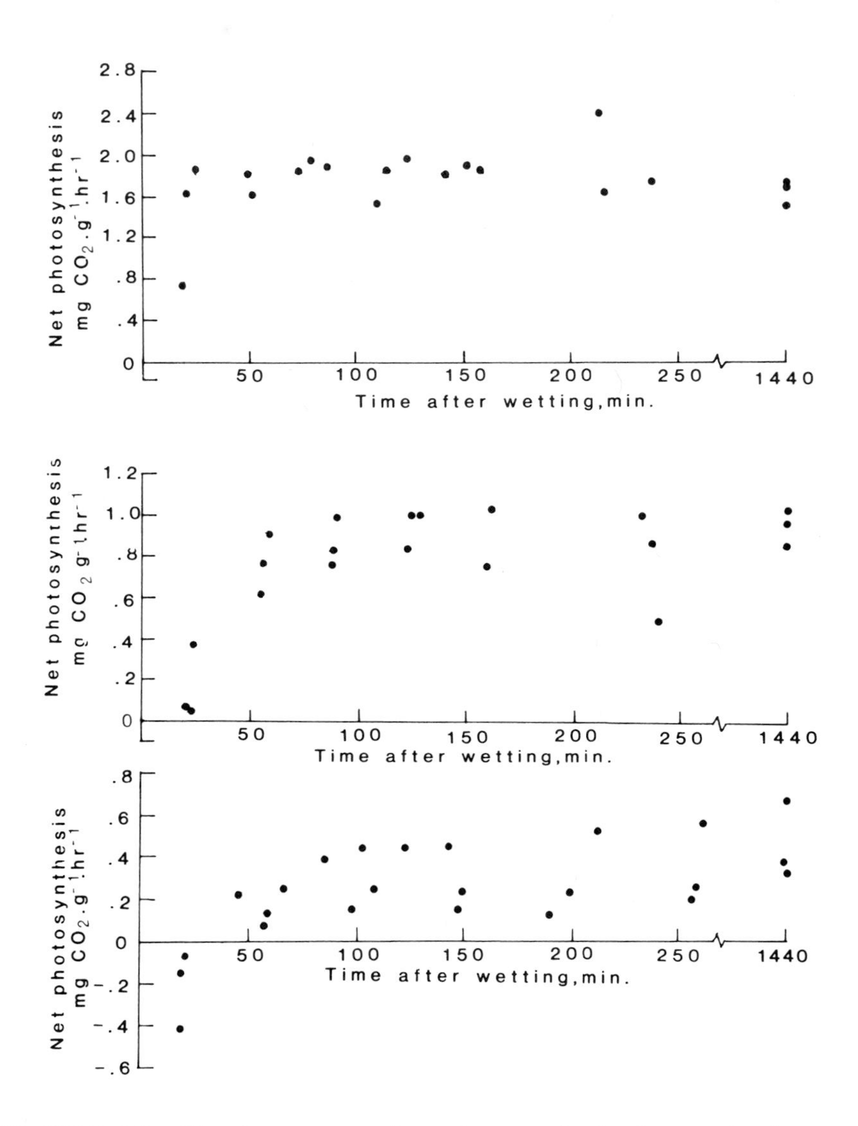

FIGURE 6.3. The time course of net photosynthesis after wetting in three subarctic lichens. The carbon dioxide exchange rates were measured in an open infrared gas analysis system. The lichens were maintained at 15° C and 75–85 percent relative water content under an irradiance of 1000 μE m^{-2}sec^{-1}. Between readings the samples were incubated at 25° C and 100 percent relative water content under 250 μE m^{-2}sec^{-1} on a 12/12 photoperiod. The species from top to bottom are: *Cetraria cuculata, Cladonia rangiferina,* and *Nephroma arcticum.* Unpublished data provided by M.J. Lechowicz and M. Groulx.

have shown a direct relationship between thallus moisture content and relative humidity (Figure 6.4), the rate of uptake being dependent on the particular morphological design exhibited by the lichen (Ried, 1960a; 1960b; 1960c; 1960d). In most studies, living and dead thalli exhibit similar water vapor absorption rates, although differences due to conformational thallus changes during heat killing have been observed (Heatwole, 1966; Cuthbert, 1934; Smith, 1962; Showman & Rudolph, 1971). Sorediate and isidiate lichen species tend to have greater absorption rates than species without soredia or isidia (Blum, 1973). However, there do not appear to be marked correlations between lichen water absorption rates and availability of water in the environment. Species from xeric habitats are not necessarily better at water accumulation than species from more mesic habitats, nor is thallus water content for optimal photosynthesis correlated with ecological conditions. However, more work is needed before accurate generalizations of this sort can be made.

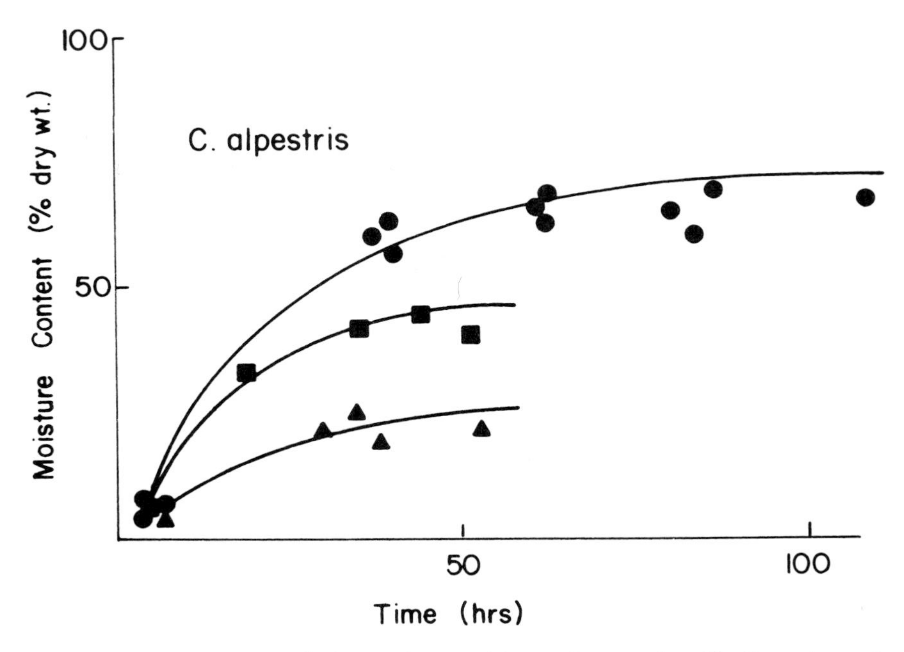

FIGURE 6.4. Rates of absorption and loss of water by *Cladina alpestris* under different conditions of relative humidity. Values near 100 percent R.H. (●); values at 75 percent R.H. (▲); values at 100 percent R.H. (■), but using thalli that had been killed by oven drying at 105° C for 12 h. Adapted from H. Heatwole. 1966. Reprinted by permission from *Mycologia* 58: 148–156. Copyright 1966, The New York Botanical Garden.

Morphological Control over Water Relations

Although lichens are not able to control water uptake and loss in the same way as vascular plants, recent studies suggest that there is more control than previously thought. Larson and Kershaw (1976) investigated the water relations of numerous populations of *Alectoria ochroleuca, A. nitidula, Cetraria nivalis* and *Cladina alpestris* from different habitats in arctic and subarctic Canada. They found evidence for morphological control over water relations in all of the species they sampled. Three control mechanisms were apparent in these lichens:

1. an increase in surface area-to-weight ratios through finer branching,
2. a change in surface characteristics,
3. a change in growth habit.

These modifications resulted in increased evaporative resistances that permitted greater photosynthetic productivity.

Larson (1981) conducted a series of experiments in a "raining wind tunnel" designed to determine if certain morphological characters (rhizines, isidia, papules, apothecia, and lamellae) affected thallus saturation rates. Thallus disks with and without various structures were placed in a wind tunnel with an attached misting device to provide a water source. Rhizines facilitated wetting in some species (e.g., *Umbilicaria* species), but not in others (e.g., *Peltigera* species). Lamellae produced by *Umbilicaria muhlenbergii* also increased saturation rates. Other structures had no influence on thallus saturation. Larson also compared normal thallus saturation patterns with those observed when either the upper or lower cortex was blocked with a water-repellent substance. Some species accumulated water more rapidly from one surface or the other; others absorbed water equally from both surfaces.

Kunkel (1980) also investigated morphological variations in ecological populations (ecads) of *Aspicilia desertorum* from distinctly different microhabitats in western Colorado. He found that individuals from cooler microhabitats tended to exhibit a fruticose growth form and those from more exposed xeric microhabitats tended toward a crustose growth form. The crustose individuals had lower mean water-holding capacities but higher mean algal volumes than fruticose individuals (Table 6.4). There was also a marked difference in the distribution of fungal biomass to various tissues. The increased thickness of the fruticose thalli was due almost entirely to an increase in medullary tissue. It was apparent from this study that more fungal biomass could be maintained per unit algal biomass in *Aspicilia desertorum* individuals from more mesic microhabitats, and the increase was accompanied by—and perhaps also caused by—an increase in water-holding capacity by the medulla.

TABLE 6.4. Morphological Characteristics of *Aspicilia desertorum* Ecads from Exposed Microhabitats (Crustose) and Cooler, More Mesic Microhabitats beneath Snow Patches (Fruticose) in Western Colorado

	Crustose		Fruticose	
	Thickness (μm)	%	*Thickness (μm)*	%
Thallus	407	100	1044	100
Cortex	67	16	114	11
Algal layer	93	23	61	6
Medulla	247	61	869	83
#Algal cells/g dry wt. ($\times 10^8$) ± S.D.	3.14 ± 0.19		1.74 ± 0.45	
Max. g water/g dry wt. ± S.D.	1.85 ± 0.36		2.54 ± 0.22	

G. Kunkel. 1980. "Microhabitat and structural variation in the *Aspicilia desertorum* group (lichenized Ascomycetes)." *Am. J. Bot.* 67:1137–1144. Reprinted with permission of American Journal of Botany.

Rundel (1974) found a correlation between morphology, distribution, and water-holding capacity in *Ramalina menziesii* that suggested a morphological control over water relations in this lichen. *Ramalina menziesii* is a fog-zone lichen in California and exhibits a low desiccation tolerance. In the summer months when precipitation is essentially absent, *R. menziesii* depends almost entirely on fog and moisture-laden sea breezes for water. Rundel suggested that the unique reticulated thallus structure of this lichen (Figure 6.5) is an adaptation to maximize water-holding capacity. Comparisons of water saturation rates and water-holding capacity of *R. menziesii* with *Letharia vulpina*, a nonreticulate pendulous species, showed no difference in rates of water absorption but higher water-holding capacities for *R. menziesii* thalli. The reticulations in the thalli of this lichen apparently trap water and hold it, thereby allowing photosynthetic periods to be extended. Morphological variations in *R. menziesii* also correlated with microenvironmental factors. Morphs exhibiting the greatest degree of reticulation were found to occur in areas beyond the direct reach of coastal fogs, but where maritime conditions produced humidities near 100 percent each morning. Morphs showing little reticulation occurred in more favorable foothill areas where heavy fogs saturated thalli almost daily; large numbers of fine reticulations are apparently less necessary in these habitats. If development of the reticulations seen in *R. menziesii* thalli are under genetic control, they are apparently modified by natural selection to maximize water-holding capacity in habitats where atmospheric moisture is most limiting.

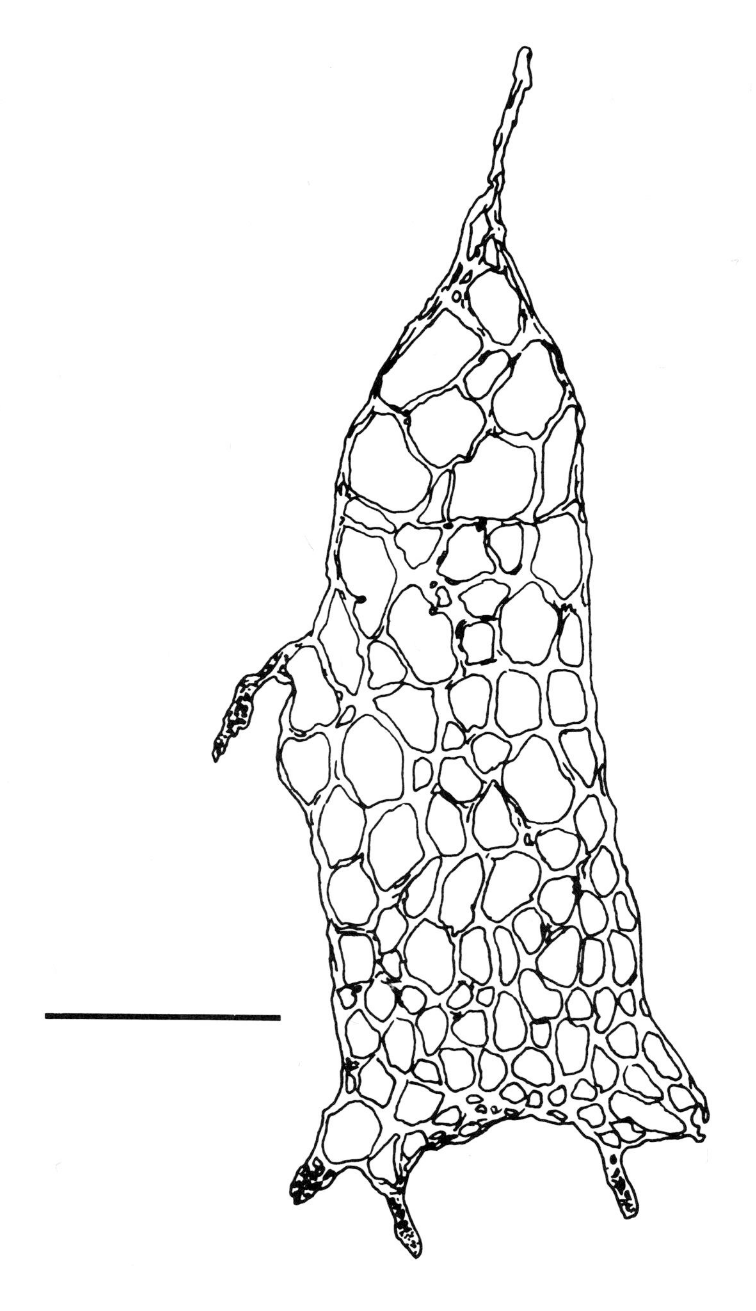

FIGURE 6.5. *Ramalina menziessi,* bar = 1 cm. Provided by M.E. Hale, Jr.

Fluctuating Water Conditions and Physiological Buffering

It has been suggested that fluctuating water conditions are necessary for the maintenance of a net carbon balance in lichens (Farrar, 1973; 1976a; 1976b). Figure 6.6 shows that when maintained under conditions of constant saturation, thalli of *Hypogymnia physodes* exhibited rapid reductions in photosynthetic activity; when subjected to wetting and drying cycles, a high rate of photosynthesis was retained (Farrar, 1973; 1976a; 1976b).

It is very likely that lichens experience such drying and rewetting cycles in nature, but to varying degrees depending on the microhabitat. There is an energetic cost to a lichen thallus in such a fluctuating environment. Smith and Molesworth (1973) discovered two distinct effects of rewetting:

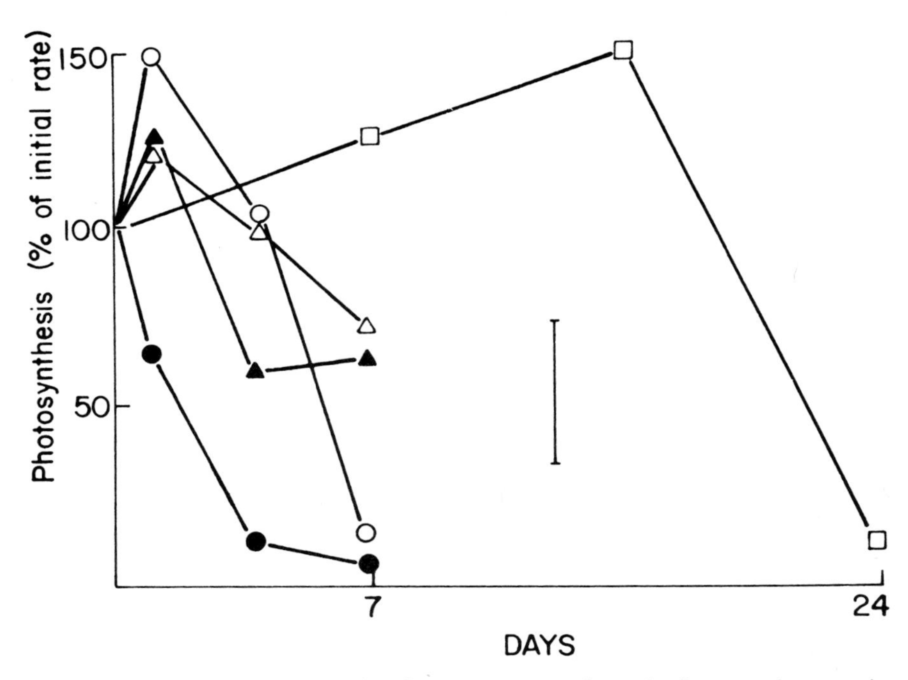

FIGURE 6.6. Photosynthesis of *Hypogymnia physodes* from various environmental regimes. The vertical bar represents the least significant difference for $p = 0.05$. Environments: soaked thalli in light at 20° C (○); soaked in dark at 20° C (●); soaked in light at 5° C (△); soaked in dark at 5° C (▲); thalli subjected to wetting and drying in a 12 h day at 5° C (□). Redrawn from J. Farrar. 1976d. "The lichen as an ecosystem: Observation and experiment." In *Lichenology: Progress and Problems* (D.H. Brown, D.L. Hawksworth, & R.H. Bailey, eds), pp. 385–406. London: Academic Press. Reprinted with permission of Academic Press.

1. an immediate nonmetabolic release of gas, mainly CO_2
2. a rapid rise in respiration to rates well above basal respiration and involving different pathways.

They also found that *Peltigera polydactyla* thalli from mesic habitats were more sensitive to these rewetting effects than thalli of *Xanthoria aureola*, which are found in more xeric habitats. It is still not clear whether resaturation represents an adaptation to fluctuating environments or a temporary loss of metabolic control during rewetting that must be tolerated by the lichen.

Lichens are apparently able to withstand periods of drying and rewetting despite the resaturation effects because they maintain relatively high concentrations of ethanol-soluble carbohydrates that can serve as respirable substrates during periods of rewetting. This has been called "physiological buffering" by Farrar (1976b).

Thus, under normal circumstances, lichen metabolism is divided into two sets of processes:

1. Growth and repair processes.
2. Respiratory and stress-resistant processes.

The soluble respiratory carbohydrates are turned over rapidly, whereas the insoluble growth and repair compounds are used as surpluses become available. This may explain the slow growth rates and low productivities of lichens in nature.

A lichen's ecological success may relate more to its ability to maintain adequate thallus moisture than to its maximum photosynthetic rate in an ideal environment. A good example of this is provided by Lechowicz and Adams (1974b), who measured the net CO_2 exchange responses of three *Cladonia* species from the Wisconsin Pine Barrens to irradiance, thallus temperature, and thallus water content. Although one of these species, *C. uncialis*, exhibited the lowest net photosynthetic rates, it was the most frequently observed lichen in the region. The investigators attributed this to the slower drying rate of *C. uncialis*, which allows it to photosynthesize longer than other *Cladonia* species under comparable drying regimes.

Rapid water losses by lichens in fluctuating environments would have a profound effect upon their energy budgets. Hoffman and Gates (1971) measured the total diffusion resistance (a combination of internal and boundary layer resistances) of *Xanthoparmelia conspersa* and numerous other plants to evaluate water loss as an energy dissipation mechanism in energy exchanges between various plants and their environments. Results (Table 6.5) indicated that cryptogams (*X. conspersa* and a liverwort, *Reboulia hemisphaerica*) had very low diffusion resistances compared to vascular plants. Comparable resistances were found only in very young seedlings of

TABLE 6.5. Minimum Diffusion Resistances Calculated for Several Vascular and Nonvascular Plant Species

Species	*Total Diffusion Resistance (sec cm^{-1})*
Helianthus annuus	
Cotyledon stage	0.26
4-leaf stage	1.54
6-leaf stage	1.81
9-leaf stage	4.13
13-leaf stage	3.85
Datura stramonium	
6-leaf stage	2.37
Pelargonium zonale	
4-leaf stage	2.76
Cucumis sativus	
5-leaf stage	3.13
Coleus blumei (Victoria)	
6-leaf stage	10.60
Kalanchoë pinnata	
6-leaf stage	13.80
Epiphyllum ackermannii	54.00
Xanthoparmelia conspersa	0.57
Reboulia hemisphaerica	0.44

G.R. Hoffman & D.M. Gates. 1970. "Transpirational water loss and energy budgets of selected plant species." *Oecol. Plant.* 6:115–131. Reprinted with permission of Oecologia Plantarum.

vascular plants (e.g., *Helianthus annuus*, cotyledon stage). It is apparent from these results that cryptogams solve their energy-exchange problems in ways very different from vascular plants. As temperatures increase, evaporative energy dissipation in a lichen thallus is very rapid, thus maintaining thallus temperatures below lethal levels while the thallus is moist (Hoffman & Gates, 1970). Once in a dry state, lichens can tolerate higher temperatures than when moist (Haynes, 1964) and reradiation and convective energy dissipation replace evaporative cooling. The ability of many lichens to begin metabolic activity rapidly when hydrated and to shut down rapidly when dry may thus be an adaptation to fluctuating conditions. It also suggests they are opportunistic; they tend to maximize production during favorable periods and reduce metabolic loss and risk of injury during unfavorable periods.

Algal Desiccation Resistance

One of the ways in which lichen fungi are supposed to cultivate their captured algae is by reducing moisture stress on the algae by protecting them from the

TABLE 6.6. Lowest Water Potentials Permitting Photosynthesis by Whole Lichen Thalli and Isolated Lichen Algae. Values are Lowest Water Potentials at which Detectable Photosynthesis Occurred. Values in Parentheses Are Relative Humidities Corresponding to the Respective Water Potentials

Lichen Species	*Whole Thallus*	*Isolated Alga*
Cladina submitis	−56 bar (96.0%)	−28 bar (98.0%)
Lepraria membranacea	−307 bar (80.0%)	−145 bar (90.0%)
Usnea substerillis	−56 bar (96.0%)	−7 bar (99.5%)
Letharia vulpina	−164 bar (88.5%)	−28 bar (98.0%)
Rock alga (unlichenized)	—	−145 bar (90.0%)

T.D. Brock. 1975. "The effect of water potential on photosynthesis in whole lichens and in their liberated algal components." *Planta* 124:13–23. Reprinted with permission of Springer-Verlag and the author.

environment. There have been few attempts to investigate this. Brock (1975) measured lichen phycobiont photosynthetic activity in both lichenized and nonlichenized (i.e., isolated from the lichen thallus) conditions. Four lichen species were used: *Lepraria membranacea, Cladina submitis, Usnea substerilis*, and *Letharia vulpina*. Brock found that algal photosynthesis could continue at lower water potentials when the alga was in the lichen thallus than when it was isolated, even though the liberated alga was not affected by its removal from the thallus and was able to photosynthesize quite well on its own (Table 6.6). Brock also found that an unidentified "rock alga" was able to photosynthesize at relatively low water potentials, but not as low as the lichenized phycobiont of *L. membranacea*. This may be part of the explanation for the relatively infrequent occurrence of free-living *Trebouxia* species in nature. Comparable measurements of desiccation resistance of *Trebouxia* species and other terrestrial algae need to be done to determine if the protection afforded *Trebouxia* cells by the lichen fungus has resulted in a requirement for lichenization for these common phycobionts.

Cellular and Molecular Effects of Water Relations

Ultimately, lichen desiccation resistance will be understood at the cellular and molecular levels. Particularly needed are studies on the desiccation resistance of metabolic systems in both the lichen alga and fungus. A recent study of desiccation resistance of lichen metabolic processes was reported by Cowan, Green, and Wilson (1979). They used a tritium label to study the anhydrobiotic metabolism of *Ramalina celastri* and *Peltigera polydactyla*. Incorporation of the tritium label into compounds from the major classes of metabolites at various humidities (Figure 6.7) demonstrated that different metabolic systems

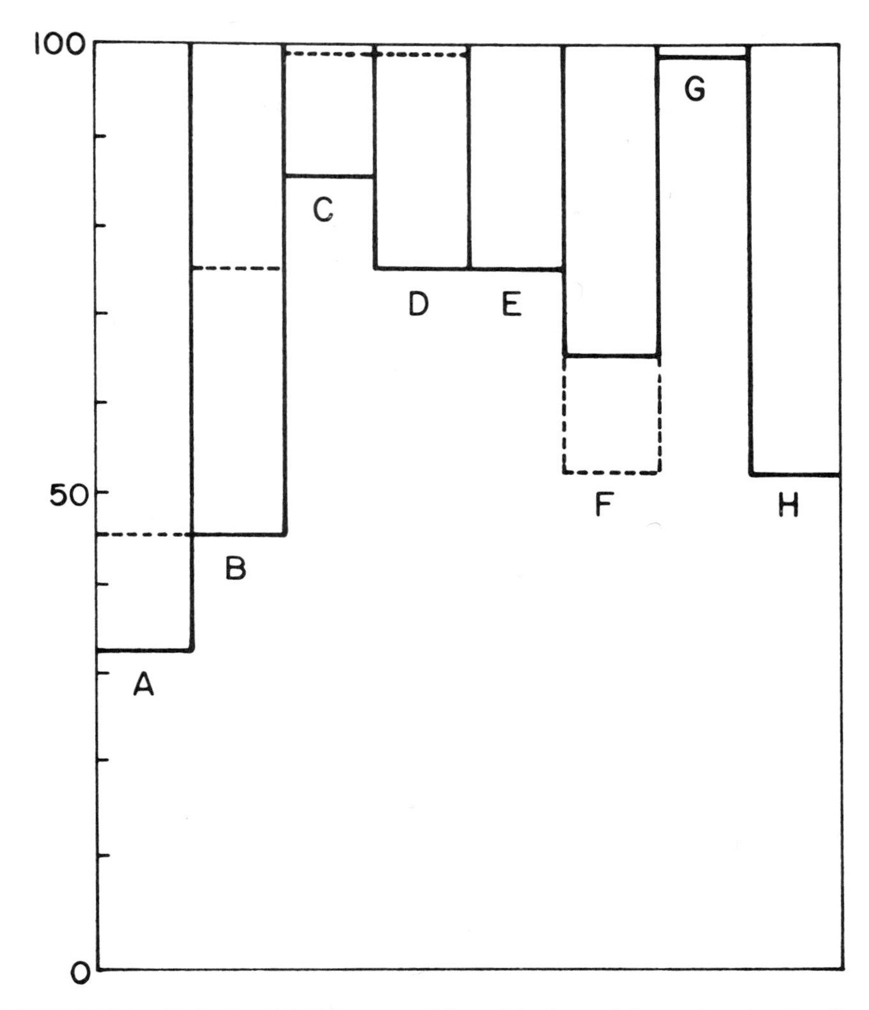

FIGURE 6.7. Relationship between tritium labeling of the major classes of metabolites and the experimental incubation humidity for *Ramalina celastri* (——) and *Peltigera polydactyla* (------). A, amino acids; B, organic acids; C, sugars; D, sugar phosphates; E, total lipids; F, sugar alcohols; G, soluble macromolecules; H, insoluble cell residues. From D.A. Cowan, T.G.A. Green, & A.T. Wilson. 1979. "Lichen metabolism. 1. The use of tritium labelled water in studies of anhydrobiotic metabolism in *Ramalina celastri* and *Peltigera polydactyla*." *New Phytol.* 82: 489–503. Redrawn with permission of Academic Press.

have different levels of resistance to desiccation. Amino acids, organic acids, and sugar alcohols were found to have incorporated the label at relatively low humidities. Apparently, enzymes associated with the tricarboxylic acid cycle and related transamination reactions, and enzymes associated with sugar alcohol metabolism are capable of functioning at fairly low water potentials. Since soluble carbohydrates are important as physiological buffers (Farrar, 1976b; Green, 1970), the ability of lichens to synthesize these compounds at low water potentials may represent an adaptation to fluctuating water conditions.

The rapid responses of lichen photosynthesis to changes in relative humidity indicate the operation of some kind of water-sensitive switch, turning the processes on or off, depending on whether conditions are favorable or unfavorable. Sigfridsson (1980) studied luminescence and fluorescence in *Cladina impexa* and *Collema flaccidum* at different levels of humidity, and found rapid changes in a number of different photosynthetic processes as humidity is varied. In a study of desiccated samples of *C. impexa* and the phycobiont *Trebouxia pyriformis*, Sigfridsson and Öquist (1980) also showed that energy distribution between photosystem II and I (PSII & PSI) is regulated by relative humidity (RH). As RH decreased, captured excitation energy was distributed preferentially to PSI instead of being reemitted as fluorescence from PSII; therefore, fluorescence was observed to decrease. These changes in energy distribution were interpreted to be caused by membrane conformational changes at different water content levels.

The actual mechanism of control over internal water potential is not known for lichens, but there have been some attempts to explain this in terms of rapid synthesis of carbohydrates and/or free amino acids in response to water stress. Feige (1973) found that the coastal marine lichen, *Lichina pygmaea*, responds to water stress caused by high salt concentrations by increasing the synthesis of the carbohydrate mannosidomannitol. In the laboratory, when the salt concentration of the incubation medium is changed, there is a corresponding change in mannosidomannitol concentration, suggesting that this compound may play an important role as an osmotic regulator in this marine lichen. In contrast, Weigel and Jäger (1979) showed no significant change in the concentration of several amino acids, especially proline, for the lichen *Pseudevernia furfuracea* in response to drought stress. The paucity of data on lichen osmoregulation at the cellular level obviously limits the degree to which generalizations can be made here.

Ascaso (1978) investigated the ultrastructural modifications induced in thalli of *Parmelina tiliacea*, *Lasallia pustulata*, and *Ramalina protecta* by changes in relative humidity (RH). At RH of 40–60 percent, the pyrenoid matrix of algal chloroplasts was found to be partially disorganized and the number of pyrenoglobuli, presumably lipid storage products (Jacobs & Ahmadjian, 1969; 1971a; 1971b) that seem to increase in damp thalli

(Rudolph & Giesy, 1966; Peveling, 1969), was drastically reduced. These effects have been observed before (Peveling, 1968; Jacobs & Ahmadjian, 1973; Ascaso & Galvan, 1976a); however, an explanation is not yet possible. Storage starch granules were also found to be numerous at RH 40–60 percent, possibly the result of net accumulation of photosynthates during periods of low fungal metabolic activity. At higher humidities (RH 60–80 percent), the number of pyrenoglobuli increased and starch inclusions were absent, reflecting a lower photosynthetic production and higher fungal metabolic activity. No ultrastructural changes were evident in the mycobiont. Other studies of ultrastructural modifications induced by ecological factors have been done by Brown and Wilson (1968), Jacobs and Ahmadjian (1971a), Peveling and Galun (1976), and Harris and Kershaw (1971). Many of these studies have generated contradictory results, so few generalizations can be made at the present time.

LIGHT

Light quantity (intensity) and quality (wavelength) directly regulate the photosynthetic activity of lichen algae and indirectly influence the metabolic activities of both the alga and the fungus by modifying thallus and substrate temperatures. Stocker (1927) reported that respiration in air-dried thalli of *Lobaria* and *Umbilicaria* species was greater in diffuse light than in darkness, despite the fact that some photosynthetic activity would be expected to occur in light. The effect was not observed in hydrated thalli. Since whole thallus gas-exchange measurements were made, it is difficult to interpret these results; it appears, however, that fungal respiration is increased in light. Evidence from one study (Pearson & Brammer, 1978) suggests that increasing light intensity retards respiratory rates in fungal medullary tissue of *Peltigera canina* and *Evernia prunastri*. This light-induced response was not associated with photosynthetic tissue, nor was it observed in other fungal tissues (cortex, rhizines, and apothecia). There was no attempt to explain why medullary respiration was reduced in light; however, it is the first suggestion that changes in light intensity may elicit responses in achlorophyllous lichen tissue, and it is apparent that further experimental work will be required to understand this interaction.

Photosynthetic Pigments

Photosynthetic pigments, particularly chlorophylls, have frequently been extracted from lichen thalli for analysis. Table 6.7 shows that the total chlorophyll content of lichen thalli (chlorophyll a + chlorophyll b) is considerably less than that found for most vascular plants (Table 6.8). The

TABLE 6.7. Selected Chlorophyll Data for Lichen Species

Species	*Total Chlorophyll (mg/g dry wt.)*	*Chl a/Chl b Ratio*	*Source*
Sticta filix	0.69	1.52	Rundel et al. (1979)
Pseudocyphellaria delisei	0.69	0.91	"
Cladina subtenuis			
Sun populations	0.46	1.20	Rundel (1972)
Shade populations	1.04	1.39	"
Cetraria chlorophylla	0.76	2.16	Beltman et al. (1980)
Evernia prunastri	0.35	2.18	"
Hypogymnia physodes	0.40	2.07	"
Parmelia acetabulum	0.37	1.85	"
P. saxatilis	0.46	1.70	"
P. subrudecta	0.29	2.22	"
P. sulcata	0.29	1.90	"
Platismatia glauca	0.51	2.64	"
Ramalina farinacea	0.45	2.00	"
R. fastigiata	0.43	2.00	"
Pseudevernia furfuracea	0.83	2.46	"
Usnea hirta	1.14	2.16	"
Xanthoria parietina	0.28	2.11	"
Peltigera canina	0.8–1.2	—	Wilhelmsen (1959)
Parmelia physodes	0.8–1.6	3.8–4.2	"
Xanthoria parietina	0.5–1.0	3.0–4.0	"
Peltigera canina			
Shade populations	1.07*		Hampton (1973)
Sun populations	0.68*		"
Haematomma ventosum	0.1–0.8		Czygan (1976)
Xanthoria parietina			
From trees	4.1		Hill & Woolhouse (1966)
From rocks	3.5		"
From rocks by sea	1.7		"

*Chlorophyll a measured only.

difference is perhaps greater than a first glance at these data would indicate, since the total chlorophyll data for lichens are based on thallus oven-dry weight and the data for vascular plants are based on leaf-fresh weight. There is a great deal of seasonal variability in lichen chlorophyll content as well. Wilhelmsen (1959) found that in winter, lichens have 0.33 to 0.50 more chlorophyll than in summer. There are also differences related to habitat.

Boardman (1977) has reviewed the literature on photosynthetic properties of sun and shade plants, and comments that shade leaves are generally thinner and contain more chlorophyll than sun leaves. Shade leaves also tend to have a higher proportion of chlorophyll b to chlorophyll a, a characteristic that

TABLE 6.8. Selected Representative Chlorophyll Values of Vascular Plant Species

Species	*Chlorophyll a + b (mg/g fresh wt.)*
Sun species	
Atriplex patula	1.8
Echinodorus berteroi	2.3
Mimulus cardinalis	1.6
Plantago lancelolata	2.2
Solidago spathulata	1.8
Shade species	
Adenocaulon bicolor	3.2
Aralia california	3.0
Disporum smithii	2.8
Trillium ovatum	3.4
Viola glabella	3.2

N.K. Boardman. 1977. Reproduced, with permission, from the Annual Review of Plant Physiology, Vol. 28.

might be associated with increased grana formation relative to stroma lamellae in the leaf chloroplasts. Rundel (1972) measured the total chlorophyll and chlorophyll a/b ratios of sun and shade populations of *Cladina subtenuis* in North Carolina. He found higher levels of total chlorophylls in shade plants, a result consistent with most vascular plant studies. However, he found slightly higher proportions of chlorophyll b to a in sun plants, the opposite of what is found in most vascular plants. Hampton (1973) found a higher concentration of chlorophyll a in shade populations of *Peltigera canina* from Michigan than in sun populations. Also, he found a higher ratio of phycoerythrin to chlorophyll in shade specimens, although the ratio of phycocyanin to chlorophyll was the same in shade and sun populations.

Rundel (1972) also measured total carotenoids in sun and shade populations of *Cladina subtenuis* (Table 6.9). He found a large difference in the ratio of total chlorophyll to total carotenoid content, with the sun plants containing much smaller concentrations of carotenoids than the shade plants. This indicated that characteristics of the algal photosynthetic apparatus change with light intensity. However, the fungus may also be involved inasmuch as carotenoids are found in isolated mycobiont cultures (Henriksson, 1963; Henriksson & Pearson, 1968). In a survey of 19 samples of lichens belonging to the family Collemataceae, Henriksson (1963) found that both the mycobionts and the phycobionts produce carotenoid pigments. These compounds are normally masked by other pigments produced by the whole lichen thallus.

A recent series of studies by Czeczuga (1979a; 1979b; 1980a; 1980b) has

TABLE 6.9. Selected Carotenoid Data for Lichen Species and Isolated Phycobionts

	Total Carotenoids (mg/g dry wt.)	*Total Chlorophyll/ Carotenoids*	*Source*
Cladina subtenuis			
Sun populations	0.24	1.91	Rundel (1972)
Shade populations	3.32	0.31	″
Roccella montagnei	1.5–2.0*	—	Murty & Subramanian (1958)
Haematomma ventosum	1.5–2.3	0.06–0.35	Czygan (1976)
Cetraria crispa	2.15	—	Czeczuga (1980b)
C. islandica	5.99	—	″
Hypogymnia physodes	6.12	—	″
H. tubulosa	20.24	—	″
Parmelia acetabulum	24.64	—	″
P. caperata	15.44	—	″
P. conspersa	18.27	—	″
P. dubia	12.32	—	″
P. olivacea	1.47	—	″
P. omphalodes	3.13	—	″
Pseudevernia furfuracea	4.28	—	″
Peltigera rufescens	42.03	—	Czeczuga (1979a)
Cladonia coniocraea	1.92	—	″
C. minor	16.28	—	″
C. pyxidata	93.27	—	″
Xanthoria parietina	16.23	—	″
Physcia aipolia	0.70	—	″
P. grisea	4.65	—	″
P. ciliaris	22.69	—	″
Usnea comosa	2.89	—	Czeczuga (1979b)
U. dasypoga	5.35	—	″
U. fulvoreagens	4.95	—	″
U. hirta	3.25	—	″
Peltigera aphthosa	8.02	—	Czeczuga (1980a)
P. canina (on ground)	7.34	—	″
P. canina (on wood)	3.48	—	″
P. erumpens	1.34	—	″
P. hazszlynski	4.73	—	″
P. horizontalis	3.31	—	″
P. leucophlebia	2.05	—	″
P. malacea	7.19	—	″
P. polydactyla	4.73	—	″
P. praetextata	3.34	—	″
P. rufescens	3.11	—	″
P. spuria	7.23	—	″

TABLE 6.9. *(Continued)*

	Total Carotenoids (mg/g dry wt.)	Total Chlorophyll/ Carotenoids	Source
Isolated phycobionts			
Trebouxia decolorans	0.27	2.9	Giudici de Nicola & Tomaselli (1961a; 1961b)
T. humicola	0.24	3.4	"
T. decolorans	0.26–0.28	3.5–3.8	Giudici de Nicola & di Benedetto (1962)
T. humicola	0.24	4.1	"
T. albulescens	0.43	2.3	"
Chlorella from *Haematomma ventosum*			
Normal culture	4.25	5.03	Czygan (1976)
N-deficient culture	3.6	0.02	"

*Average carotene content based on air dry weight. Approximately half was determined from melting point to be predominantly β-carotene.

revealed the presence of numerous carotenoid compounds in lichens from various families (Usneaceae, Peltigeraceae, Parmeliaceae, Cladoniaceae, Pertusariaceae, Teloschistaceae, and Physciaceae). Total carotenoid data are listed in Table 6.9. Because both alga and fungus can produce carotenoids and the quantities appear to vary considerably with changes in environment, the taxonomic significance of patterns of carotenoid production is difficult to assess at this time. These preliminary studies do demonstrate, however, that carotenoids represent a diverse set of lichen compounds worthy of further study.

Pigment Content of Isolated Phycobionts

Algal cells within a lichen thallus generally show a marked reduction in photosynthetic pigment when compared to those in culture or free-living situations (des Abbayes, 1951). This is particularly true of *Trentepohlia* and *Gloeocapsa* species. Ahmadjian (1966b) attributes this to a reduced nutrient supply in the lichen thallus. Some representative chlorophyll data obtained for isolated lichen phycobionts are listed in Table 6.10. Giudici de Nicola and Tomaselli (1961a; 1961b) and later Giudici Di Nicola and Di Benedetto (1962) isolated pigments from *Trebouxia decolorans* cultures obtained from the lichens, *Buellia punctata* and *Xanthoria parietina*, and compared them with similar pigments extracted from *T. humicola*, a free-living light-insensitive species of *Trebouxia*. Both species had the same pigments, but the lichen phycobionts contained more chlorophyll and carotenoids and less β–carotene

TABLE 6.10. Chlorophyll Contents of Isolated Lichen Phycobionts

Species	*Total Chlorophyll (mg/g dry wt.)*	*Chl a/Chl b Ratio*	*Source*
Trebouxia decolorans	0.78	2.9	Giudici de Nicola & Tomaselli (1961b)
Trebouxia humicola	0.82	2.8	"
Trebouxia decolorans	0.78–0.92	1.2–1.3	Giudici de Nicola & di Benedetto (1962)
T. humicola	0.82	1.3	"
T. albulescens	1.15	1.3	"
Trebouxia from *Acarospora fuscata*	6.42	3.1	Kinraide & Ahmadjian (1970)
Trebouxia from *Cladonia boryi*	20.1	2.4	"
Chlorella from *Haematomma ventosum*			
Normal culture	21.37	2.47	Czygan (1976)
N-deficient culture	0.08	1.66	"

than the free-living form. The chlorophyll/carotenoid ratio was also found to be much higher in the free-living form, an observation that may explain the light insensitivity of this species. Czygan (1967) isolated the *Chlorella* phycobiont of *Haematomma ventosum* and tested cultures maintained under various degrees of N-starvation for numerous carotenoids as well as chlorophylls. He found a decrease in total chlorophyll in N-deficient cultures and a reduced chlorophyll a/b ratio. A reduction in total carotenoids was also observed; however, the concentration of secondary carotenoids like astaxanthin-ester, canthaxanthin, and others increased under N-deficient conditions. The explanation for this observation is not yet altogether clear, but it appears that production of secondary pigments may be a general physiological response to extreme environmental conditions.

Pigment Extraction

Photosynthetic pigment extraction procedures for lichens are more complicated than for vascular plants. Extracts of lichen pigments often contain acidic lichen compounds as well, and the presence of these compounds causes degradation of chlorophylls to phaeophytins (Brown & Hooker, 1977; Brown, 1980). Since chlorophyll b has a greater stability than chlorophyll a during acidification, chlorophyll a/b ratios would obviously be subject to considerable experimental error depending on the extraction technique used. A method has been proposed (Brown & Hooker, 1977) that involves washing thalli that contain acidic compounds with absolute acetone before pigment extraction. The use of this method should prevent errors in pigment analysis caused by acidification of the extract.

Fungal Control over Light Penetration to Algae

The variance in photopigment concentration exhibited by various populations of lichens is probably greater than that exhibited by vascular plant leaves, since photopigment concentration in lichens is not genetically determined but rather is due to the density of algae maintained in the thallus. Despite the fact that lichen algae appear to exhibit structural and physiological adaptations to varying light intensities, however, many lichen algae are extremely light sensitive (Fox, 1966; Ahmadjian, 1967b). This suggests that the frequency of lichen occurrence along light gradients depends on the ability of the lichen fungus to maintain algal densities sufficient to provide a net carbon gain. One way the fungus can do this is by modifying the amount of light reaching the algal layer. Rundel (1969) found that usnic acid concentration in thalli of *Cladina subtenuis* was directly related to the percentage of full sunlight hitting the plants along a light gradient in Duke Forest, North Carolina. Since cortical pigment concentration in lichens is known to influence the opacity of the fungal tissue overlying the algal cells (Ertl, 1951), this result suggested that the fungus could control the amount of light reaching the algae.

Rundel was not able to determine if the correlation between usnic acid production and light intensity was due to phenotypic plasticity or to ecotypic variation in *Cladina subtenuis* populations. Fahselt (1981) maintained thalli of *Cladina stellaris* and *C. rangiferina* under controlled growth chamber conditions at various light intensities and found that cortical substances (usnic acid and atranorin) did not vary with light intensity after one year, suggesting that concentration is independent of environmental light conditions. However, she found decreases in medullary substance concentration (fumarprotocetraric and perlatolic acids) in thalli maintained at low light intensities, a result she attributed to differences in the rates of synthesis of these compounds at different light levels. Decreases in concentration therefore represent losses of compounds during the experimental interval.

Sun and Shade Lichens

As Kershaw and MacFarlane (1980) have remarked, light has rarely been studied as an ecological factor in lichen ecophysiology, but those few studies that have been done suggest that lichens exhibit morphological and physiological responses to variations in light intensity that limit their distributions in the habitat. Most studies of lichen-light interactions involve a comparison between sun and shade populations or species in the field. Butin (1954) found that sun plants of *Peltigera praetextata* had higher light compensation points and net photosynthetic efficiencies at higher light intensities than shade plants. Rundel (1972) found that shade plants of *Cladina subtenuis* exhibited a higher net photosynthetic rate at light saturation than sun plants, but found no differences in the light compensation point or dark respiration rate of the two populations. Lechowicz and Adams (1973) found no significant differences in the light dependence of net photosynthesis of sun and shade populations of *Cladina mitis*. Kershaw and MacFarlane (1980) observed a number of physiological responses to light in three *Peltigera* species. They found that *P. canina* acclimated rapidly to low or high light levels and could therefore maintain a given rate of photosynthesis despite changes in canopy coverage. *Peltigera scabrosa* also showed a response to low light, even when air dried. *Peltigera aphthosa* thalli collected from dense shade were very sensitive to light; populations collected from open habitats were not as sensitive. In all species, it appeared that continuous light conditions were detrimental to lichens and ultimately resulted in the death of the phycobiont.

The problem with some of these studies is that they frequently do not take into account the density of algal cells in lichens from different light intensities. Observed differences in gas-exchange rates along light gradients may not reflect differences in whole plant response but rather differences in the ratio of photosynthetic to nonphotosynthetic tissue. Plummer and Gray (1972) measured phycobiont densities and growth of three *Cladonia* species from granitic rock outcrops in the Georgia piedmont. For all three species, they observed the greatest algal densities in thallus tips and on the south- and west-facing portions of thallus clumps. They also found higher densities in May than in February and suggested that these differences would contribute significantly to seasonal growth patterns in thallus clumps. Harris (1971) studied the relationship between distribution and ecophysiology of three corticolous *Parmelia* (*sensu lato*) species in South Devon, England. He found that light compensation points and rates of net carbon assimilation were correlated with habitat (tops and bases of trees), time of year, and number of algal cells per square centimeter surface area of thallus. Algal cell counts were found to be higher in treetop thalli than treebase thalli of both *Hypogymnia physodes* and *Pseudoparmelia caperata* (Table 6.11). However, light compensation points were different for the sun and shade ecotypes of these two species. Sun plants of *P. caperata* always had a lower compensation point than shade plants (Table 6.12 and Figure 6.8); this is not unexpected since sun plants also had higher algal cell counts than shade plants and thus higher photosynthetic capabilities at all light intensities. Sun plants of *H. physodes*, however, had higher compensation points than shade plants despite the fact that sun plants had higher algal counts. This is because shade plants have

TABLE 6.11. Algal Cell Counts per cm^2 of Thallus for Three Lichens Collected from the Tops and Bottoms of Oak Trees in South Devon, England

Species	*Collection Site on Trees*	*Cell Number*
Hypogymnia physodes	Top	33.3 ± 4.17 (6)*
	Bottom	18.8 ± 6.03 (4)
Pseudoparmelia caperata	Top	34.2 ± 8.91 (8)
	Bottom	24.0 ± 6.95 (10)
Parmelia sulcata	Top	34.2 ± 0.78 (2)

*Mean cell number ± S.D. Sample size in parentheses.

G.P. Harris. 1971. "The ecology of corticolous lichens. II. The relation between physiology and the environment." *J. Ecol.* 59:441–452. Reprinted with permission of Blackwell Scientific Publications.

TABLE 6.12. The 24-Hour Light Compensation Values (in ft.-candles) for Three Lichen Species from Oak and Birch Trees in South Devon, England

	Pseudoparmelia caperata		*Hypogymnia physodes*		*Parmelia sulcata*
	Top	Bottom	Top	Bottom	Top
September	160	320	410	200	310
January	230	250	470	270	620
March	800	1250	850	690	1520

G.P. Harris, 1971. "The ecology of corticolous lichens. II. The relation between physiology and the environment." *J. Ecol.* 59: 441–452. Reprinted with permission of Blackwell Scientific Publications.

higher photosynthetic efficiencies at low light intensities (Figure 6.8), an observation that suggests an adaptation to low light intensities. Whether this is due to changes in algal or fungal physiology or both remains to be determined. Whatever the mechanism responsible for this response in *H. physodes,* it apparently does not operate in *P. caperata.*

TEMPERATURE

Lichens have long been considered exceptionally temperature-tolerant organisms. Kappen (1973) reviewed the vast literature on lichen responses to extreme environments, and elsewhere in this book endolithic lichen biology in polar and hot desert regions is discussed.

It is difficult to study lichen responses to temperature alone since temperature is so strongly coupled with water and light conditions. However,

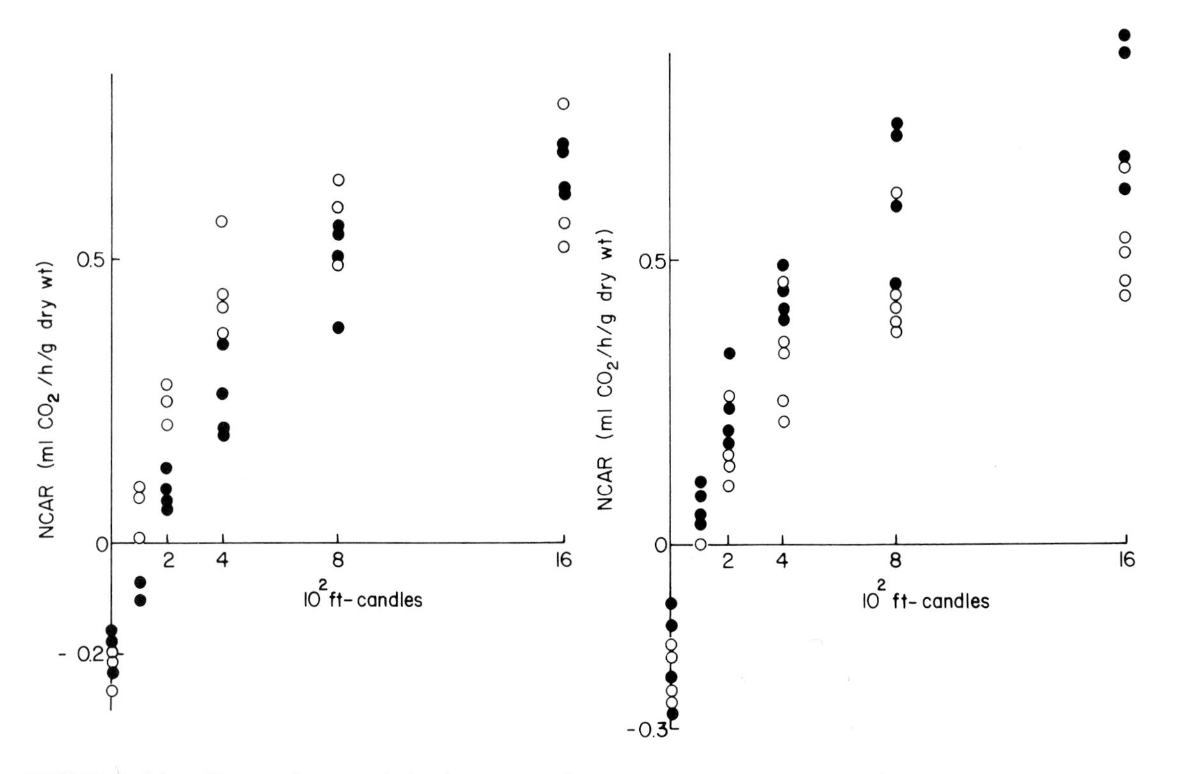

FIGURE 6.8. Net carbon assimilation rate of *Hypogymnia physodes* (left) and *Pseudoparmelia caperata* (right) at various light intensities; top form (●); bottom form (○). From G.P. Harris. 1971. "The ecology of corticolous lichens. II. The relation between physiology and the environment." *J. Ecol.* 59: 441–452. Redrawn with permission of Blackwell Scientific Publications.

temperature optima for various processes in lichens are frequently measured and generally reflect the range of temperature conditions encountered by the lichens in their habitats. Kershaw and Smith (1978), for example, studied the control of seasonal rates of respiration and net photosynthesis in *Stereocaulon paschale* in subarctic Canada by moisture, light, and temperature. They found that a relatively high temperature optimum (20° – 30° C) for net photosynthesis combined with a relatively high light requirement (1000 μE m^{-2} sec^{-1}) correlated well with the open nature of the spruce-lichen woodland where *S. paschale* occurs and also helps to explain the disappearance of this lichen when the canopy closes later in succession. They also observed a marked sensitivity to thermal stress in thalli of *S. paschale*, even when the thalli were dry. This thermal sensitivity may explain the delayed entry of *S. paschale* into the successional sequence in this system.

Because lichens are so often considered to be resistant to temperature changes when dry, little attention has been focused on thermal sensitivity of air-dried thalli. However, MacFarlane and Kershaw (1978) have found that air-dried thalli of *Peltigera canina* populations from mesic and xeric habitats show different responses to temperature that correlate with their natural habitats. Responses were determined by measuring nitrogenase activity, net photosynthesis, and respiration in air-dried thalli after storage at various temperatures. These results, along with similar results for other lichen species (MacFarlane & Kershaw, 1980; Tegler & Kershaw, 1981) showed that thermal sensitivity is a characteristic even of dry thalli and may thus be extremely important in regulating lichen distributions in nature.

The reason that lichen thermal sensitivity has not been fully appreciated by lichenologists is perhaps due to the difference in sensitivity exhibited by lichen mycobionts and phycobionts. MacFarlane and Kershaw (1980) found that net photosynthesis was more sensitive to thermal stress than nitrogenase activity and much more sensitive than respiration. This suggested to them that the lichen mycobiont has a greater thermal stress tolerance than the phycobiont, and may explain why such extreme temperatures were required by Lange (1953; 1954) to measure respiratory stress responses in lichens. A reevaluation of stress tolerance in lichen physiological processes is required to fully appreciate temperature as an ecological factor.

Thallus Color

Lichen thallus color would be expected to influence thallus temperature and water relations and thus regulate rates of photosynthesis and respiration. In a very interesting study, Kershaw (1975a) tested this idea experimentally by painting thalli of the light-colored *Thamnolia vermicularis* black and comparing temperatures of painted and unpainted thalli at different light intensities. He found an increasing temperature difference between painted and normal air-dried thalli with increasing illumination. Then, he measured thallus temperatures of *Alectoria nitidula* (dark brown) and *A. ochroleuca* (pale yellow) during drying under natural field conditions and obtained similar color-related responses. He suggested that thallus color may explain the distribution and ecophysiology of *Alectoria nitidula*, a dark-colored lichen found on exposed ridge tops in the Canadian subarctic, and *Cladina stellaris*, a light-colored lichen found on lower slopes. During the summer, *A. nitidula* would be expected to have a carbon deficit due to respiratory losses at high temperature and short, infrequent intervals of physiological activity whereas *C. stellaris* would dry more slowly and be metabolically active longer and more frequently. In the winter, exposed dark thalli of *A. nitidula* would attain thallus temperatures suitable for frequent intervals of photosynthetic activity whereas the light thalli of *C. stellaris* are buried under several meters of snow where low temperatures and light levels result in little photosynthetic activity. It is evident from this study that thallus color can help to explain lichen distributional and ecophysiologic patterns in nature.

NUTRIENTS

All plants require mineral elements for normal physiological activity. These include H, B, C, O, Na, Mg, P, S, Cl, K, Ca, Mn, Fe, Cu, Zn, and Mo—all have been observed in lichen thalli (Nieboer, Richardson, & Tomassini, 1978). We would not expect the requirements for these essential nutrient elements to be the same for all lichens, nor would they be the same for lichen mycobionts and phycobionts. Of all elements found in lichen thalli, perhaps nitrogen has been studied more than any other. As a component of all enzymes and many other important compounds (chlorophyll, nucleic acids, nucleotides, coenzymes, and vitamins), nitrogen has long been appreciated as an essential element in plant nutrition. The fact that many lichen thalli contain blue-green algae capable of N_2-fixation makes nitrogen metabolism an area of particular interest to lichenologists.

Other essential elements have also been studied in lichens, though to a much lesser extent. As an essential component of ATP, nucleic acids, phosphorylated sugars, and phospholipids, phosphorus is important in fungal and algal metabolism, as are potassium and calcium, which are involved in osmoregulation and cell permeability. Sulfur is an important constituent in some proteins, and magnesium is an essential component of the chlorophyll molecule. Other elements are necessary in various enzyme systems.

Farrar (1976d) has posed two important questions relating to lichen nutrient accumulation and physiology. First, how do the methods of nutrient accumulation and utilization by lichens correlate with habitat? Second, how do algae obtain necessary nutrients when they are surrounded by fungal hyphae known to be extremely efficient element accumulators? To these questions, I would add the following:

1. How different are the element requirements of lichen algae and fungi?

2. How do nutrient deficiencies influence physiological activities of the whole lichen thallus and the isolated symbionts?

3. To what extent do additions of essential elements increase lichen growth in the laboratory and in the field; to what extent does this fertilization destabilize the lichen symbiosis?

Nitrogen

Nitrogen metabolism has been reviewed by Millbank and Kershaw (1973) and Millbank (1976). Most lichens are found in habitats where nutrients, particularly nitrogen, are limiting. However, only a relatively small proportion of lichen species contain N_2-fixing blue-green algae. Furthermore, there is no apparent correlation between the distribution patterns of blue-green alga-containing lichens and habitat availability of nitrogen (Stewart, 1966; Farrar, 1976d; Smith, 1975). Some lichens have been shown to secrete extracellular enzymes that break down nitrogenous substrates (Galinou, 1956; Moissejeva, 1961). It may therefore be possible for many lichens that do not contain blue-green algae to obtain nitrogen from decaying organic material.

Lichens that contain blue-green algae exhibit active N_2-fixation, usually measured by assaying for nitrogenase activity with an acetylene reduction technique. There are strong correlations between total nitrogen content of lichen thalli and presence of N_2-fixing, blue-green algae (Green et al., 1980), and evidence is accumulating that lichens can contribute significant amounts of fixed N_2 to certain forested ecosystems (Pike et al., 1972; Denison, 1973; Forman, 1975).

It was once believed that the process of N_2-fixation went on in vegetative cells as well as in the heterocysts of the blue-green algae, because rates of N_2-fixation were higher than could be accounted for by heterocyst activity alone (Griffiths, Greenwood, & Millbank, 1972). Subsequent studies (Millbank, 1972; Hitch & Millbank, 1975) indicated that the rates of N_2-fixation are not as high as they were originally thought to be and could be accounted for solely by heterocyst activity.

It was also once believed that symbiosis facilitated N_2-fixation in the lichen thallus because of low O_2 tensions around the heterocysts and the frequently observed decrease in the rate of N_2-fixation upon removal of the blue-green algae from the lichen thallus. It has recently been shown (Millbank, 1977) that the environment around the N_2-fixing heterocysts of *Nostoc* from *Peltigera* species is not constantly anaerobic and that O_2 evolution frequently occurs in light. Also, the decrease in N_2-fixation rates observed when *Nostoc* algae are removed from the thallus have been accounted for by disruption of *Nostoc* filaments during isolation.

Rates of N_2-fixation are frequently higher in lichen cephalodia containing *Nostoc* than in lichen thalli containing *Nostoc*. This observation can be explained on the basis of differences in heterocyst frequency in the two areas of the thallus (Table 6.13). Stewart and Rowell (1977) measured N_2-fixation in *Peltigera aphthosa*, which has the green phycobiont, *Coccomyxa*, as a thallus symbiont and *Nostoc* in cephalodia. They found a heterocyst frequency of 21 percent for *Nostoc* in the cephalodia, whereas the heterocyst frequency of free-living *Nostoc* is 4–5 percent. It is not clear why there is a higher frequency of heterocysts in cephalodia than in free-living or lichenized blue-green algae. Stewart and Rowell (1977) argued that cephalodia blue-green can maintain a greater proportion of nonphotosynthetic heterocysts relative to photosynthetic cells because the primary green phycobiont is able to supply the mycobiont with sufficient amounts of photosynthate to compensate for the reduced photosynthate production in the cephalodia. Another explanation

TABLE 6.13. The Percentage of Heterocysts in Blue-Green Phycobiont Filaments

	Cephalodia		
Lichen	*Internal*	*External*	*Thallus*
Leptogium burgessii	—	—	4.5
Lobaria amplissima	21.6	3*	—
L. laetevirens	30.4	—	—
L. pulmonaria	35.6	—	—
L. scrobiculata	—	—	3.9
Nephroma laevigatum	—	—	4.1
Peltigera aphthosa var. *variolosa*	21.1	—	—
P. canina	—	—	4.9
P. polydactyla	—	—	5.8
P. praetextata	—	—	4.4
P. venosa	—	—	7.8
Sticta limbata	—	—	4.9
S. sylvatica	—	—	3.5

* *Dendriscocaulon*like morphotype.

J. Millbank. 1976. "Aspects of nitrogen metabolism in lichens." In *Lichenology: Progress and Problems* (D.H. Brown, D.L. Hawksworth, & R.H. Bailey, eds.) pp. 441–455. London: Academic Press. Reprinted with permission of Academic Press and the author.

offered by Hitch and Millbank (1975) is that increased cephalodial heterocyst frequency may be stimulated by production of photosynthate by the green phycobiont. Regardless of the mechanism, however, levels of N_2-fixation are apparently increased as a result of lichenization, especially in cephalodia.

A number of studies have focused on the flow of fixed N_2 from phycobiont to mycobiont in N_2-fixing lichens. It has long been known that substantial amounts to fixed N_2 move in the form of ammonia from alga to fungus in the whole lichen thallus; however, this transport ceases when the alga is removed from the thallus (Millbank & Kershaw, 1973). Stewart and Rowell (1977) have proposed an explanation for this observed reduction in ammonia transport from *Nostoc* when fungal contact is terminated (Rowell & Stewart, 1981). In free-living *Nostoc*, ammonia is convered to glutamine through the action of the enzyme, glutamine synthetase. In symbiosis, however, there is a marked reduction in the activity of this enzyme (Table 6.14). Thus, the ammonia normally converted to glutamine is made available for release by the alga (Figure 6.9). The cause of the observed enzyme suppression—if this is what it is—in the lichen thallus is not known, but the presence of the fungus is apparently required.

Glucose is required for high N_2-fixation rates (Smith, 1960a; 1960b), and since rates increase with light intensity, one might infer that glucose released from the lichen phycobiont during photosynthesis stimulates ammonia release. However, Rai, Rowell, and Stewart (1981) have found that *Peltigera aphthosa* thalli containing *Nostoc* in cephalodia may also fix N_2 at high rates and for prolonged intervals in darkness when photosynthate concentrations are much reduced. They provide evidence suggesting that fungal CO_2 fixation via phospho-enol pyruvate (PEP) carboxylase along with stored polyglucose in *Nostoc* cells may be important in sustaining N_2-fixation in darkness. These observations may also explain the slow cell growth of *Nostoc* in cephalodia compared to free-living *Nostoc*.

TABLE 6.14. Activity of the Enzyme Glutamine Synthetase in Symbiotic *Nostoc* from *Peltigera canina* and the Cephalodia of *Peltigera aphthosa,* and Free-living *Nostoc* Isolated from *P. canina* and Maintained in Culture

	Enzyme Activity (nmol mg^{-1} protein min^{-1})
Peltigera canina	
Symbiotic *Nostoc*	4.1
Isolated *Nostoc*	72.4
Peltigera aphthosa	
Cephalodial *Nostoc*	4.5

W.D.P. Stewart and P. Rowell. 1977. Reprinted by permission from *Nature,* Vol. 265, pp. 371–372.

NOSTOC in isolation

$N_2 \longrightarrow NH_4 \xrightarrow[\text{synthetase}]{\text{glutamine}} \text{glutamine}$

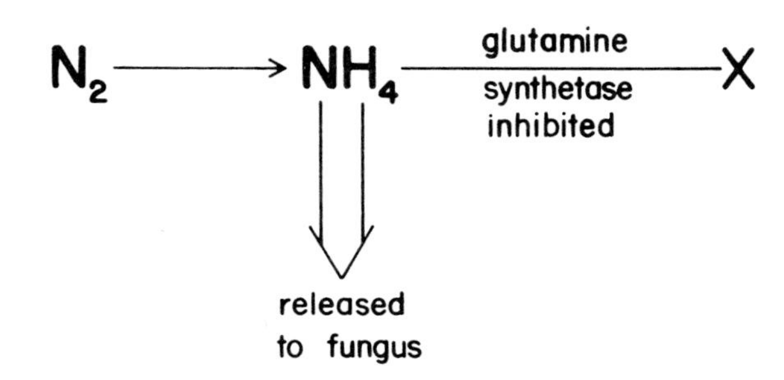

FIGURE 6.9. Nitrogen assimilation pathways in free-living and symbiotic *Nostoc*. Adapted from D.C. Smith. 1978. Reprinted by permission from *Mycologia,* 70: 915–934. Copyright 1978, D.C. Smith and The New York Botanical Garden.

Most non-lichenized fungi use ammonium preferentially over nitrate nitrogen, and there is no evidence available to suggest that lichenized fungi are different in this respect. Farrar (1976d) has hypothesized that this preference for ammonium by the fungus allows the alga to obtain nitrogen in the form of nitrate since the fungus does not appear to accumulate nitrate efficiently. Kershaw and Millbank (1970) found that adding a dilute KNO_3 solution to *Peltigera aphthosa* resulted in healthier thalli and suggested that the green phycobiont *Coccomyxa* cells received only 3 percent of their volume-corrected proportional share of the N_2 fixed in the cephalodia containing *Nostoc*. Therefore, even in lichen thalli containing N_2-fixing, blue-green algae in cephalodia, the green phycobionts appear to be nitrogen limited.

Other Elements

Lichen elemental concentrations and accumulation processes have been reviewed by Tuominen and Jaakkola (1973), Brown, (1976) and Nieboer, Richardson, and Tomassini (1978). The role that elemental accumulation

TABLE 6.15. Characteristic Levels of Selected Nutrient Elements in Lichens

Element	*μg/g dry wt.*
N	6000–50,000
K	500–5000
P	200–2000
Ca	200–40,000
Mg	100–1000
Na	50–1000
S	50–2000
Fe	50–1600
Mn	10–130
Zn	20–500
Cu	$<$1–50
Mo	0–3

Modified from E. Nieboer, D.H.S. Richardson, & F.D. Tomassini. 1978. "Mineral uptake and release by lichens: An overview." *Bryologist* 226–246. Reprinted with permission of the American Bryological and Lichenological Society.

plays in rock weathering and pedogenesis has been discussed by Syers and Iskandar (1973), and the importance of lichens in ecosystem mineral cycles has been reviewed by Pike (1978).

Element concentrations in lichen species examined to date are extremely variable (Table 6.15), and probably reflect availability of nutrients in the habitat. Little information of general applicability is available to suggest causes for species-specific variations in element concentration, although there appears to be a correlation between Ca content, oxalic acid production, and frequency of occurrence of certain lichen species on calcareous substrates (Mitchell, Birnie, & Syers, 1966; Syers, Birnie, & Mitchell, 1967). Other species also appear to exhibit habitat preferences that reflect mineral requirements. For example, many species occur on rocks near bird rookeries (Du Rietz, 1932) or on whalebones rich in phosphate (Almborn, 1948). Others occur on substrates influenced by salt spray in coastal areas (Fletcher, 1976). Many of these distributional patterns are discussed by Barkman (1958). The extent to which these patterns reflect mineral requirements and not tolerance of high element concentrations cannot be determined as yet because so little is known about the mineral requirements of lichens.

The experimental study of lichen mineral nutrition is in a developmental stage, and much laboratory work is required before nutritional requirements of lichens are well known (Fletcher, 1976). Some growth studies have been done using nutrient additions. Over a two-year period, Hakulinen (1966) poured a solution containing bird excrement on lichens growing on rocks two meters above the shoreline in a coastal area in Finland and observed consistently higher growth rates for these lichens than for those receiving only rainfall. However, the effect of adding water along with the nutrients was not controlled. Jones and Platt (1969) moistened thalli of *Xanthoparmelia conspersa* under laboratory conditions with either distilled water or a nutrient solution and observed significant increases in growth for thalli to which nutrients had been added, suggesting that nutrients are limiting for many lichens in nature. The separate effects of various nutrient elements on lichen growth and physiological processes need to be determined in the laboratory before much progress can be made in this area.

Accumulation Processes

Element accumulation processes have been the subject of numerous studies, and reviews are available (Brown, 1976; Nieboer, Richardson, & Tomassini, 1978). Mechanisms most commonly studied are extracellular ion-exchange processes (Puckett et al., 1973b). However, the mechanisms responsible for the internal (cellular) accumulation of essential nutrients by lichen fungi and algae are not as well known. It is assumed that accumulation of K^{+}, Mg^{+}, and Ca^{++} occurs passively along an electrochemical potential gradient, as in other plants and animals (Higinbotham, 1973). Accumulation of other divalent ions like Mn^{++}, Zn^{++}, and Cu^{++} is probably also passive and perhaps is aided by ion-exchange in the cell (Higinbotham, 1973). Major anions like NO_3^-, $H_2PO_4^-$, and SO_4^{--} are probably accumulated against an electrochemical potential gradient, as in other organisms (Higinbotham, 1973; Nissen, 1974).

Despite their low water solubility (Iskandar and Syers, 1971), lichen secondary compounds, particularly those with electron donor groups in the *ortho* position, are known to form chelates with divalent cations (Ascaso and Galvan, 1976b; Syers, 1969; Williams & Rudolph, 1974; Iskandar and Syers, 1972). This has led some workers to suggest that these substances might be important in binding ions within the lichen thallus. Lichen substances suitable for cation chelation are found in varying concentrations in lichen thalli. However, no comparative field studies of cation uptake capability have been done using thalli differing either qualitatively or quantitatively in lichen substance concentration. There have also been studies showing no effect of lichen substances on cation accumulation. In two separate experiments, Brown (1976) actually observed an increase in total cation uptake in lichen thalli from which lichen substances were removed with acetone. Thus, it presently appears that lichen substances can function as chelators, but their role in cation binding is probably less important than other mechanisms.

Cellular Uptake and Loss

Farrar (1976c) experimented with phosphate uptake by *Hypogymnia physodes* and observed active uptake from very dilute solutions. He also found that labelled phosphate became incorporated into an insoluble fraction, which he suggested was polyphosphate sequestered mainly in the fungus. Using Toluidine Blue-stained sections of *Collema leucocapum* and *Peltigera dolichorrhiza*, Chilvers, Ling-Lee, and Ashford (1978) found granules that appeared to be mostly polyphosphate in both the fungus and the alga, although the granules are larger and more numerous in the fungus (Figure 6.10). Electron microscope studies have also revealed polyphosphate bodies in lichen algae, but not lichen fungi (Paran, Ben-Shaul, & Galun, 1971; Jacobs & Ahmadjian, 1973). Farrar (1976c) has commented on the ecological significance of stored phosphorus supplies in the lichen thallus, a situation that parallels that of mycorrhizal fungi (Harley, 1969). Based on calculations of phosphate concentration in rainwater, rates of phosphate uptake by *H. physodes*, frequency of precipitation, and lichen growth rates, Farrar (1976b) estimated that it was unlikely that phosphorus limited the growth of this lichen in the field.

Losses of both organic and inorganic solutes have been observed when dry thalli of *Hypogymnia physodes* are rehydrated (Farrar & Smith, 1976). It has not been determined whether loss of material occurs as a result of desiccation stress or rewetting. Buck and Brown (1979) found that desiccation caused a significant loss of intracellular potassium and magnesium in desiccation-intolerant lichen species. Species from more xeric habitats exhibited less cation loss when desiccated. A Desiccation Resistance Index devised for bryophytes and based on the ratio of soluble intracellular potassium at different relative humidities (Brown & Buck, 1979) effectively classified lichen species according to habitat water availability. Those species best able to tolerate desiccation stress were found to recover most rapidly when rehydrated.

Experiments by Feige (1977) on the update of labelled sulfate and phosphate in the isolated *Coccomyxa* phycobiont of *Peltigera aphthosa* suggested that uptake was inhibited in the presence of the mycobiont and thallus homogenate but was increased when the algal cells were completely isolated from the thallus. Interference with algal phosphate uptake may be caused by high rates of fungal accumulation. However, mycobionts apparently absorb sulfate at very low rates. Preliminary experiments with a protein inhibitor (cycloheximide) suggested that the normalization of active ion uptake by isolated *Coccomyxa* cells is perhaps dependent upon the synthesis of a protein carrier.

Importance of Lichens in Mineral Cycling Processes

Pike (1978) has discussed the role lichens play in ecosystem material flow processes. In particular, he was interested in determining the standing crop of the mineral capital in lichens, the rate and degree to which minerals are turned over in ecosystems relative to non-lichen components and the sources of

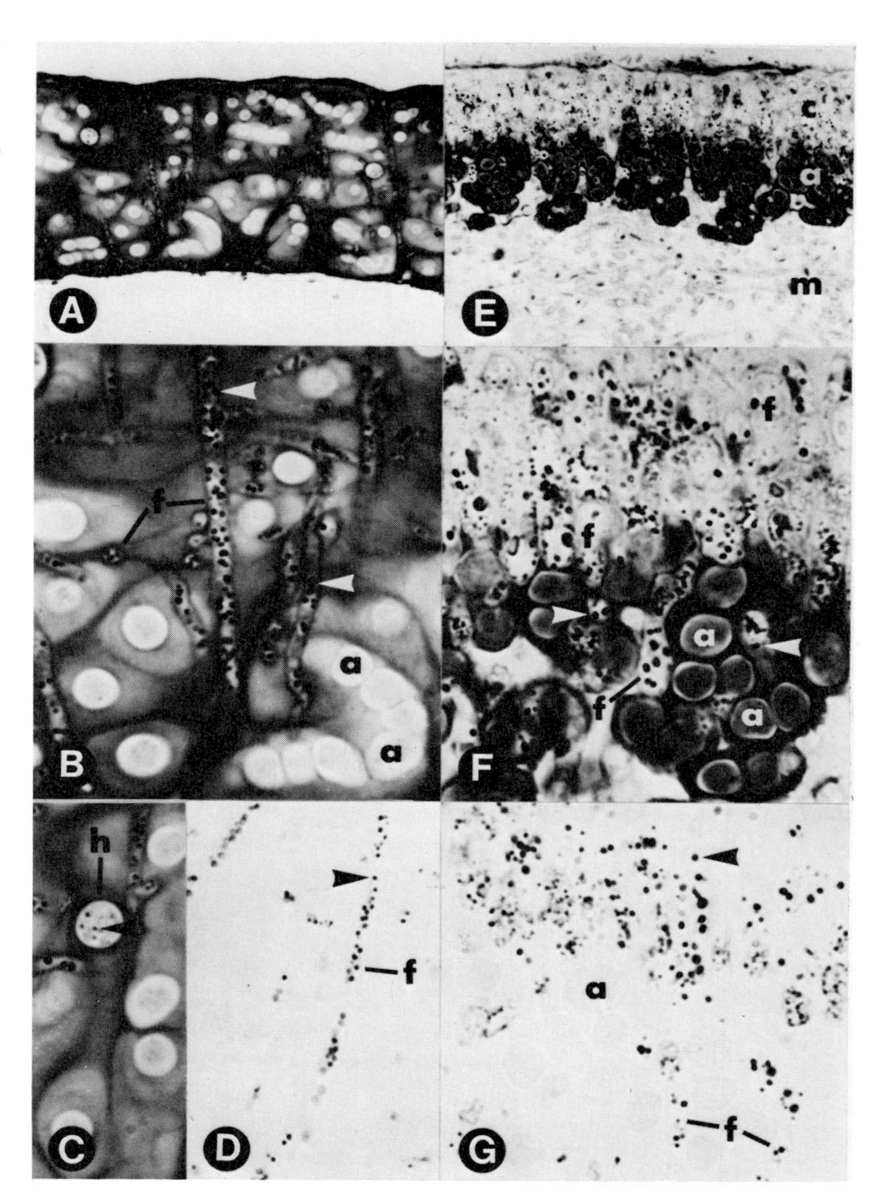

FIGURE 6.10. Microphotography of vertical sections through two lichen species. Nos. 1–4, *Collema leucocapum;* Nos. 5–7, *Peltigera dolichorrhiza.* a, algal layer or algal cell; c, cortex; m, medulla; h, heterocyst; f, fungal hypha. Arrows point to granules. (a) *Collema leucocapum* stained with Toluidine Blue at pH 4.4 (×450); (b) *Collema leucocapum* stained to show numerous metachromatic granules distributed throughout the fungal hyphae (×1500); (c) Another region of above showing granules in an algal heterocyst (×1500); (d) *Collema leucocapum* section stained with Toluidine Blue at pH 1.0. Only

minerals utilized by lichens. In forested systems that have been extensively studied, vascular plants make up most of the plant biomass. A comparison of lichen biomass with the rapidly cyclable foliage component of these systems (Table 6.16) suggests that lichens account for around 6 percent of the readily cyclable biomass. Nitrogen appears to account for a relatively large component of the lichen mineral capital, with K and P exhibiting intermediate values and Ca and Mg low values. In Douglas fir forests, mineral turnover rates for lichens are in the same proportions as the bulk mineral capitals. In oak woodlands, higher relative turnover rates are seen for total lichen biomass and lichen Ca and Mg. Turnover for Ca and Mg may not be as great if throughfall is included. In northern hardwood forests, lichen biomass is assumed to turn over every five years, and the quantity of minerals cycled through lichen biomass is approximately 1 percent of that cycled through leaf litter.

Release of minerals from lichens occurs through decomposition, leaching, and grazing activity of lichen herbivores. Decomposition rates measured for lichen material are relatively rapid, ranging from three months in oak woodland to one year in Douglas fir and balsam fir forests (Pike, 1978). Lang, Reiners, and Heier (1976) found that lichens submerged in water released Ca, Mg, and K into the solution, suggesting that nutrient fluxes could be due, at least in part, to leaching losses. However, these losses are very small compared to lichen mineral capital.

Very little work has been done on nutrient fluxes due to lichen herbivore grazing. Although this mechanism of nutrient transfer is probably of little significance in forested ecosystems where lichens and lichen herbivores make up only a small proportion of the total biomass, it may be more important in polar regions where lichens are dominant and nutrient fluxes are more strongly regulated by lichen biomass turnover. The potential for nutrient transfer between lichens and lichen herbivores was demonstrated by Lawrey (1980b), who used atomic absorption spectrophotometric and energy-dispersive, microanalytic techniques to measure calcium accumulation by *Xanthoparmelia conspersa* and subsequent transfer to herbivores on two Potomac River islands in Maryland. On Plummers Island, *X. conspersa* thalli contained significantly higher concentrations of calcium and other elements

---

the metachromatic staining of the granules remains (×1500); (e) *Peltigera dolichorrhiza* (×450); (f) High power micrograph of above showing metachromatic granules in hyphae of lower cortex and algal layer (×1500); (g) High power micrograph of a similar region of *Peltigera dolichorrhiza* section stained with Toluidine Blue at pH 1.0 (×1500). From G.A. Chilvers, M. Ling-Lee, & A.E. Ashford. 1978. "Polyphosphate granules in the fungi of two lichens." *New Phytol.* 81: 571–574. Reprinted with permission of Academic Press and the author.

TABLE 6.16. Ratios of Nutrient Capital or Turnover for Lichens to Nutrient Capital or Turnover for Other Ecosystem Components

			Element				
Ecosystem	*Ratio*	*Biomass*	*N*	*P*	*K*	*Ca*	*Mg*
Douglas fir forest	Lichen capital / Foliage capital	0.06	0.11	0.06	0.06	0.02	0.02
Douglas fir forest	Lichen turnover / Total litterfall and throughfall	0.06	0.19	0.09	0.06	0.01	0.01
Oak woodland	Lichen turnover / Leaf litter	0.15	0.15	—	—	0.09	0.07
Northern hardwoods	Lichen capital ÷ 5 / Leaf litter	0.007	0.011	0.017	0.010	0.005	—

L.H. Pike. 1978. "The importance of epiphytic lichens in mineral cycling." *Bryologist* 81:247–257. Reprinted with permission of the American Bryological and Lichenological Society.

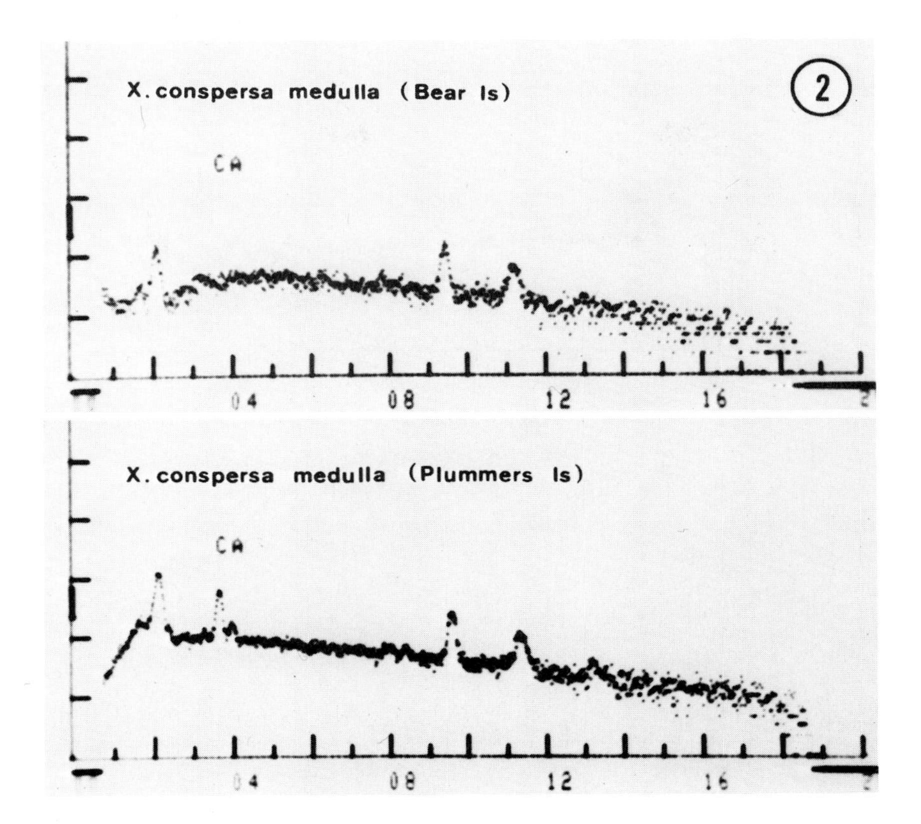

FIGURE 6.11. X-ray spectra obtained from medullary probes of *Xanthoparmelia conspersa* sampled from Bear Island and Plummers Island, Maryland. Calcium peaks are indicated. Aluminum and gold peaks are due to specimen preparation. Ordinate units are counts per channel (log scale); abscissa units are X-ray energy units from 0 to 20 KeV. From J.D. Lawrey. 1980b. Reprinted by permission from *Mycologia*. 72: 586–594. Copyright, 1980, J.D. Lawrey and The New York Botanical Garden.

than on Bear Island (Figure 6.11). Plummers Island is located below a heavily traveled highway and probably receives elements from particulate automobile exhaust and road deicing agents. Bear Island is located at a distance several kilometers from pollution sources. Lichen herbivores collected from *X. conspersa* thalli were analyzed for Ca content using energy-dispersive, X-ray microanalysis, a nondestructive method that does not require large samples. X-ray energy spectra obtained for herbivores sampled from both islands

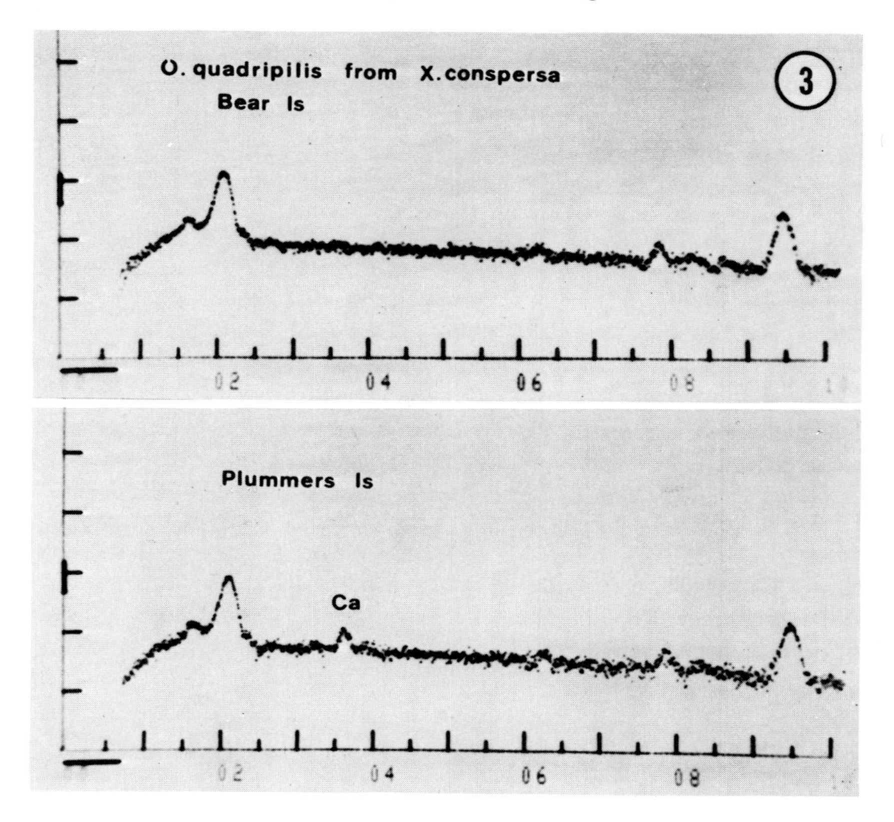

FIGURE 6.12. X-ray spectra obtained from probes of the lichen herbivore *Oribata quadripilis* sampled from *Xanthoparmelia conspersa* on Bear Island and Plummers Island, Maryland. Calcium peaks are indicated. Aluminum and gold peaks are due to specimen preparation. Ordinate and abscissa units as in Figure 6.11. From J.D. Lawrey. 1980b. Reprinted by persmission from *Mycologia.* 72: 586–594. Copyright 1980, J.D. Lawrey and the New York Botanical Garden.

(Figure 6.12) show that Ca accumulated by lichens on Plummers Island can be transferred to lichen herbivores.

CARBOHYDRATE FLOW BETWEEN SYMBIONTS

As Smith (1976) has commented, lichen associations are particularly well suited to the study of some of the general properties of biotrophic symbioses. In these sorts of associations, nutrients move between living cells of the symbionts and there is little destruction of one symbiont by the other. Evident over evolutionary time as well as ontogenetic time, there is also the progressive development of mechanisms that facilitate nutrient flow and balanced growth of the symbionts.

As a result of lichenization, lichen fungi and algae show a number of morphological and physiological changes that differentiate the whole lichen thallus from its isolated associates. Among these are:

1. production of bizarre fungal secondary compounds;
2. fungal tissue differentiation;
3. changes in rates of algal cell division;
4. increased rates of photosynthate release by phycobionts;
5. increased rates of ammonia release by lichenized blue-green algae.

The extent to which these changes are regulated by the fungus and/or alga is only partly understood.

Due mainly to the work done by D.C. Smith, his students, and colleagues, a great deal is known about the flow of complex metabolites between fungus and alga. Most of these studies involved the use of radioactive tracers and lichen disks incubated on various liquid media for various short intervals of time. The first use of this technique in lichen physiology by Smith (1961) demonstrated that photosynthetically fixed carbon was transferred from alga to fungus. Subsequent studies have shown that this flow is massive and the type of carbohydrate transferred is dependent upon the alga. Blue-green algae have been found to release glucose; green algae release a sugar alcohol. Once absorbed by the fungus, this carbohydrate is converted to another carbohydrate, frequently mannitol or arabitol. If the alga is isolated from the fungus, the massive flow of carbohydrate rapidly ceases and algal metabolism is characterized instead by synthesis of ethanol-insoluble compounds—presumably structural or storage. Figure 6.13 illustrates these properties of photosynthate transfer in lichens. What is presently not very well understood is the degree of control the fungus has over algal metabolism and carbohydrate release, although a number of interesting recent studies have greatly increased our understanding of this.

Identity of the Transfer Carbohydrate

Hill (1976) has summarized the studies done on carbohydrate transfer between alga and fungus. These studies demonstrated that glucose is released by blue-green phycobionts; ribitol is released by *Trebouxia, Myrmecia*, and

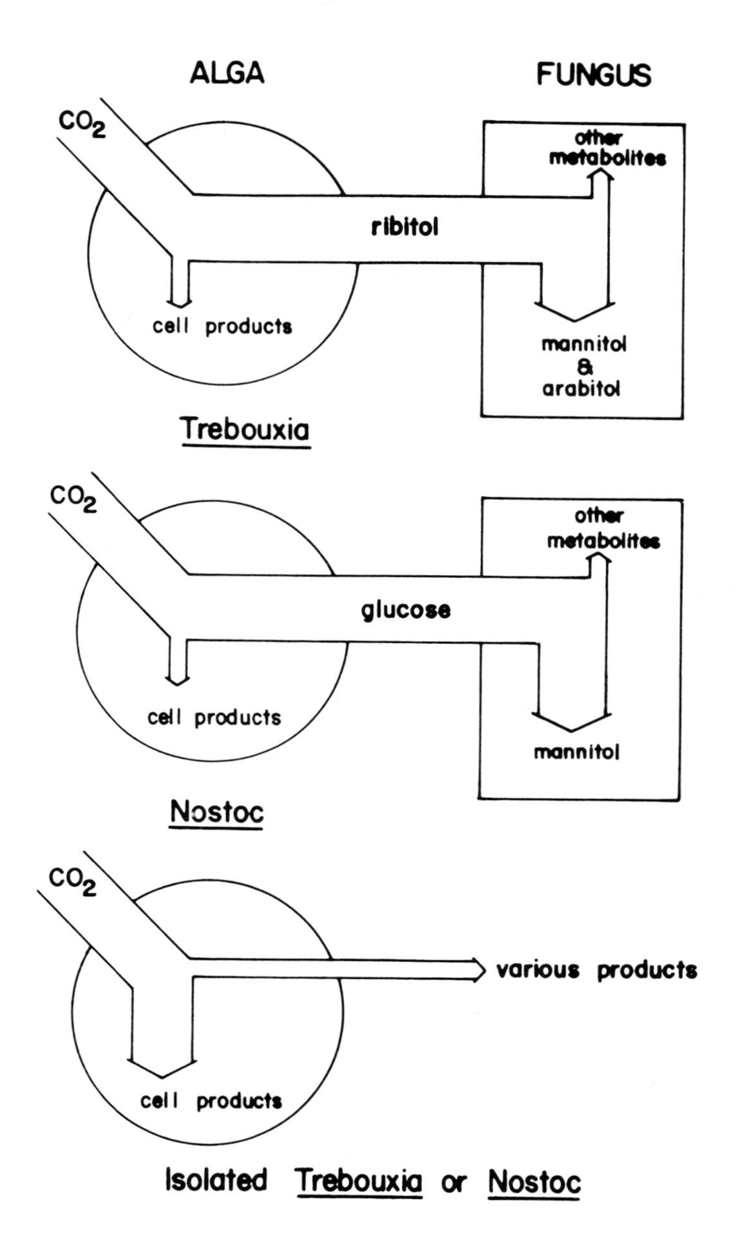

FIGURE 6.13. Biotrophic movement of photosynthate from alga (left) to fungus (right) and in isolated phycobionts (bottom). Adapted from D.C. Smith. 1978. Reprinted by permission from *Mycologia* 70: 915–934. Copyright 1978, D.C. Smith and the New York Botanical Garden.

Coccomyxa; sorbitol is released by *Hyalococcus* and *Stichococcus*; and *Trentepohlia* releases erythritol. One of the interesting problems in lichen physiology is the mechanism involved in the massive release of a specific carbohydrate by algae in specific associations (Smith, 1978). In addition, the extent to which the alga, the fungus, or both are responsible for this needs to be determined.

The identity of these transfer carbohydrates has been determined by a number of techniques. The *inhibition technique* was first used by Drew and Smith (1967b). When they incubated *Peltigera polydactyla* disks on a labelled carbon source in light, transfer of labelled glucose from alga to fungus was rapid and easily detectable; however, when they added unlabelled glucose to the incubation solution, transfer of labelled carbon between symbionts ceased, and instead was pumped into the solution. They hypothesized that the unlabelled glucose in solution was competing for fungal carbohydrate uptake sites and inhibited uptake of labelled glucose produced by the alga. Inhibition was not observed when other carbohydrates (fructose, galactose, mannose, and mannitol) were added to the medium. This suggested that the fungal uptake sites were specific to the carbohydrate released by the alga.

The inhibition technique has proved useful in identifying the transfer carbohydrate in lichens, but there are problems with the technique. The presence of carbohydrate in the incubation solution probably affects the release of glucose by the alga (Drew & Smith, 1967b; Hill & Smith, 1972); it probably also reduces the rate of glucose uptake by the fungus. Quantitative studies of carbohydrate transfer using the inhibition technique therefore should probably be designed using appropriate carbohydrate analogues rather than compounds identical to the transfer carbohydrates (Chambers, Morris, & Smith, 1976).

A more direct *autoradiographic technique* that avoids many of these problems has been developed by Tapper (1981a, 1981b). This technique does not require disruption of the symbionts and allows a detailed examination of the spatial distribution of the ^{14}C label in the lichen thallus. Basically, it involves freeze-sectioning and freeze-drying thallus disks exposed to $H^{14}CO_3$ and coating sections with a nuclear emulsion film for exposure. After development of the exposed films, the slides can be viewed under the light microscope; areas of high concentration of the label are clearly evident (Figure 6.14). Using this method, Tapper (1981a, 1981b) found that the quantity of photosynthetically fixed carbon transferred from *Trebouxia* to the mycobiont of *Cladonia convoluta* was much higher than that measured previously with different techniques.

Vertical and Horizontal Transfer of Photosynthates

Vertical transfer between the algal layer and the medulla was first shown by Smith (1961) for *Peltigera polydactyla*. It was the first evidence of transfer of

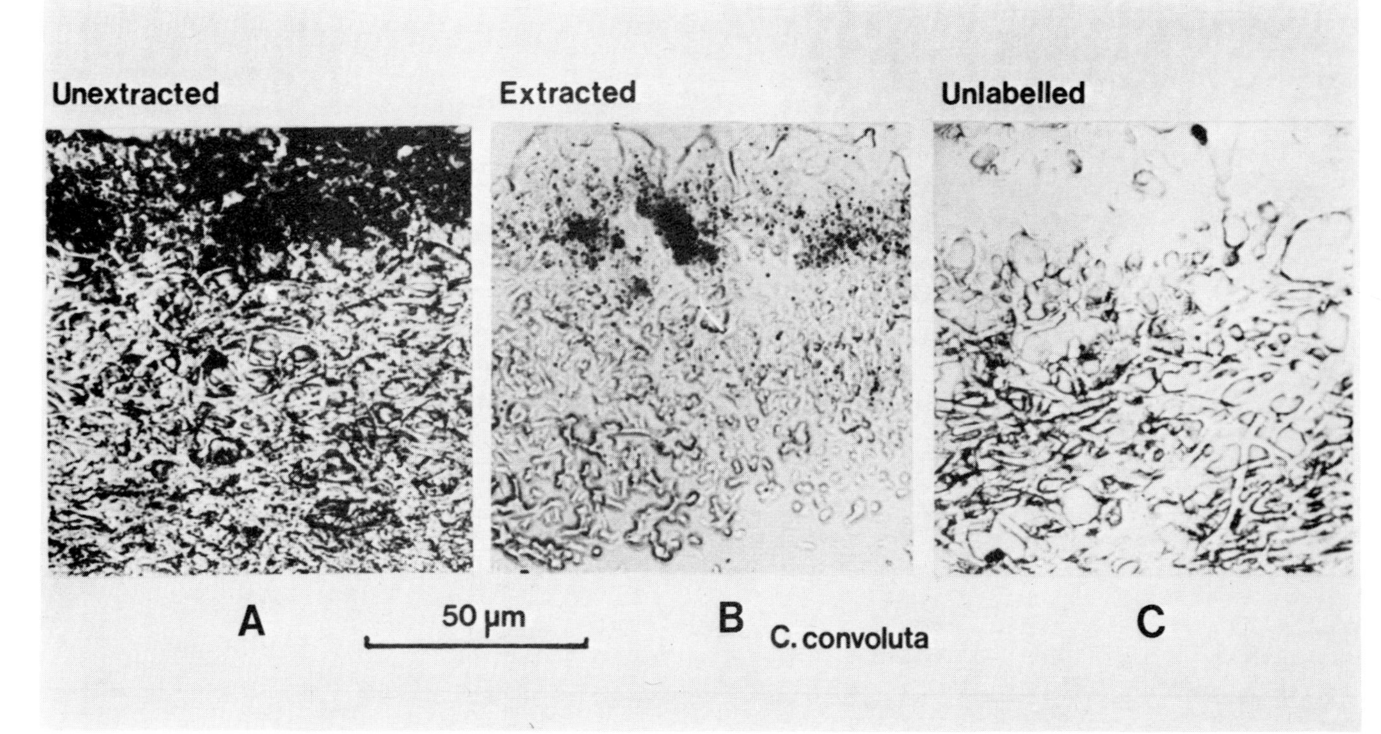

FIGURE 6.14. Autoradiographs of *Cladonia convoluta* following a pulse with $H^{14}CO_3^-$; (a) Untreated control section following chase; (b) Section extracted with EtOH following pulse; (c) Unlabelled section. From R. Tapper. 1981a. "Direct measurement of translocation of carbohydrate in the lichen *Cladonia convoluta* by quantitative autoradiography." *New Phytol.* 89: 429–437. Reprinted with permission of Academic Press.

algal photosynthetic products to the lichen fungus. The transfer required about two to four hours. Horizontal movement of photosynthates has not been investigated. Clearly, there is a need to understand the extent to which photosynthate can be transferred from place to place in the thallus. This information will permit a better understanding of the energetics of lichen growth and reproductive patterns. It will also be interesting to determine if different growth forms (e.g., thread-shaped fruticose lichens like *Usnea* species) translocate more efficiently than others (Hill, 1976).

The total amount of carbohydrate transferred from alga to fungus is considerable. Hill and Smith (1972) found that the amount of glucose transferred in *P. polydactyla* was 0.7–1.1 mg g^{-1} (dry wt) h^{-1}. They also found that 30–70 percent of the total CO_2 fixed was released by the alga. Massive transfer of carbohydrates is characteristic of all lichens studied except those with *Trentepohlia* phycobionts (Smith, 1974). The reason for this is not clear, but there may be a phylogenetic significance to this observation since many *Trentepohlia*-containing lichens are looser, more primitive associations. Whether or not the amount of transfer observed is sufficient to sustain the fungus is not presently known. Other carbohydrate sources available to the fungus are products of cell lysis (Hill, 1976) and organic products of decomposition present in the substratum (Smith, 1978). Smith has even suggested that the slow growth rates of lichen fungi in nature could probably be accounted for solely by environmental sources of carbon. Additional sources from algae may thus be required to physiologically buffer lichen thalli from environmental fluctuations (Farrar, 1976b). Studies on the energetics of lichens, both theoretical and empirical, are needed to resolve some of these problems.

Carbohydrate Utilization by the Fungus

In thalli of *Peltigera polydactyla,* which has a *Nostoc* phycobiont, glucose absorbed by the fungus is converted to mannitol (Drew & Smith, 1967a). In *Xanthoria aureola*, a lichen containing *Trebouxia*, the transfer carbohydrate is ribitol and the fungus converts this into mainly mannitol and some arabitol (Richardson & Smith, 1968). There is no information available to suggest a biosynthetic pathway for any of these conversions, but in nonlichenized fungi, mannitol may be formed either from fructose or fructose-6-phosphate via mannitol phosphate (Lewis & Smith, 1967).

Stored fungal carbohydrates may function as metabolic reserves. In darkness, the mannitol and arabitol content of *Xanthoria aureola* was found to decline rapidly, completely depleting the arabitol pool, which apparently functions as a short-term carbohydrate reserve (Richardson & Smith, 1966). During resaturation respiration losses in *Peltigera polydactyla*, mannitol content was also observed to decline (Smith & Molesworth, 1973). The rate at which absorbed carbohydrates are converted to cell structural components in the fungus would obviously depend on the total availability and composition of carbohydrate pools. However, not much direct evidence is available.

Environmental Control of Photosynthate Movement from Phycobiont to Mycobiont

Tysiaczny and Kershaw (1979) developed a method to monitor the temperature and moisture conditions that control the movement of labelled glucose from the phycobiont of the lichen *Peltigera praetextata* to the mycobiont. Using the ratio of labelled mannitol to labelled photosynthetic products as an index of glucose transport between the symbionts, they discovered that temperature had little effect on transport, but the level of hydration was extremely important. The investigators emphasized that this method produces results that are useful only as relative measures of glucose transport because there are respiratory losses by both symbionts prior to and during the transfer. However, it is extremely useful in evaluating the environmental control over complex whole thallus processes.

In a study using the same technique, MacFarlane and Kershaw (1982) observed similar results with *Peltigera polydactyla, P. rufescens,* and *Collema furfuraceum.* Temperature had little effect in controlling glucose movement between symbionts in the three species, but thallus hydration was found to be very important. In *P. polydactyla,* most glucose transfer occurred at, or close to, full thallus saturation; below 300 percent thallus moisture, glucose movement declined sharply. In *C. furfuraceum* and *P. rufescens,* however, transport continued at constantly declining rates with decreasing levels of thallus hydration. These results are consistent with the characteristic habits of the three species: *P. rufescens* and *C. furfuraceum* were collected from relatively xeric habitats, whereas *P. polydactyla* was collected from a more mesic woodland habitat.

These investigations indicate that the continuous, massive flow of carbohydrates observed in most laboratory studies of photosynthate movement between lichen symbionts (Bednar & Smith, 1966; Drew & Smith, 1967a; Richardson, Hill, & Smith, 1968; Richardson & Smith, 1968; Chambers, Morris, & Smith, 1976) is a direct result of the use of thallus disks on aqueous solutions of $NaH^{14}CO_3$, the standard experimental method used in most work in whole thallus physiology. Obviously, the use of this method results in data on efflux rates that occur only at full thallus saturation, a very rare condition for many lichens in nature. Moreover, the range of physiological responses observed when environmental conditions change is essential to understanding the true pattern of carbohydrate balance between the symbionts under natural conditions. Indeed, the results obtained when thallus hydration is varied support the widely held hypothesis that alternate wetting and drying cycles are required for maintenance of a balanced symbiotic association (Farrar, 1973; 1976b; Pearson, 1970; Dibben, 1971; Harris & Kershaw, 1971).

CONTROL OF PHOTOSYNTHATE RELEASE

The fact that the presence of the lichen fungus influences the carbohydrate release by the alga is demonstrated when the alga is removed from the thallus. Almost immediately after isolation, the alga exhibits a sharp decrease in carbohydrate release. This reduction in carbohydrate release is accompanied by an increase in the amount of photosynthate incorporated into ethanol-insoluble—presumably structural—cell components. After a short period of time, the physiological characteristics of freshly isolated lichen algae are identical to those of isolated phycobionts maintained for long periods of time in culture. One of the more persistent unsolved mysteries in lichen physiology is the reason for these observed changes in algal behavior upon isolation from the lichen thallus. It appears that the fungus somehow regulates carbohydrate flow from the alga; when the fungus is removed, this flow ceases. The question is: How does the fungus control the alga?

Smith (1974; 1975) has suggested that there are three categories of potential mechanisms to explain photosynthate release by the alga and the fungal control over this release:

1. *chemical*—release of some compound or factor by the fungus that alters the behavior of the alga;

2. *physical*—alteration of membrane potential or some other property that affects algal transport;

3. *second order effects*—internal metabolic changes in the alga that influence the production of mobile carbohydrates, transfer to release sites, or synthesis of carrier compounds.

Chemical Factors

Symbioses between algae and invertebrates (Taylor, 1970; 1973; Smith, 1973) are similar to lichen symbioses in that carbohydrate release by the alga is massive and generally involves a specific compound. Unlike lichens, where carbohydrate transfer is intercellular, symbiotic algae are often found inside the cells of invertebrates. In many invertebrate-alga and invertebrate-chloroplast symbioses, a chemical factor is produced by the animal that stimulates carbohydrate release by the algal cell or the chloroplast (Smith, 1974). The identity of these factors has not been determined. Numerous attempts to stimulate carbohydrate release by isolated lichen algae using homogenates of animals containing these factors have been entirely unsuccessful.

In an attempt to identify the presence of specific lichen factors, Green (1970) added lichen thallus homogenates and thallus washes to isolated lichen algae to induce carbohydrate release. He was unable to show significant increases in carbohydrate release using these homogenates, nor was he successful in altering algal release behavior with numerous other compounds (benzylaminopurine, kinetin, coconut milk, and a range of sugars and sugar alcohols).

Lichen acids have often been considered as carbohydrate release factors (Follmann, 1960; Follmann & Villagrán, 1965). The low water solubility of lichen substances would generally argue against their role as release factors; however, they have been shown to influence the protoplast permeability of numerous higher plant cells (Follmann and Nakagava, 1963). Furthermore, an aqueous solution of a sodium salt of usnic acid was found to increase the permeability of *Trebouxia* cells isolated from *Acarospora fuscata* and *Cladonia boryi* (Kinraide & Ahmadjian, 1970). Green (1970) has attributed many of these effects of lichen acids to pH reduction in the medium. When the medium is buffered properly, the effects of lichen acids on membrane permeability are considerably diminished.

pH

Green (1970) found that the amount of ribitol released by *Trebouxia* cells isolated from *Xanthoria aureola* was dependent upon the pH of the medium. At pH values below 5, ribitol release was considerable; at pH 6, however, very little ribitol was released. The fungus may be able to reduce the pH of the algal environment—perhaps with lichen acids—and thus facilitate carbohydrate release. It is not known, however, how long enhanced release can be maintained as a result of pH reductions. Also, there appear to be quite different pH optima for carbohydrate release of different phycobionts. Green (1970) found that glucose release by *Nostoc* was greatest at pH 5.9. It is possible that reduced release of glucose by *Nostoc* at low pH is associated with increased incorporation of fixed carbon into insoluble compounds (Hill, 1972), one of which has been identified as a storage compound, α-1, 4-glucan (Hill, 1976). No effect of pH on release by *Nostoc* was observed at all if isolated cells were allowed to age in distilled water. It appears from the evidence available, therefore, that pH effects by themselves are insufficient to account for the observed control over algal carbohydrate release.

Dilution

In the lichen thallus, algal cells would be expected to continue to release photosynthate so long as the fungus continued to absorb it. However, when the algae are removed from the thallus, the fungus can no longer absorb the released carbohydrates. Therefore, isolated algae would be expected to release photosynthate to the medium until an equilibrium is established between the internal and external carbohydrate concentrations. In a test of this hypothesis, Green (1970) found that by changing the carbohydrate dilution factor, the amount of photosynthate released could be changed, but not enough to explain the level of carbohydrate release characteristic of intact thalli.

Membrane Potentials

Smith (1974) argued that a high membrane potential of the lichen fungus may be able to alter the membrane potential of the alga to facilitate photosynthate flow from alga to fungus. There is some indirect evidence to support this idea. The membrane potential of the alga *Chlorella* has been measured at −40 millivolts (Barber, 1968) and that of the ascomycete *Neurospora* at −200 millivolts (Slayman, 1965). If lichen fungi and algae have similar membrane potentials, it is apparent that a significant electrical potential difference could actually exist between the two; the lichen fungus would therefore be able to alter algal membrane behavior and cause photosynthates to flow out of algal cells. Even though there is no direct contact between lichen algal and fungal membranes due to the presence of thick fungal and algal cell walls, charges may be carried along various protein, polysaccharide, or glycoprotein bridges in the cell walls.

This argument now seems less compelling because of some experiments done by Chambers, Morris, and Smith (1976). Carbohydrate transfer was observed between symbionts of *Peltigera polydactyla* in the presence and absence of digitonin. Digitonin abolishes the ability of the fungus to absorb glucose and increases the permeability of the fungal cell membrane so that substantial amounts of mannitol are released to the medium. Digitonin has no detectable effect on *Nostoc*, the phycobiont of *P. polydactyla*. In the presence of digitonin, there is a large sustained release of glucose into the incubation medium despite the fact that the fungus is not absorbing glucose (Table 6.17).

TABLE 6.17. Effect of Digitonin on the Release of Fixed ^{14}C from *Peltigera* Disks

Treatment	*% Fixed ^{14}C Released to Medium*
0.01% (w/v) digitonin, then $NaH^{14}CO_3$ added	54.3 (as glucose)
water, then $NaH^{14}CO_3$ added	0.4

S. Chambers, M. Morris, & D.C. Smith. 1976. "Lichen physiology XV. The effect of digitonin and other treatments on biotrophic transport of glucose from alga to fungus in *Peltigera polydactyla*." *New Phytol.* 82:489–500. Reprinted with permission of Academic Press.

Obviously, digitonin would also totally disrupt the membrane potential of the fungus, thereby eliminating this effect as a possible explanation for the release. So even if it is determined that the fungus is able to alter the alga's membrane potential, it appears not to be the explanation for algal carbohydrate release in the intact thallus.

The presence of haustorial contact between alga and fungus has also frequently been observed in electron micrographs (Peveling, 1973), and some workers have suggested that this represents a mechanism for the mass transfer of carbohydrates between symbionts. However, autoradiographic studies of carbohydrate transfer between symbionts have shown that transfer is not restricted to haustorial regions but rather occurs over the entire algal cell (Jacobs & Ahmadjian, 1971b; Hessler & Peveling, 1978). Collins and Farrar (1978) summarized the information presently available on haustorial contacts between lichen symbionts and found no clear pattern suggesting that haustoria are required for biotrophic nutrition. They suggested that diffusion, rather than haustorial contact, is a better explanation for mass transfer of carbohydrates between lichen symbionts.

Control of Algal Cell Division

Although algal cell division is often observed in lichen electron micrographs, it is assumed that cell division is more rapid in isolated phycobionts than in intact thalli simply because isolated algae grow more rapidly than lichenized algae. If algal cells are prevented from dividing, either by physical envelopment by fungal hyphae or by production of some inhibitor, then the photosynthate not used by the alga to manufacture new cells could be released to the medium or to the fungus. Since isolated lichen algae stop releasing carbohydrates and instead begin producing internal structural compounds—and probably also begin dividing more rapidly although this has not been determined—this explanation may be worth pursuing. However, no evidence for fungal control over algal cell division in the lichen thallus is presently available. The digitonin experiments described above (Chambers, Morris, & Smith, 1976) also strongly suggest the absence of any requirement for fungal membrane integrity for carbohydrate release. Fungal control over algal cell division, if it occurs at all, would have to be passive.

Facilitated Diffusion and Other Explanations

Chambers, Morris, & Smith (1976) suggested four ways in which carbohydrate efflux might be induced and sustained in the lichen thallus:

1. The fungus somehow modifies the environment of the alga so that algal carbohydrate metabolism is changed; then, high concentrations of carbohydrate accumulate at transport sites, and efflux occurs by facilitated diffusion.

2. The fungus induces formation of a carbohydrate transport system in the alga that shuts down when the association is terminated.

3. An existing algal transport system is modified by the fungus in such a way that carbohydrate efflux is stimulated.

4. There is some as yet unknown property of cells or membranes that prevents efflux of metabolites, and the fungus is somehow able to override this property.

Using the inhibition technique, Hill and Smith (1972) found that darkness reduced carbohydrate release in *Peltigera polydactyla* disks; release resumed when the thallus disks were placed in light (Figure 6.15). One possible explanation for this observation is that light energy is required to supply ATP from photophosphorylation for active secretion of carbohydrate by the alga. This idea was tested experimentally (Swindlehurst, Jordan, & Hill, reported in Hill, 1976) using 3-(3,4-dichlorophenyl)-1, 1-dimethyl urea (DCMU), a

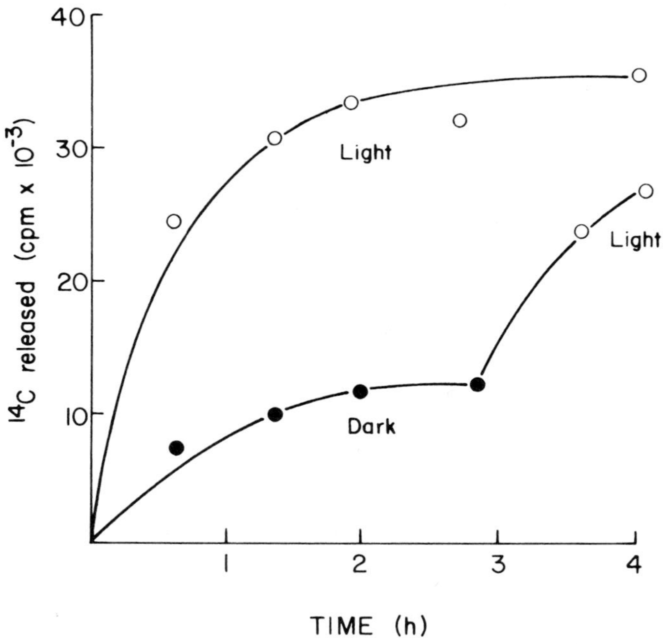

FIGURE 6.15. Effects of light and dark on release of ^{14}C by *Peltigera polydactyla* thallus disks into the growth medium during "inhibition" after pulse of $H^{14}CO_3^-$ in the light. From D.J. Hill. 1976. "The physiology of lichen symbiosis." In *Lichenology: Progress and Problems* (D.H. Brown, D.L. Hawksworth, & R.H. Bailey, eds.), pp. 457–496. London: Academic Press. Redrawn with permission of Academic Press.

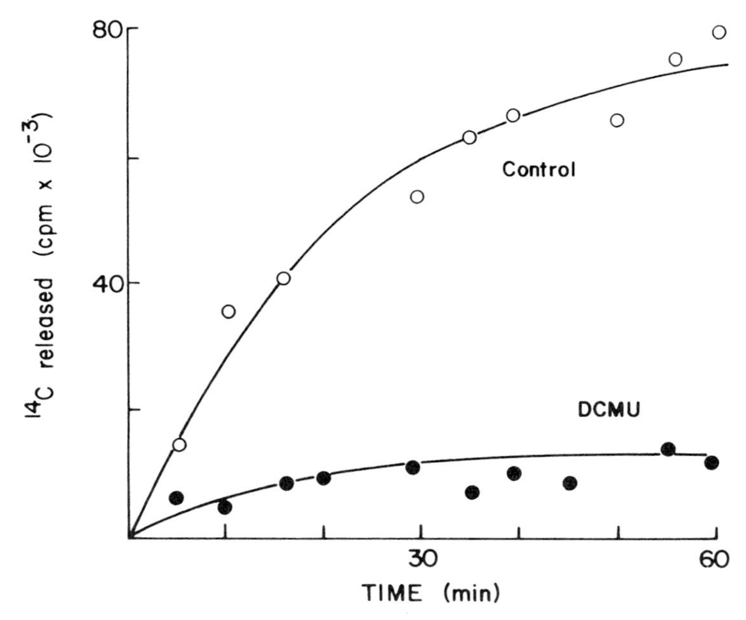

FIGURE 6.16. The effect of DCMU on the release of ^{14}C into the incubation medium by *Peltigera polydactyla.* From D.J. Hill. 1976. "The physiology of lichen symbiosis." In *Lichenology: Progress and Problems* (D.H. Brown, D.L. Hawksworth, and R.H. Bailey, eds.), pp. 457–496. London: Academic Press. Redrawn with permission of Academic Press.

compound that inhibits noncyclic photophosphorylation and NADPH formation but not cyclic photophosphorylation. The effects of DCMU on carbohydrate release were very similar to those of darkness (Figure 6.16); carbohydrate release was reduced in the presence of DCMU. Release may thus be an active process, but more work needs to be done to determine this unequivocally.

If energy is found to be required for transport, it must be determined if it is provided solely by the alga or if the fungus is somehow involved. Smith (1974) suggested the use of specific inhibitors to investigate this problem. A number of inhibitors have been tried, including cycloheximide, chloramphenicol, nystatin, griseofulvin, dinitrophenol, sodium azide, phloridzin, ouabain, and N-acetyl glucosamine, and no effect on transport was observed for any of these (Smith, 1975). However, pretreatment of *Peltigera polydactyla* thallus disks with sorbose resulted in reduced carbohydrate transfer between alga and fungus (Smith, 1975; Chambers, Morris, & Smith, 1976). There was a reduction of fixed carbon released to the medium in inhibition experiments. These results have suggested that sorbose enters the algal cell and competes with glucose for the carrier involved in glucose efflux. This is further evidence of a carrier-mediated, facilitated diffusion efflux system in lichens; such carrier systems in yeasts are known to have an affinity for sorbose (Cirillo, 1961). This appears to be the best explanation for carbon transfer between lichen symbionts at the present time. However, as Smith (1975) has remarked, our understanding of the nature and mechanism of fungal control over photosynthate efflux is still quite limited.

It is obvious that a precise understanding of biotrophic transport in lichens will require additional work in a number of areas, particularly in algal and fungal membrane functions. Nevertheless, biotrophic transport mechanisms are best known in lichenized fungi, and it is likely that the model for biotrophic associations will continue to be the lichen association. Numerous similarities to other fungal biotrophic associations have already been identified (Smith, 1978). The fact that at least 25 percent of all "real" fungi (i.e., exclusive of lichenized fungi) are biotrophic suggests that the study of transfer processes in lichens has a great deal of relevance to fungi generally (Smith, 1978).

7 Growth and Demography

Lichen growth is notoriously slow, a result of the interaction of numerous environmental and physiological factors. Although a number of workers have investigated the environmental factors thought to influence lichen growth (rainfall, sunlight, and nutrients), the biological basis for lichen growth has not been adequately explained. Despite all that we have learned about the various physiological processes of lichen symbionts, both in the intact thallus and in isolation, we do not yet know specifically how these processes result in thallus growth.

As a consequence of all this, we must rely on theoretical models that assume certain anatomical and physiological attributes. A number of these models have been published in recent years, and all provide quite plausible hypotheses for testing. The extent to which they actually explain lichen growth will require considerable experimental study, however.

This chapter will briefly evaluate the methods used to measure lichen growth and discuss the more important environmental factors controlling growth. Theoretical treatments will be considered briefly along with the underlying assumptions each makes about lichen structure and function. Then, because lichen growth can be considered at two distinct levels, the individual thallus and the entire colony, a consideration of population growth, age structure, and other demographic characteristics will be provided. Finally, there will be a brief discussion of lichenometry, the use of lichen growth data to determine the age of colonizable substrates.

A few excellent review articles on lichen growth have appeared recently (Hale, 1973a; Armstrong, 1976; Topham, 1977). Since most of the older papers on growth will not be considered in this discussion, a serious search of the growth literature should begin with these reviews.

GROWTH

Growth Measurement Methods

One- and Two-Dimensional Growth. Lichen growth rates are most frequently reported as thallus radius or surface area increase with time. Such measurements can be obtained directly using photographic techniques or indirectly by measuring the size of thalli growing on datable substrates (e.g., buildings, tombstones, etc.). Early workers made fairly accurate measurements by establishing a fixed basepoint on the substrate and measuring changes in the distance from this point to the thallus growing edge. Hale (1970) modified this technique by establishing fixed points on older portions of the thallus and measuring changes in the distance between these points and the growing edge. This reduces errors in radial growth estimates caused by expansion of the central parts of the thallus.

The most reliable techniques are photographic because they provide permanent records and can be enlarged to permit very precise growth measurements for relatively short intervals of time in thalli of various sizes, including extremely small juvenile thalli. Examples of two photographic records of *Pseudoparmelia baltimorensis* from Maryland studied by Lawrey and Hale (1977) are provided to illustrate the technique (Figures 7.1 & 7.2). In the first case (Figure 7.1), a large thallus of *P. baltimorensis* overgrows a juvenile thallus of the same species. In the second (Figure 7.2), the rapid early growth of a small thallus of *P. baltimorensis* is shown.

Despite the popularity of photographic techniques for growth assessment and their obvious superiority to other direct methods (Hale, 1973a), there are potential sources of error of which investigators should be aware. Hooker and Brown (1977) discussed these problems and compared the accuracy of two direct methods, a photographic method and an unusually accurate measurement method developed by Armstrong (1973). They demonstrated how magnification and parallax errors in measurement were the result of camera angle and suggested several ways to eliminate these errors. Once they are eliminated, lobe growth accuracies to 0.1 millimeter were obtained, well within the range of accuracy demonstrated by Armstrong (1973).

Three-Dimensional Growth. Difficulties in assessing growth in three dimensions have limited the number of these studies in the literature. For this reason, the use of lichen surface area measurements to estimate changes in volume or dry weight is common; unfortunately, it is also quite unreliable

FIGURE 7.1. Thalli of *Pseudoparmelia baltimorensis* photographed at different times on Plummers Island, Maryland; (a) March 20, 1966; (b) August 21, 1966; (c) December 4, 1966; (d) February 26, 1967; (e) December 24, 1968; (f) June 17, 1969. From J.D. Lawrey & M.E. Hale, Jr. 1977. "Natural history of Plummers Island, Maryland. XXIII. Studies on lichen growth rate at Plummers Island, Maryland." *Proc. Biol. Soc. Washington* 90: 698–725. Reprinted with permission of the Biological Society of Washington and the authors.

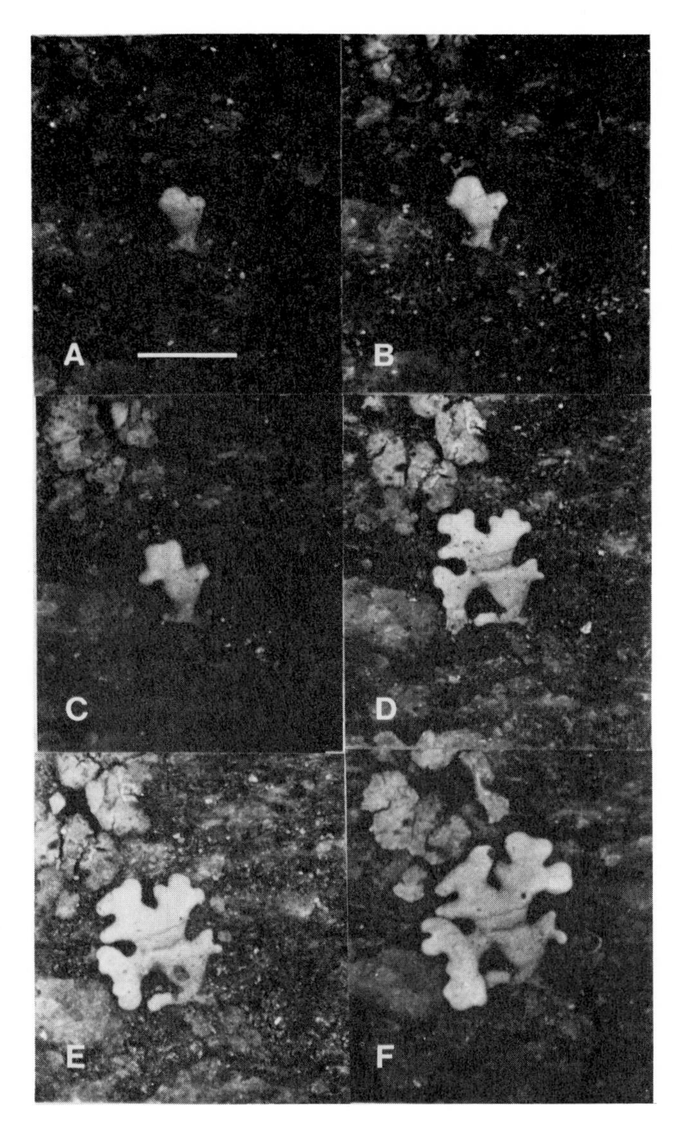

FIGURE 7.2. Successive photographs of minute thallus of *Pseudoparmelia baltimorensis* taken at Plummers Island, Maryland. All photographs same scale; bar = 1 mm. (a) March 15, 1969; (b) May 15, 1969; (c) June 17, 1969; (d) February 1, 1970; (e) April 12, 1970; (f) June 21, 1970. From J.D. Lawrey & M.E. Hale, Jr. 1977. "Natural history of Plummers Island, Maryland. XXIII. Studies on lichen growth rate at Plummers Island, Maryland." *Proc. Biol. Soc. Washington* 90: 698–725. Reprinted with permission of the Biological Society of Washington and the authors.

(Farrar, 1974). Interest in this subject will probably increase in the future, however, since a thorough physiological understanding of lichen growth will ultimately require accurate methods of measuring changes in lichen biomass in the field and the laboratory.

Field studies of lichen biomass accretion are limited to fruticose species or large, chunky foliose species that can be easily removed from the substrate. In subarctic Finland, Kärenlampi (1971) measured relative growth rates of a number of fruticose species by placing clumps of thalli in plastic boxes on the forest floor. The same samples could then be removed, air-dried, and weighed at various time intervals to estimate growth rate. Several potential sources of error were discussed, among these:

1. An unnatural microclimate inside the boxes.

2. The fact that continuous air drying is necessary for accurate weighing, and this appears to be detrimental.

3. The plastic boxes collect water and nutrients, thereby influencing growth.

These problems may be more serious in some studies than others. For example, Miller (1966, cited in Hale, 1973a) had problems with thallus disintegration in a study of foliose lichen growth using Kärenlampi's direct technique.

In an interesting field study of *Lobaria oregana* growth in the Pacific Northwest, Rhoades (1977) also made biomass measurements, but in his study, biomass was measured at the end of the growth interval and the relationship between surface area and biomass was established by regression analysis. Although less direct than Kärenlampi's (1971) method, this method avoids experimental disturbances of thalli and thereby allows a more accurate assessment of natural biomass accretion rates. Such an approach could also be taken with foliose lichens not easily removed from the substrate, since thalli would only have to be removed once for weighing.

Harris and Kershaw (1971) described a method for measuring changes in lichen biomass under laboratory growth chamber conditions. Small thalli of *Parmelia sulcata* and *Hypogymnia physodes* were placed on wetted filter paper in the chambers and the air supply to the chambers was filtered. To estimate growth rates, experimental material was air-dried and weighed at 28-day intervals for four months. During the experimental period, both lichens grew continuously under conditions of fluctuating moisture and light. Daily watering and/or continuous illumination resulted in death within three months. Other laboratory investigations (Pearson, 1970; Pearson & Benson, 1977) give roughly the same results. Obviously, the need to air-dry material before and after exposure to growth chamber conditions limits the reliability of correlations between the measured growth and the environmental conditions in the chambers. Such experimental error cannot be controlled as yet; however, these studies demonstrate that successful experiments involving whole lichen thalli can be done in laboratory growth chambers so long as reliable results can be obtained in a reasonable interval of time.

Marginal Zonation or Growth Rings. Some lichens produce distinct concentric marginal rings of alternating light and dark tissue that appear to represent annual increments (Hale, 1973a). If these patterns are truly annual, then estimates of annual growth rate and population age distribution can be determined by measuring and counting rings.

Species of *Pertusaria, Lecanora, Ochrolechia, Buellia, Xanthoria,* and *Caloplaca* are known to produce rings. Hooker (1980) has identified several zonation patterns in Antarctic crustose species. In *Buellia russa,* the zonation is confined to the darker prothallus and appears to represent one year's growth. In other species, however, each ring may represent as many as five years' growth. During a period of 16 months representing two growing seasons, *Buellia coniops* thalli formed only one ring; in *Caloplaca cirrochrooides* no new ring was observed.

There may also be distinct thallus ring patterns caused by unusually severe weather conditions; Hooker (1980) called these climatic rings. For example, he observed thalli of *Xanthoria elegans* that appeared to have annual rings. Further investigation revealed that *X. elegans* thalli from other sites did not have rings. Hooker hypothesized that the ring effect was caused by a short period of stunted lobe growth during unfavorable climatic conditions.

It is obvious that this aspect of lichen growth warrants further study. Unfortunately, few lichens develop rings, climatic, annual, or otherwise, so most lichen growth investigations will have to rely on other methods.

Growth Phases

Lichen thalli of different sizes grow at different rates, suggesting that lichens pass through distinct growth phases. Beschel's (1958) study of lichens on grave markers in Switzerland provided indirect evidence for at least three phases:

1. a juvenile stage during which very small thalli grow relatively slowly,

2. a "great period" of rapid growth in which small thalli become established in the habitat,

3. a period of linear growth until maturity is reached.

Some species also appeared to exhibit reduced growth rates at the onset of thallus disintegration, suggesting that lichens have a maximum size beyond

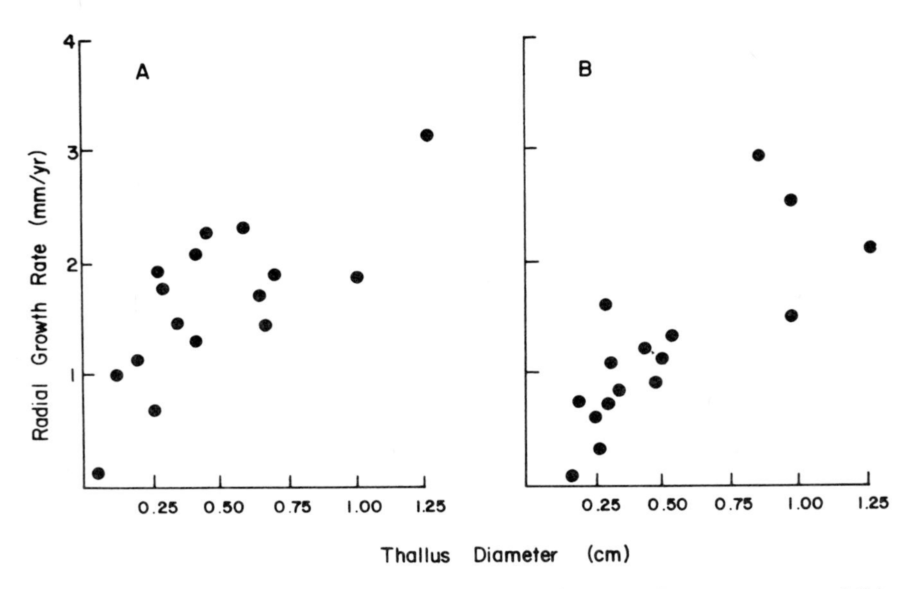

FIGURE 7.3. Radial growth rate of (a) *Xanthoparmelia conspersa* and (b) *Melanelia glabratula* in relation to thallus diameter. From R.A. Armstrong. 1974. "Growth phases in the life of a lichen thallus." *New Phytol.* 73: 913–918. Redrawn with permission of Academic Press and the author.

which radial growth slows or ceases and the thallus center dies and falls away. Hale (1973a) reported similar growth patterns for *Xanthoparmelia conspersa* in Virginia.

Armstrong (1974; 1976) and Topham (1977) have discussed growth phases and concluded that many lichens, if not most, exhibit an exponential radial growth phase as juveniles, then a linear phase until thallus disintegration takes place. The evidence for this two-stage pattern comes primarily from the work of Armstrong, who has thoroughly studied growth of saxicolous foliose species on Ordovician slate at various sites in Wales. In one report (Armstrong, 1974), he compared growth of small thalli—less than 1.25 cm diameter—and senescent thalli—those with fragmenting centers—of *Xanthoparmelia conspersa* and *Melanelia glabratula* ssp. *fuliginosa* using two growth measures, a radial rate (mm/yr) and a relative rate ($cm^2/cm^2/yr$). He observed a rapid increase in radial growth in juvenile thalli of both species (Figure 7.3) consistent with the hypothesis that early growth is exponential; relative growth ($cm^2/cm^2/yr$) was also rapid until a diameter of 2–3 millimeters was reached, whereupon growth of both species declined. Radial growth of mature thalli is related to thallus diameter in an approximately linear fashion, a result that confirmed previous experiments (Armstrong, 1973). A compari-

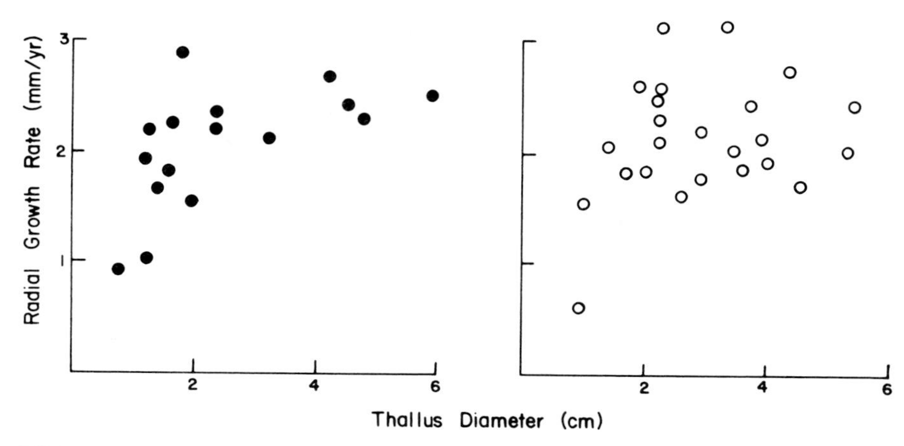

FIGURE 7.4. Growth rate of *Melanelia glabratula*. ● = nonfragmenting thalli; ○ = fragmenting thalli. From R.A. Armstrong. 1974. "Growth phases in the life of a lichen thallus." *New Phytol.* 73:913–918. Redrawn with permission of Academic Press and the author.

son of fragmenting and nonfragmenting (senescent) thalli revealed no significant differences in radial growth rate (Figure 7.4), suggesting that the thallus center of these particular species contributed little to radial growth of mature thalli.

Whether this pattern is a general one for lichens is not clear as yet. As Armstrong (1974) has mentioned, it is difficult to test many of the emerging lichen growth hypotheses using previously published data because few investigators have accurately measured growth of very small and very large—including senescent—thalli.

Seasonal Growth and Environmental Correlates

Despite the fact that seasonal lichen growth has been actively investigated (Phillips, 1963; Rydzak, 1961; Hale, 1970; Armstrong, 1973; 1975; Showman, 1976; Lawrey & Hale, 1977; Fisher & Proctor, 1978), only a few generalizations can presently be made. This is because numerous environmental factors (light, moisture, temperature, substrate, and nutrients) interact in complex ways to influence growth.

An example of the influence of interacting environmental factors on seasonal growth is provided by Armstrong (1975), who measured radial growth of *Melanelia glabratula* ssp. *fuliginosa* for a year on rock surfaces in Wales facing south and northwest. Thalli on northwest-facing rocks grew more than those on south-facing rocks in spring and fall months; south-facing thalli grew the most in the winter (Figure 7.5). Armstrong suggested that these differences reflected changes in optimal growth conditions on the two surfaces during the year. Optimal growth is assumed to be most closely coupled with temperature and moisture conditions. When very wet or very dry, thalli were expected to have low net carbon assimilation rates (NCAR). In the spring and fall when temperatures were moderate to high and rainfall was low, lichens experienced wide fluctuations in water content. However, northwest-facing rocks dried more slowly, allowing a higher NCAR for northwestern lichens than southern lichens. In the winter when temperatures were low to moderate and rainfall was high, lichens were frequently saturated and NCAR was low. Since south-facing rocks dried more rapidly than northwest-facing rocks, a higher winter lichen NCAR could be maintained in southern exposures.

These results emphasized the importance of thallus moisture content in controlling lichen growth rate. The majority of available correlation analyses, regardless of the level of sophistication, indicate that rainfall, cloud cover, relative humidity, fog frequency, and other factors that contribute directly or indirectly to thallus wetting and drying are most closely coupled with growth.

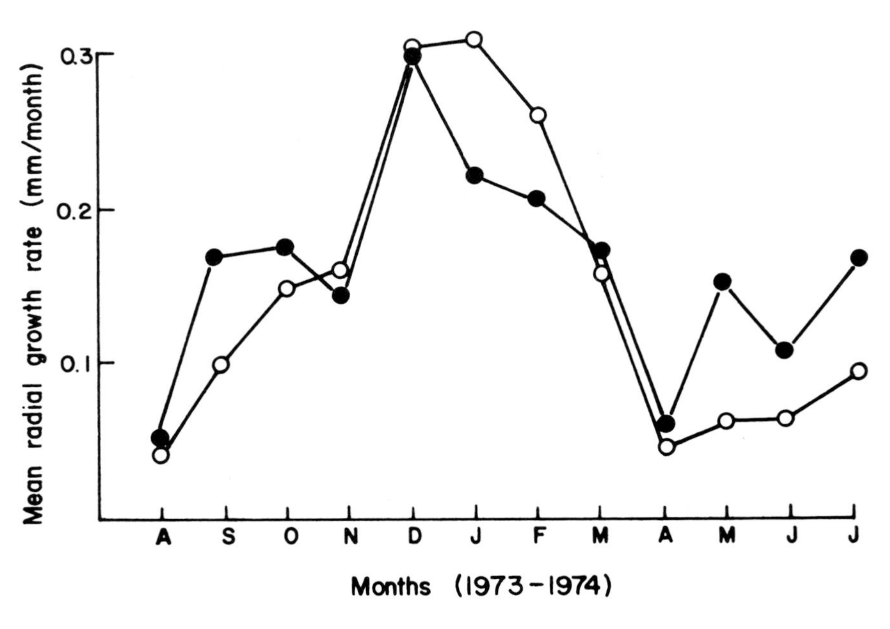

FIGURE 7.5. Seasonal growth of thalli from two populations of *Melanelia glabratula;* (●) northwest face; (○) south face. Redrawn from R.A. Armstrong. 1975. "The influence of aspect on the pattern of seasonal growth in the lichen *Parmelia glabratula* ssp. *fuliginosa* (Fr. ex Duby) Laund." *New Phytol.* 75:245–251. Redrawn with permission of Academic Press and the author.

Other environmental factors have been less frequently studied. Nutrient element availabilities are of unquestionable importance. Hakulinen (1966) and Jones and Platt (1969) found that fertilization of lichen thalli significantly enhanced growth rates of a number of lichens. However, as Smith (1975) has commented, nutrient additions do not cause particularly marked growth stimulations; growth is never as rapid as in higher plants. We know very little about the specific ways nutrients contribute to lichen growth. We know that each lichen symbiont requires nutrients and may obtain them from the other, but we know practically nothing about the mechanisms involved. Most needed are studies of:

1. nutrient requirements of symbionts,
2. localization, turnover and transfer of nutrients within lichen thalli.

Farrar has suggested that, "next to reproduction, this is the area of lichen biology most in need of clarification" (1976d, p. 398).

Relationships between Growth and Reproduction

The frequently measured exponential growth phase of juvenile thalli represents a period during which assimilative capacity increases with age, at least for a short time. It is reasonable to assume that as a thallus passes through this exponential phase to a linear phase, its assimilative capacity reaches some sort of equilibrium determined by an as yet poorly understood interaction between the environment and physiological constraints. A number of lichen growth models have been developed to account for these growth patterns with reasonable biological interpretations. These will be discussed in the section that follows.

Topham (1977) has suggested that the attainment of linear growth by a thallus may be accompanied by production of reproductive tissues, since the assimilative zone responsible for thallus growth remains approximately constant in size, releasing the remaining center portion of the thallus for sexual or asexual reproduction. Onset of linear growth can therefore be considered an indicator of reproductive maturity. A number of predictions can be derived from this very interesting hypothesis:

1. A species-specific minimum size must be attained before reproduction can begin.

2. The assimilative zone, a band of tissue extending from the thallus edge inwards, will remain approximately constant in area after a thallus has attained a linear growth phase.

3. Reproductive effort will be positively correlated with the size and assimilation capacity of the central zone of the thallus.

4. The time required to reach linear growth will be correlated with the level of competition or severity of the physical environment.

At present, these predictions cannot be tested because so little is known about the relative assimilative capacities of central versus marginal portions of the thallus or the movement to assimilate from place to place within the thallus.

Few measurements of reproductive effort (RE) in lichens have been made. This is clearly an area requiring more thorough investigation. Fahselt (1976) measured growth rate in vegetative and reproductive tissue of *Xanthoparmelia cumberlandia* in a transplant experiment in southern Ontario. Using a diamond saw, she subdivided a single thallus still attached to its rock substrate into several portions and relocated samples at different sites. After one year, vegetative growth rates of transplanted thalli were not significantly different from controls at the original location; however, reproductive effort appeared to have been stimulated at the southernmost transplant locations. Increased initiation of new apothecia was observed as well as a significant expansion of previously formed apothecia, especially in the younger portions of the transplanted thalli. Fahselt gave several possible explanations for these results. The southern transplant locations may have been more favorable for apothecial development because the climate is generally milder. There is also a possibility that air quality was involved. Fahselt's results strongly support the premise that lichen reproductive effort is greater in the most environmentally favorable environments. They also show that lichen growth rate and RE are not necessarily related. Reproductive characteristics of lichens may therefore be more sensitive to environmental fluctuations (both natural and anthropogenic) than vegetative characteristics.

Theoretical Treatments of Lichen Growth

There have been a number of interesting recent attempts to explain lichen growth in a biologically meaningful way through the use of theoretical models (Woolhouse, 1968; Proctor, 1977; Armstrong, 1976; Topham, 1977; Fisher & Proctor, 1978; Aplin & Hill, 1979; Childress & Keller, 1980). Some of these will be considered briefly in this section. Hill (1981) has written an excellent review of lichen growth models and the various assumptions implicit in each. The main conclusions that emerge from this review are:

1. We know very little about the actual biological mechanisms of lichen growth.
2. Better focused experiments are needed before many of the theories already published can be properly evaluated.

Some of the more troublesome questions that need to be answered are listed in Table 7.1.

TABLE 7.1. Questions Concerning the Biological Basis of Lichen Growth

1. Is there a general relationship between photosynthetic rate and growth? Are there similar relationships between growth and other lichen attributes (reproductive characteristics, secondary compound production, etc.)?

2. Is photosynthate produced in all thallus portions equally or are there differences due to tissue age?

3. To what extent can carbohydrates move from place to place in lichen thalli?

4. In what form is the transferable carbohydrate? By what process does it move?

5. How are growth in thallus surface area (apical growth) and growth in thallus thickness (vertical growth) different?

6. How do growth rates correlate with thallus longevity? Is early senescence correlated with rapid growth rate? What are the mechanisms responsible for senescence and longevity?

7. To what extent is algal cell turnover different in thallus portions of different age? To what extent is cell turnover frequency related to thallus longevity?

8. Is slow lichen growth a consequence of life in physiologically stressful habitats? Are there important trade-offs between lichen growth and reproductive effort in habitats that differ markedly in the level of stress they induce?

Partial data adapted from several sources, primarily D.J. Hill. 1981. "The growth of lichens with special reference to the modelling of circular thalli." *Lichenologist* 13:265–287, and P.S. Aplin & D.J. Hill. 1979. "Growth analysis of circular lichen thalli." *J. Theoret. Biol.* 78:347–363.

Theoretical treatments of lichen growth generally assume a circular thallus structure. In one of the earliest of these models, Woolhouse (1968) suggested that the relative growth rate concept frequently applied to higher plants should also be used to describe lichen growth. He argued that the growth potential of lichens depends on the thallus surface area available for light interception; the larger the surface area, the greater the potential growth rate. This explains the general observation that large thalli grow more rapidly than very small thalli. However, it does not explain the more or less linear radial growth of intermediate and large thalli observed by Armstrong (1973; 1974) and others.

A modification of Woolhouse's (1968) concept that attempts to account for these observations was developed by Proctor (1977), who suggested that growth is not dependent on the entire thallus surface area, but is proportional to the area in an annulus of constant width at the growing edge. This hypothesis stems from a suggestion made by Beschel (1961) and further developed by Armstrong (1975) to the effect that the thallus center makes very little contribution to radial growth, and relative growth rates are proportional to the peripheral area rather than the surface area of the thallus. Proctor tested his model with *Buellia* (=*Diploicia*) *canescens* growth data and obtained a relatively good correspondence between observed and expected values.

There is some indirect evidence to support the concept of a peripheral growth ring in lichens. Fungal and algal cells are apparently most metabolically active at growing tips (Boissière, 1972; 1979; 1982; Galun, Paran, & Ben-Shaul, 1970a; Greenhalgh & Anglesea, 1979); radial growth is also apparently restricted to an "elongation zone" within a few millimeters of the thallus periphery (Boissière, 1972; Hale, 1970; Fisher & Proctor, 1978). However, as Hill (1981) has accurately pointed out, there is little direct structural or functional evidence to presume *a priori* the existence of a marginal zone of constant width in circular thalli.

A more functional approach to lichen growth modeling has been taken by Aplin and Hill (1979), Childress and Keller (1980), and Hill (1981). The model of Aplin and Hill initially contained five constants representing biologically important processes thought to influence growth:

1. rate of net photosynthesis,
2. proportion of photosynthate contributing to radial growth,
3. proportion of photosynthate contributing to thallus weight,
4. movement of photosynthate to growing edges,
5. rate of growth in relation to photosynthate concentration.

They reduced their initial model to the following simpler equation describing the change in radius width (r) with time (t), in which α is a distance constant that defines the extent to which centrally produced carbohydrates contribute to radial growth and the constant β is the potential rate of increase irrespective of the translocation of photosynthate. This would imply that α is almost totally an expression of fungal growth whereas β is a function of algal photosynthetic rate.

$$\frac{dr}{dt} = \frac{\alpha r}{\beta(r + 2)}$$

Hill (1981) has since modified this growth equation to make the constants conform to current usage. The new equation is

$$\frac{dr}{dt} = \frac{\alpha s r}{r + 2s}$$

in which α is the growth rate constant (equivalent to $1/\beta$ of Aplin & Hill (1979)) and s is the distance constant defining the width of the peripheral thallus ring contributing to lichen growth (equivalent to α of Aplin & Hill (1979)).

Because they reflect the actual processes that contribute to growth, the values of α and s must be accurately estimated. Hill (1981) provided a method of determining α and s from photographic growth data. Radial growth (Δ r) is measured for thalli of various sizes during a time interval (Δ t), and Δ r is plotted on a graph against Δ ln r. This relationship should be linear. The value of Δ r when Δ ln r = 0 (the intercept on the Δ r-axis) and the value for Δ ln r when Δ r = 0 (the intercept on the Δ ln r-axis) are both estimated and the values of α and s can be obtained as follows:

$$\alpha = \frac{2}{\Delta t} \Delta r \text{ (when } \Delta \ln r = 0)$$

$$s = \frac{\ln r \text{ (when } \Delta r = 0)}{\alpha \Delta t}$$

In calculating these growth constants, there is an accuracy problem because the points on the Δ r/Δ ln r graph are sometimes widely scattered and the intercepts a matter of some judgment. Hill (1981) discussed these problems and provided a list of constants obtained using a number of different methods (Table 7.2). Although they vary considerably, the constants are all within the same order of magnitude and allow interesting biological comparisons. For example, the *Parmelia* species generally have a larger distance constant than *Diploicia canescens*, suggesting that photosynthate is translocated farther from the growing tips in *Parmelia* than in *D. canescens*.

DEMOGRAPHY

Demography is the study of life history statistics, especially age-specific birth, growth, and death rates. An individual's life history characteristics reflect the selective forces that have influenced its evolutionary development, and demographic studies provide the raw data needed to characterize life histories (Solbrig & Solbrig, 1979).

Demographic theory developed primarily for human and animal populations cannot be applied uncritically to plants because the concepts of individual birth and death are far less relevant (Harper & White, 1974). This is especially true for clonal plants like lichens because growth can be considered

TABLE 7.2. Values for Growth Constants Obtained from Various Sources

Species	*Rate Constant* (yr^{-1})	*Distance Constant* (*cm*)[a]	*Reference*[b]
Diploicia canescens	$RGR_O = 2.0$	$s = 0.08$	Proctor (1977)
D. canescens	$RGR_O = 1.5$	$s = 0.075$	Proctor (1977)
D. canescens	$RGR_O = 1.3$	$s = 0.13$	Proctor (1977)
D. canescens	$1/\beta = 1.27$	$\alpha = 0.130$	Aplin & Hill (1979)
D. canescens	$1/\beta = 2.00$	$\alpha = 0.088$	Aplin & Hill (1979)
D. canescens	$1/\beta = 1.61$	$\alpha = 0.079$	Aplin & Hill (1979)
D. canescens	$1/\beta = 2.13$	$\alpha = 0.133$	Aplin & Hill (1979)
D. canescens	$1/\beta = 1.61$	$\alpha = 0.144$	Aplin & Hill (1979)
D. canescens	$1/\beta = 2.22$	$\alpha = 0.058$	Aplin & Hill (1979)
Lecanora muralis	$\alpha = 0.50$	$\omega = 0.7$	Farrar (unpubl.)
Lobaria oregana	$\alpha = 1.45$	$\omega = 0.5$	Farrar (unpubl.)
Pseudoparmelia baltimorensis	$1/\beta = 1.26$	$\alpha = 0.57$	Hill (unpubl.)
Xanthoparmelia conspersa	$\alpha = 0.8$	$\omega = 0.23$	Farrar (unpubl.)
X. conspersa	$1/\beta = 1.02$	$\alpha = 0.73$	Aplin & Hill (1979)
X. conspersa	$1/\beta = 0.88$	$\alpha = 1.00$	Aplin & Hill (1979)
X. conspersa	$1/\beta = 1.10$	$\alpha = 0.73$	Aplin & Hill (1979)
Melanelia glabratula	$1/\beta = 1.43$	$\alpha = 0.197$	Aplin & Hill (1979)
M. glabratula	$\alpha = 0.7$	$\omega = 0.31$	Farrar (unpubl.)
Parmelia saxatilis	$\alpha = 0.4$	$\omega = 0.20$	Farrar (unpubl.)
Physcia caesia	$\alpha = 0.8$	$\omega = 0.50$	Farrar (unpubl.)
Phaeophyscia orbicularis	$\alpha = 0.32$	$\omega = 0.34$	Farrar (unpubl.)
Rhizocarpon geographicum	$1/\beta = 0.27$	$\alpha = 0.194$	Proctor (unpubl.)

[a]The distance constants are not all equivalent because they are derived differently; ω and s are assumed to be discrete real distances of the width of an imaginary annulus whereas α is a hypothetical distance that relates to the translocation of photosynthate (Hill, 1981).

[b]The actual data used are identified in Hill's (1981) paper.

D.J. Hill. 1981. "The growth of lichens with special reference to the modelling of circular thalli." *Lichenologist* 13:265–287. Reprinted with permission of Academic Press.

at two different levels of organization, the individual and whole colony. As we have seen in previous chapter sections, individual thallus growth has frequently been measured. However, studies of lichen colony or population growth have only rarely been conducted. For this reason, little is known about lichen demography.

The distinction between individual and colony growth is frequently blurred in lichen investigations because it is difficult to decide what an individual lichen really is. In some groups (e.g., *Umbilicaria* species) individuals are easily distinguished because each possesses a unique structure (in the case of *Umbilicaria*, an umbilicus) without which it cannot exist independently. For most lichens, however, one cannot easily determine when a lichen colony is composed of genetically distinct individuals of separate origin (genets of Harper & White, 1974) and when it is simply a large thallus that has become subdivided into smaller genetically identical parts (ramets of Harper & White, 1974).

Size Frequency Analyses

Although age-specific statistics are rarely determined in lichen studies, size frequencies have been measured in a number of lichenometric investigations (Benedict, 1967; Andersen & Sollid, 1971; Lindsay, 1973; Denton & Karlén, 1973; Matthews, 1973). These measurements provide information about the history of a lichen population and the substrate on which it is found. A population with numerous small thalli and no thalli larger than a certain size would be expected on relatively young substrates, and a population with representatives of all size classes would be expected on older undisturbed substrates. Furthermore, the frequency of individuals in each size category should be proportional to the rate of thallus colonization and inversely proportional to the size-specific thallus growth rate (Farrar, 1974).

Topham (1977) has suggested that lichen population size frequencies can be used to estimate colonization rates. Data from lichenometric studies that generally involve *Rhizocarpon geographicum* indicate an exponential colonization rate, suggesting that individual thallus mortality is relatively low. However, age-specific mortality rates have not been determined directly for lichen populations. Farrar (1974) observed an exponential colonization rate for *Physcia caesia* on an asbestos roof in Berkshire, England. In the same habitat, however, *Lecanora muralis* thalli showed an initially rapid colonization rate followed by a decline, suggesting a gradual replacement of *L. muralis* by *P. caesia* over successional time (Figure 7.6). Brightman (1959) also measured size frequencies of *L. muralis* in another location in England and observed a relatively constant colonization rate, a result that implies no secondary production of young thalli by older thalli (Topham, 1977).

Age-Specific Statistics

Additional lichen demographic studies will have to be done to accurately measure:

1. age-specific mortality rates,
2. age-specific rates of thallus division (cloning) and fusion.

The most direct way to determine these rates is by tracking a large number of individual thalli from year to year for long intervals of time and recording the number of individuals that subdivide or die each year. Also, the number and position of new colonists must be accurately recorded.

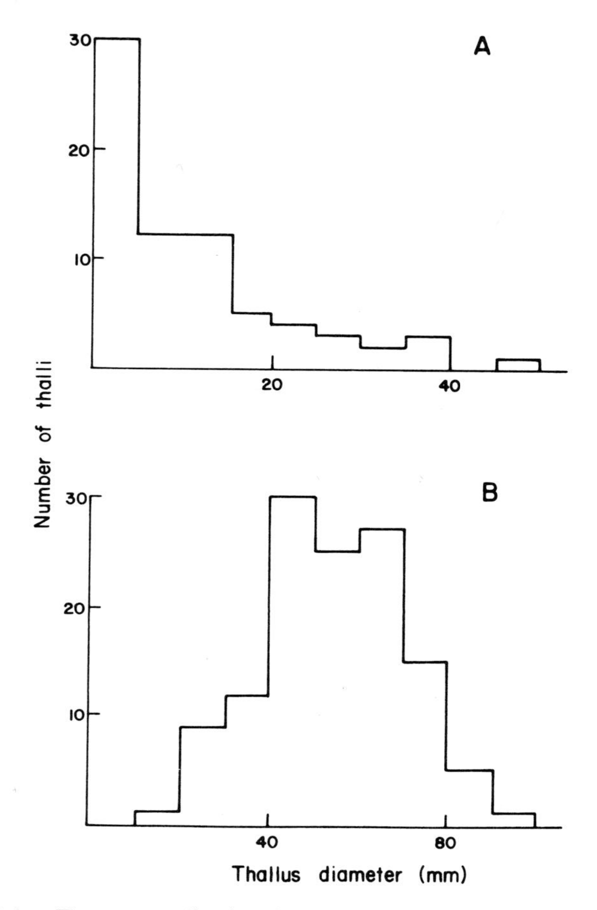

FIGURE 7.6. Frequency distribution of (a) *Physcia caesia* and (b) *Lecanc muralis* of various sizes on a 30-year-old asbestos roof in Berkshire, Englar From J.F. Farrar. 1974. "A method for investigating lichen growth rates a succession." *Lichenologist* 6:151–155. Redrawn with permission of Acader Press.

In one of the few investigations of this sort done to date, Yarrant (1975) measured the population growth of *Cladina stellaris* by following progress of individual podetia from year to year. In northern Ontai populations of *C. stellaris* were photographed from 1968 to 1974 in a num of previously burned areas of known age. Thallus birth and death rates w estimated at each site as accurately as possible by noting the emergence disappearance of thalli in successive photographs, although one's abilit distinguish between individuals of this particular species is extremely diffic

TABLE 7.3. Demographic Summary of *Cetraria oakesiana* and *C. pinastri* Populations from Spruce Knob, West Virginia

	Size Categories (cm)			
	<1	*1–2*	*2–3*	*>3*
C. oakesiana	667* (69.33)	205 (21.31)	76 (7.90)	14 (1.45)
C. pinastri	1096 (89.25)	114 (9.28)	14 (1.14)	4 (0.32)

*Numbers are cumulative summaries of individuals per size category from a total of 98 colonies (percent frequencies in parentheses).

because of its characteristic clonal growth behavior. Yarranton found that population growth was logistic with a typical carrying capacity estimated at 500 grams per square meter 30 years after establishment. Individual growth rates varied considerably, but were generally lower in the oldest and densest plots. Reproduction occurred almost entirely by cloning; very few nonclonal individuals became established during succession. Some podetia were found to disappear, especially in older plots; overgrowth by the moss, *Pleurozium schreberi*, was largely responsible. Age-specific mortality rates were not calculated, but it is likely that the highest mortality is associated with the smallest thalli and these were not visible in the photographs.

Demography and Ecology

In addition to characterizing the population ecology of a single lichen species, demographic statistics can be used to emphasize ecologically important differences between species. As an example, consider the species *Cetraria oakesiana* and *C. pinastri*. Though chemically quite distinct, these species are morphologically similar; each produces narrow, strap-shaped thallus lobes and marginal soralia. Their geographic distributions overlap in alpine habitats in portions of the Appalachian Mountains of the eastern USA where they can frequently be found together on both rocks and trees. Since each species reproduces vegetatively (apothecia are rarely produced), individuals are best viewed as diffuse and highly subdivided colonies. A preliminary assessment of the demographies of these two species was done in the summer of 1982 in a red spruce-dominated alpine habitat at Spruce Knob, West Virginia (altitude 1482 m) by Lawrey (unpublished). The colonies of each species had distinctly different size frequency patterns (Table 7.3). *Cetraria pinastri* colonies were dominated by small thalli with very few over three centimeters in diameter; *C. oakesiana* colonies were generally smaller with a greater proportion of intermediate- and larger-sized thalli.

One explanation for these demographic differences is that a distinct colonization strategy is employed by each species. A few large thalli of *Cetraria oakesiana* are apparently able to cover as much substrate area as numerous small thalli of *C. pinastri*. Since neither species can maximize both thallus number and thallus size, there has to be an energetic trade-off between the two. The observed demographic patterns reflect how these trade-offs differ for each species.

Further support for this idea comes from an analysis of the size/weight ratios of each species. Thalli of *Cetraria pinastri* are generally lighter than thalli of *C. oakesiana* of the same size (Figure 7.7). The reasons for these

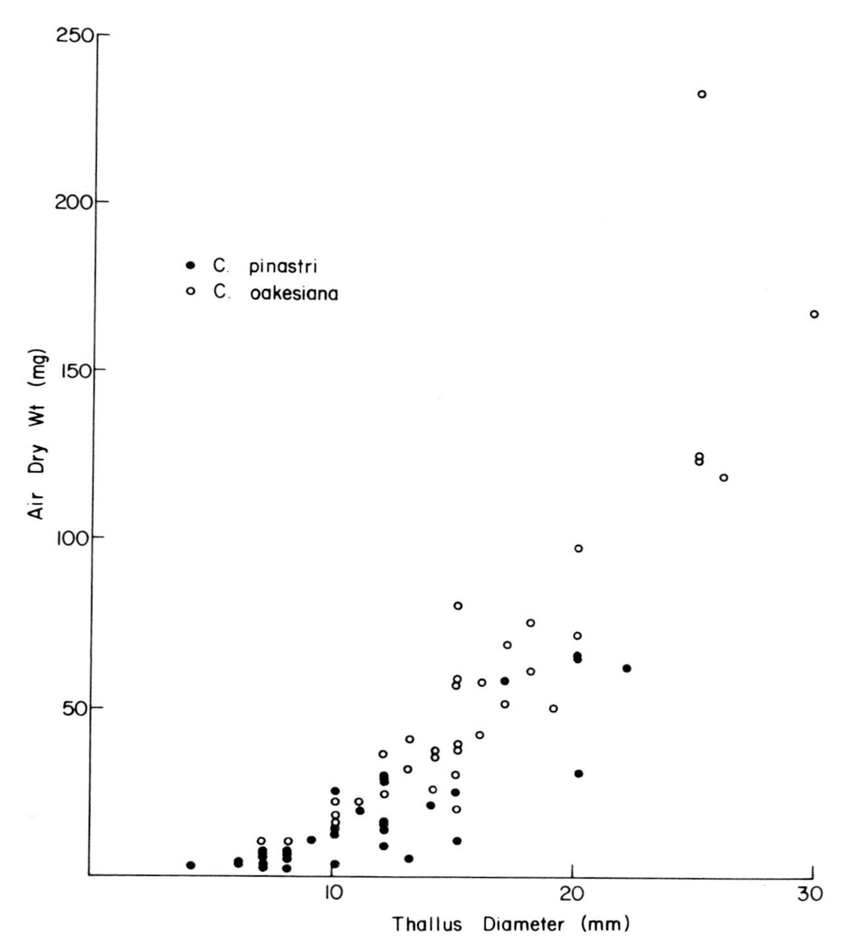

FIGURE 7.7. Diameter-weight relationships for *Cetraria pinastri* (●) and *C. oakesiana* (○) from Spruce Knob, West Virginia.

differences are not known, but may reflect different energetic costs associated with chemical production, nutrient acquisition, or some other ecophysiological process. One hypothesis that appears to be part of the explanation is that small thalli of *C. pinastri* contain higher concentrations of vulpinic and pinastric acids than large thalli; however, this does not explain why large thalli of *C. oakesiana* are favored.

Demographic analysis of lichen colonies or populations deserves increased attention by lichenologists. When combined with investigations of growth, reproduction, physiological activity or chemistry, the analysis of population structure, particularly age structure, answers some old questions and asks new questions of ecological and evolutionary importance.

LICHENOMETRY

The use of lichens as dating tools is called lichenometry. The Austrian botanist, Dr. Roland E. Beschel, was most responsible for the development of the principles and techniques of lichenometry. His original papers (1950; 1961) and other more recent reviews (Andrews & Webber, 1969; Webber & Andrews, 1973; and other papers in a special issue of *Arctic and Alpine Research*, vol. 5, no. 4, 1973, dedicated to the memory of R.E. Beschel) provide historical details.

Lichenometry is most frequently used to estimate the age of glacial deposits, but other applications range from dating stone monuments on Easter Island (Follmann, 1961; 1965; Richardson, 1975) and limestone walls and pavements in Ireland (Trudgill, Crabtree, & Walker, 1979) to estimating earthquake frequency in central Asia (Nikonov & Shebalina, 1979).

This section will provide a general evaluation of the technique, including the assumptions implicit in most lichenometric studies.

The Technique

The technique is a relatively straightforward one that assumes a recently deposited or constructed colonizable surface will, within a certain determinable period of time, support lichens that will thereafter grow at determinable rates. Assuming no subsequent perturbations, the original lichen colonists will grow larger each year. The sizes of the largest thalli on the substrate should therefore relate in some fashion to the amount of time since the substrate first became colonizable. The size frequency of the entire population should also provide evidence about ongoing colonization rates and allow an investigator to identify periods of significant interruption in colonization and/or thallus growth patterns caused by environmental disturbances subsequent to substrate establishment.

Obviously, lichenometric measurements can only be done with species that occur in the region. However, if there are a number of species available to choose from, it is best to use those that grow slowly and persist for long periods of time without exhibiting the fragmentation or central thallus disintegration typical of many foliose species. Crustose forms, primarily *Rhizocarpon* species, have been studied most frequently. According to Beschel (1961), large thalli of *R. geographicum* range in age from 600 to 1300 years old in the Alps and from 1000 to 4500 years old in West Greenland, making them among the longest-lived organisms. An updated summary of *R. geographicum* growth rates and age estimates from a number of investigations is provided by Webber and Andrews (1973).

Once an appropriate species has been chosen, a growth curve that accurately relates thallus age to size must be produced. This is not always easy to do. As has been shown in previous chapter sections, lichen growth rates are not constant but vary as thalli progress from one growth phase to another. It was Beschel who first called attention to this fact; indeed, his study of lichens on dated grave markers in the Alps (Beschel, 1958) first demonstrated the existence of growth phases in lichens. In the absence of dated substrates, growth curves must be estimated.

Problems with Lichenometry

Beschel recognized the various problems inherent in his technique. Webber and Andrews (1973) mention many of these; they include:

1. biological problems that have to do with the physiological basis for growth, symbiosis, competition, and mode of dispersal,

2. environmental problems that relate to lichen-substrate interactions, microclimatic influence over growth, and long-term climatic variations,

3. sampling problems that relate to species identification, measurement accuracy, and reliability of growth curves.

Jochimsen (1973) argued that these problems are serious enough to call into question the entire procedure. Webber and Andrews (1973) do not consider them insurmountable, however, and cite a number of successful applications of the technique. They regard lichenometry as an important research tool that will increase in accuracy as investigators learn more about the environmental and physiological factors that control lichen growth.

8 Lichen Secondary Chemistry

Lichens produce a rich array of secondary compounds, many of which are not found in other organisms. They are generally colorless, water-insoluble extracellular products of fungal origin, although the lichen alga apparently participates in their synthesis in an as yet unknown way (for a discussion of the possible interaction of symbionts in lichen compound synthesis, see Chapter 4).

The chemistry of lichen substances has been studied for well over 100 years. Early work focused on isolation and chemical characterization of compounds; by the time the German chemist Zopf published his monumental work *Die Flechtenstoffe* in 1907, over 200 compounds were known. Later work by Asahina and students, summarized in Asahina and Shibata's book, *Chemistry of Lichen Substances* (English translation published in 1954), added new compounds and provided details on the structure of many.

In 1977, C.F. Culberson, W.L. Culberson, and A. Johnson (1977a) listed over 500 primary and secondary compounds from over 5000 species, and numerous additional discoveries are reported every year. This truly impressive body of phytochemical information makes lichens one of the best-studied groups of organisms as far as secondary chemistry is concerned.

This chapter will discuss the most common chemical categories of lichen compounds and provide representative structural formulae. It is not possible in the space allowed to list all compounds or provide detailed discussion of their chemical, physical, or biosynthetic properties. Nor will the taxonomic significance of these compounds be discussed. Information on these subjects is summarized in several reviews (Asahina & Shibata, 1954; C.F. Culberson,

1969; 1970; W.L. Culberson, 1970; W.L. Culberson & C.F. Culberson, 1970; Huneck, 1968; 1971; 1973; Mosbach, 1973; Hawksworth, 1973a; 1976; C.F. Culberson, W.L. Culberson, & A. Johnson, 1977a; C.F. Culberson & W.L. Culberson, 1977).

SURVEY OF LICHEN SECONDARY COMPOUNDS

The most commonly used classification of lichen secondary compounds is that of C.F. Culberson, W.L. Culberson, and A. Johnson (1977a), derived from that of Asahina and Shibata (1954) and Shibata (1965) and based on biogenetic origin. Three pathways account for the origin of most lichen compounds: the acetate-polymalonate, mevalonic acid, and shikimic acid pathways.

Products of the Acetate-Polymalonate Pathway

Higher Aliphatic Acids and Esters. Numerous fatty acids belonging to various structural groups are produced by lichenized fungi (Figure 8.1). They are optically active and similar in structure to nonlichen fungal fatty acids. They do not react with color reagents, but can be separated using thin layer chromatography (TLC); they must be visualized by wetting TLC plates in water and observing opaque spots as the plates dry (C.F. Culberson, 1972b; Santesson, 1973).

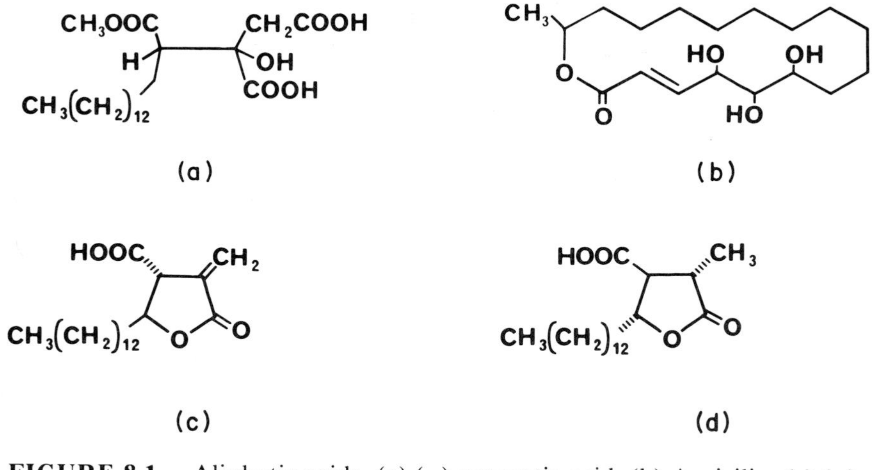

CH_3OOC CH_2COOH H OH COOH $CH_3(CH_2)_{12}$

(a)

CH_3 O O HO OH HO

(b)

HOOC CH_2 $CH_3(CH_2)_{12}$ O O

(c)

HOOC CH_3 $CH_3(CH_2)_{12}$ O O

(d)

FIGURE 8.1. Aliphatic acids. (a) (−)-caperatic acid; (b) Aspicilin; (c) (−)-protolichesterinic acid; (d) (+)-roccellaric acid.

Phenolic Carboxylic Acid Derivatives. By far the largest number of lichen secondary products are included in this group. They arise most commonly by joining together two (or occasionally three) phenolic units; however, some simple aromatic units similar to nonlichenized fungal phenolic compounds are found in lichens.

The following groups are most characteristic:

1. depsides and tridepsides,
2. depsidones,
3. the depsone derivative, picrolichenic acid,
4. dibenzofurans,
5. usnic acids and related compounds.

The depsides, tridepsides, and depsidones are formed by esterification of two types of phenolic units: the orcinol-type units (an example of one is orsellinic acid) and the β-orcinol–type units that have an extra C_1 substituent at the 3 position (Figure 8.2). These phenolic precursors are thought to be produced by an orsellinic acid-type cyclization process (Figure 8.3). Compounds formed from the orcinol- and β-orcinol–type units are usually considered separately in chemical classifications.

Precursors of the dibenzofurans are also formed by the orsellinic acid-type cyclization; however, they are joined together by different processes than depsides and depsidones. Precursors of the usnic acids are formed by a phloroglucinol-type cyclization (Figure 8.3). All of these biosynthetic differences are used to classify the various compounds and appear to be of phylogenetic importance as well.

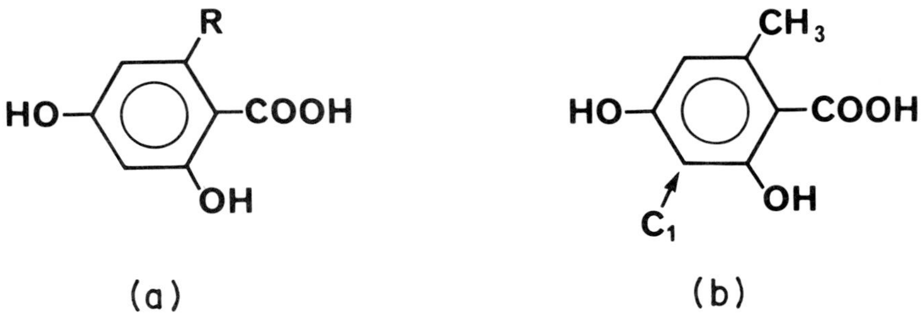

R
HO— —COOH
OH
(a)

CH_3
HO— —COOH
OH
C_1
(b)

FIGURE 8.2. Basic phenolic units derived from the acetate-polymalonate pathway. (a) Orcinol-type unit; (b) β-orcinol-type unit. From C.F. Culberson. 1969. In *Chemical and Botanical Guide to Lichen Products,* p. 32. Chapel Hill: University of North Carolina Press. Reprinted with permission of the author.

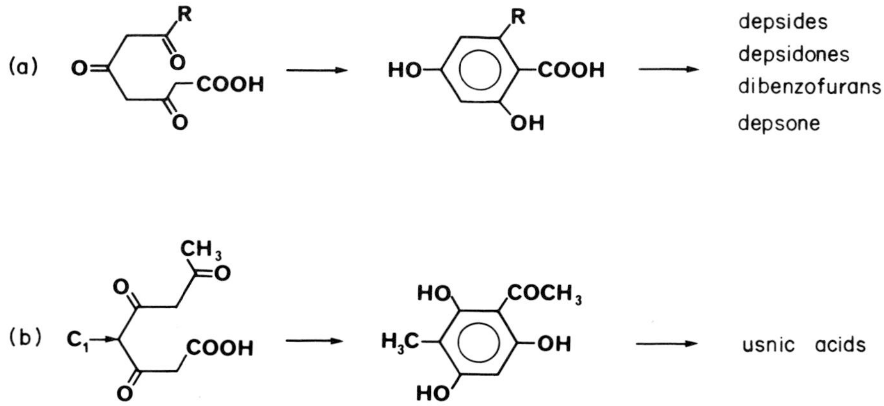

FIGURE 8.3. Comparison of (a) the orsellinic acid-type cyclization with (b) the phloroglucinol-type cyclization. From C.F. Culberson. 1969. In *Chemical and Botanical Guide to Lichen Products,* p. 31. Chapel Hill: University of North Carolina Press. Reprinted with permission of the author.

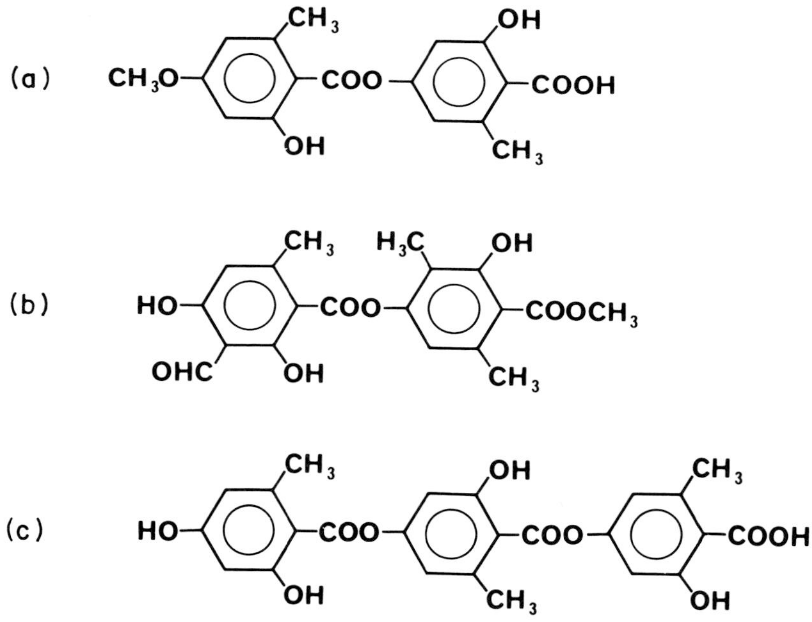

FIGURE 8.4. Lichen depsides. (a) Evernic acid, a *para*-depside of the orcinol series; (b) Atranorin, a *para*-depside of the β-orcinol series; (c) Gyrophoric acid, a tridepside of the orcinol series.

Depsides. The depsides make up the largest group of lichen substances (Figure 8.4). Most are didepsides with two orcinol or β-orcinol carboxylic acids in an ester linkage. A few tridepsides are known, including gyrophoric acid, commonly found in the Umbilicariaceae; also there is evidence that *Peltigera aphthosa* produces a tetradepside, aphthosin (Bachelor & King, 1970), although there is some doubt of this (Maass, 1975b).

Depsides are colorless crystalline substances that react variously with $FeCl_3$, KOH, $Ca(OCl)_2$ and *p*-phenylenediamine. These color reactions are used frequently in taxonomic determinations.

Although they are common lichen substances, depsides are only rarely found in free-living fungi. Umezawa et al. (1974) reported production of lecanoric acid by a *Pyricularia* species and Yamamoto, Nishimura, and Kiriyama (1976) isolated 4-*O*-demethylbarbatic acid from *Aspergillus terreus*.

Depsidones. Depsidones form the second largest group of lichen substances (Figure 8.5). They are believed to arise by oxidative cyclization of depsides; therefore, although they are a distinct class of lichen compounds, biosynthetically they are considered depside derivatives.

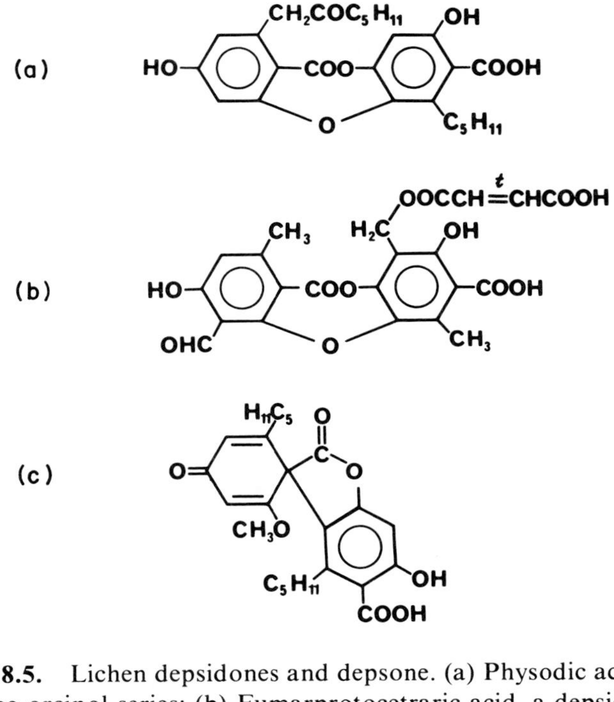

FIGURE 8.5. Lichen depsidones and depsone. (a) Physodic acid, a depsidone of the orcinol series; (b) Fumarprotocetraric acid, a depsidone of the β-orcinol series; (c) Picrolichenic acid, a depsone.

Depsone. Picrolichenic acid, the bitter compound produced by *Pertusaria amara*, is the only known depsone from lichens (Figure 8.5). It is considered to be closely related to the depsidones and is probably formed from the depside, dihydropicrolichenic acid, by an intramolecular oxidation (Mosbach, 1973).

Dibenzofurans. These rather rare compounds are essentially unique to the lichens. True dibenzofurans (Figure 8.6) are condensation products of orsellinic acid and homologues. They apparently all have antibacterial activity (Shibata & Miura, 1949; Vartia, 1973).

Usnic Acids. Unlike the dibenzofurans, the structurally similar usnic acids (Figure 8.6) are based on acetylphloroglucinol units. They are practically ubiquitous yellow-pigmented, cortical compounds that have marked antibacterial properties. They are all optically active and the isomers are more or less genus-specific. In *Cladonia*, both (+) and (−) usnic acids occur, but they have never been found in the same plant.

Chromones, Xanthones, Anthraquinones, and Naphthaquinones. Other products of the acetate-polymalonate pathway include the chromones, xanthones, anthraquinones, and naphthaquinones (Figure 8.7). These are not unique to the lichens; many have been found in free-living fungi, and some are produced by flowering plants. Some also show antibacterial activity.

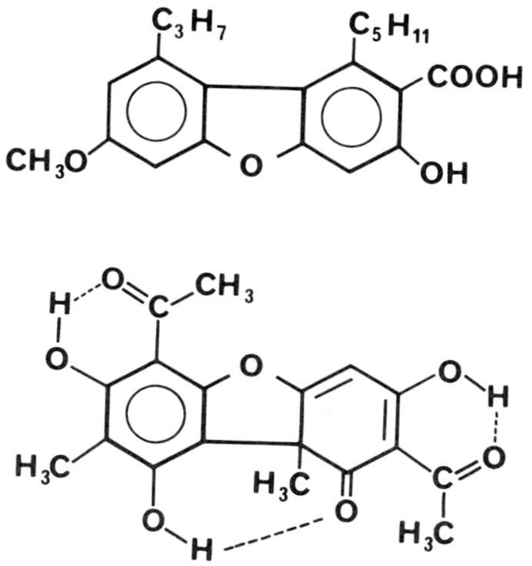

FIGURE 8.6. Didymic acid, a dibenzofuran (above); (—)-usnic acid (below).

Chromones are colorless or pale yellow compounds; few are presently known in lichens. Xanthones are all derived from norlichexanthone, and a number are chlorinated.

A number of lichen anthraquinones are known. These are orange and red pigments that react purple with KOH. The most widespread is parietin, an orange pigment produced by members of the Teloschistaceae. It is also found in free-living fungi and higher plants.

The naphthaquinone rhodocladonic acid is the familiar red pigment in the apothecia of *Cladonia* species in the subgenus *Cocciferae*. The common British soldiers lichen, *Cladonia cristatella*, is an example.

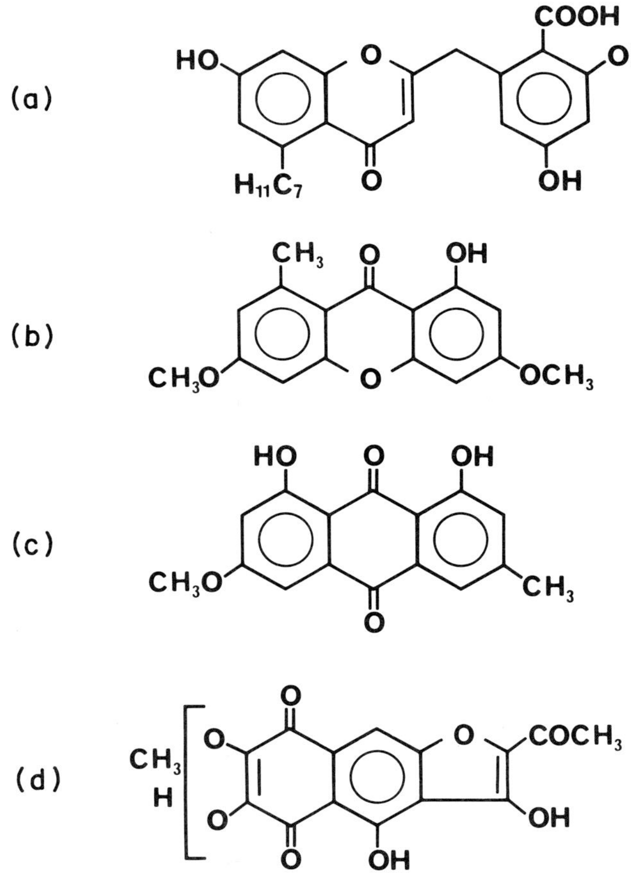

FIGURE 8.7. Chromones, xanthones, anthraquinones, and naphthaquinones. (a) Siphulin; (b) Lichexanthone; (c) Parietin; (d) Rhodocladonic acid.

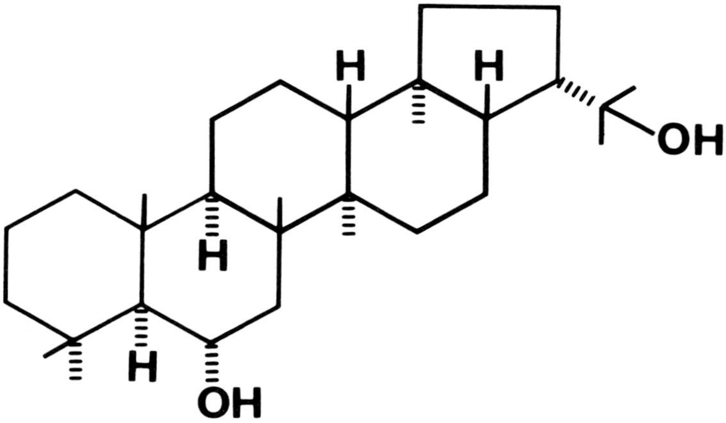

FIGURE 8.8. Zeorin, a lichen terpene.

Products of the Mevalonic Acid Pathway: Carotenoids, Sterols, and Terpenoids

Products of the mevalonic acid pathway include the familiar carotenoid pigments found in numerous plant groups. Some of these (e.g., lycopene, phytoene, and phytofluene) have been found only in lichen mycobionts.

A number of sterols are also known from lichens (ergosterol, β-sitosterol, fungisterol, and vitamin D_2). Most of these are widely distributed in other plant groups.

The majority of lichen terpenoids are triterpenes, colorless neutral pentacyclic compounds with relatively high melting points ($>$ 200° C) (Huneck, 1973). One of the most common of these, zeorin (Figure 8.8), has been reported from the isolated mycobiont of *Anaptychia hypoleuca* (Ejiri & Shibata, 1974), demonstrating that participation of the phycobiont in zeorin synthesis is not essential.

Products of the Shikimic Acid Pathway: Terphenylquinones and Pulvinic Acid Derivatives

In lichens, the pulvinic acid derivatives outnumber the known terphenylquinones, although the reverse is true in nonlichenized basidiomycetes. The biosynthesis of both of these groups goes through shikimic acid and phenylalanine in contrast to the aromatic units of the acetate-polymalonate-derived compounds.

The terphenylquinones are red or purple pigments found in the lower

(a)

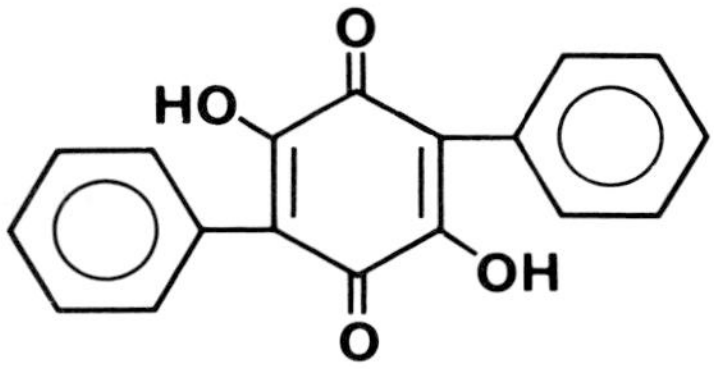

(b)

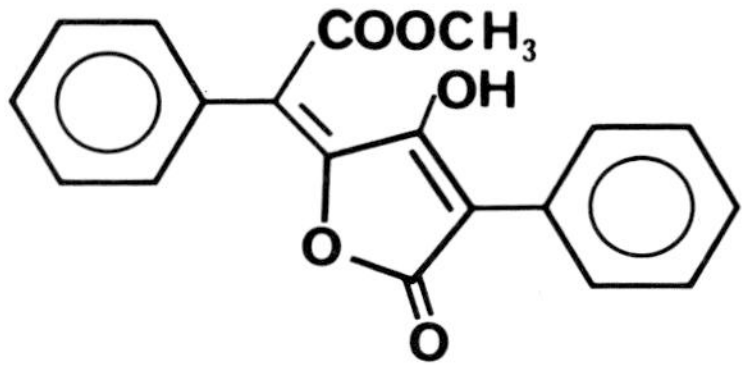

FIGURE 8.9. Products of the shikimic acid pathway. (a) Polyporic acid; (b) Vulpinic acid.

cortex and rhizines of some species of *Lobaria, Sticta*, and *Pseudocyphellaria*. Polyporic acid (Figure 8.9) also occurs in the basidiomycete genera *Polyporus* and *Peniophora*; thelephoric acid occurs in *Thelophora, Hydnum, Cantharellus*, and *Polystictus*.

The pulvinic acid derivatives (e.g., vulpinic acid in Figure 8.9) are yellow, orange, red, or violet pigments commonly found in members of the Stictaceae, although species from a variety of lichen groups are known to produce them. The fact that many of these lichens frequently contain N_2-fixing blue-green algae has suggested to some (C.F. Culberson, 1969; Rundel, 1978) that high nitrogen supplies are required for the synthesis of pulvinic acid derivatives.

IDENTIFICATION OF LICHEN SUBSTANCES

Color Reactions

Certain chemical reagents applied to lichen thalli cause color reactions indicating the presence of certain types of secondary compounds. Nylander is generally credited with the first application of color tests in lichen taxonomy. Iodine (designated I in identification keys) was first used (Nylander, 1865); later, bleaching powder ($Ca(OCl)_2$, generally designated C) and potassium hydroxide (KOH, designated K) were found to be useful in distinguishing species and identifying sterile material (Nylander, 1866a; 1866b). A short time later, Leighton (1867) found that application of K and C together gave color reactions not seen when each was used alone. The last commonly used reagent, *p*-phenylenediamine (P), was discovered by Asahina (1934).

Most lichen identification keys rely on tests with these reagents. Although identification of specific lichen compounds is generally not possible using color tests, it is possible to identify categories of chemically related compounds that react similarly to reagents (Table 8.1). Hale (1974) and Santesson (1973) have discussed these tests and a few additional ones that have been used in lichen taxonomy.

TABLE 8.1. Color Reactions of Unpigmented Lichen Substances

Reaction	*Compound*
P+, K+	*Depsides:* alectorialic acid, atranorin, baeomycesic acid, barbatolic acid, chloratranorin, decarboxythamnolic acid, haemathamnolic acid, nephroarctin, thamnolic acid
	Depsidones: constictic acid, fumarprotocetraric acid, norstictic acid, physodalic acid, protocetraric acid, salazinic acid, stictic acid, virensic acid
P+, K−	Pannarin, psoromic acid, fumarprotocetraric acid, protocetraric acid, virensic acid
P−, K+, C+	Cryptochlorophaeic acid, hiascic acid, hypothamnolic acid (K+ violet), merochlorophaeic acid, paludosic acid, ramalinolic acid, scrobiculin
P−, K−, C+ red	Anziaic acid, 4-*0*-demethylbarbatic acid, erythrin, ethyl orsellinate, gyrophoric acid, lecanoric acid, methyl 3,5-dichlorolecanorate, methyl-β-orsellinate, montagnetol, oliveoric acid, siphulin
P−, K−, C+ green	Didymic acid, pannaric acid, porphyrilic acid, strepsilin
P−, K−, C+ blue	Diploschistesic acid
P−, K−, C−, KC+	Alectoronic acid, α-collatolic acid, glomelliferic acid, lobaric acid, 4-*0*-methylphysodic acid, picrolichenic acid, microphyllinic acid, norlobaridone, physodic acid

J. Santesson. 1973. "Identification and isolation of lichen substances." In *The Lichens* (V. Ahmadjian & M.E. Hale, Jr., eds.), pp. 633–652. New York: Academic Press. Reprinted with permission of Academic Press and the author.

Microcrystallization

In 1936, Asahina introduced a relatively simple, inexpensive technique for identifying specific lichen compounds. The method involved extracting compounds from small pieces of thalli and dissolving, heating, and recrystallizing these compounds in special reagents. Each compound forms crystals with a unique shape and color. They can be identified by comparing crystals observed under a light microscope to published photographs; several sets of these photographs are available in the literature (Hale, 1967; Taylor, 1967; Thomson, 1967).

The technique has the obvious advantage over spot tests of providing positive identification of specific compounds rather than categories of compounds. Also, the sensitivity of microcrystallization is high enough to allow identification of very small amounts of a compound, although chromatographic techniques are more sensitive. One disadvantage is that some compounds, particularly aliphatic acids and terpenes, do not form suitable crystals in the standard reagents.

Chromatography

Both paper (PC) and thin layer chromatography (TLC) have been used extensively for the identification of lichen compounds. Gas-liquid chromatography (GLC) has been used in a few instances to detect triterpenes (Ikekawa et al., 1965; Shibata, Furuya, & Iizuka, 1965; Yosioka, Nakanishi, & Kitagawa, 1969), aliphatic acids (Bloomer, Eder, & Hoffman, 1970a; 1970b), anthraquinones (Furuya, Shibata, & Iizuka, 1966), low molecular weight carbohydrates (Nishikawa, Michishita, & Kurono, 1973) and usnic acid (Fahselt, 1975). C.F. Culberson (1972a) has also found GLC to be a useful separation technique for lichen fatty acids, but only after they have been converted to their trimethylsilyl derivatives. The thermal lability and low volatility of most lichen compounds limits the usefulness of GLC. High-performance liquid chromatography (HPLC) of lichen compounds, first described by C.F. Culberson (1972a), appears to be the separation technique of the future in lichen chemical investigations.

Paper Chromatography. Paper chromatography was first used in lichen chemical studies by Wachtmeister (1952) and Mitsuno (1953); Wachtmeister (1959) also published a review of the technique. Despite early interest in PC, however, most lichenologists have switched to thin layer chromatographic methods.

Thin Layer Chromatography. The first TLC separation of lichen compounds was reported by Stahl and Schorn (1961), and numerous lichenologists have since used the technique in various chemical investigations. Santesson (1973) has identified many of these earlier TLC studies.

The TLC separation technique makes use of plates coated with an adsorbent compound, almost always silica gel, although polyamide coatings have been used occasionally. These plates may be prepared in the investigator's laboratory; however, numerous types of precoated TLC plates are commercially available. Lichen compounds are extracted in an organic solvent and the extracts are spotted on the plates. The plates are then developed in one of several solvent systems.

C.F. Culberson (1972b) has described three frequently used standard solvent systems modified from earlier versions (C.F. Culberson & H. Kristinsson, 1970); she has also provided the R_f values of hundreds of compounds chromatographed in these solvents. These standard solvent systems are:

A. benzene - dioxane - acetic acid (180:45:5, 230 ml),
B. hexane - diethyl ether - formic acid (130:80:20, 230 ml),
C. toluene - acetic acid (200:30, 230 ml).

Some representative R_f values of lichen compounds chromatographed in these solvents are provided in Table 8.2. They are generally reported in relation to R_f values of cochromatographed controls containing norstictic acid and atranorin. These controls are used to indicate deviations caused by changes in solvent concentration, contamination, or temperature fluctuations.

Once developed, TLC plates are dried and treated so that the spots can be seen. If the plates are initially treated with a fluorescence indicator, then spots will be apparent under UV light. Some lichen compounds are also naturally fluorescent. Spraying the plate with 10 percent sulfuric acid and heating at 110° C for a few minutes (C.F. Culberson and H. Kristinsson, 1970) also causes the spots to become visible; some of these compounds acquire a characteristic color that aids in identification. Fatty acids can be visualized by spraying the plates with water (C.F. Culberson, 1972b; Santesson, 1973).

TLC methods will undoubtedly continue to be used in lichen chemical investigations. They are fast, efficient, accurate, and inexpensive. Recent developments in lichen TLC analysis include the use of two-dimensional methods to resolve mixtures of compounds difficult to resolve by standard methods (Maass, 1975a; 1975b; 1975c; 1975d; C.F. Culberson & A. Johnson, 1976), and the development of a new solvent system that can be used to separate and identify β-orcinol depsidones not easily resolved using standard solvents (C.F. Culberson, W.L. Culberson, & A. Johnson, 1981).

TABLE 8.2. TLC Data for Selected Lichen Products Chromatographed in Three Solvent Systems

	$R_f \times 100$ values (R_f of X / R_f of N, R_f of A*)		
Compound	*Solvent A*	*Solvent B*	*Solvent C*
Fumarprotocetraric acid	1/41, 73	25/28, 75	7/29, 77
Protocetraric acid	3/40, 74	19/30, 77	4/28, 78
Caperatic acid	4/41, 78	27/29, 76	6/34, 81
Salazinic acid	10/39, 72	7/30, 74	4/28, 77
Squamatic acid	10/42, 76	25/30, 74	25/29, 78
Pannaric acid	8/42, 77	28/30, 74	9/28, 79
Pulvinic acid	4/42, 77	36/27, 76	7/29, 80
Haemoventosin	33/42, 73	2/27, 78	17/27, 79
Galbinic acid	29/41, 75	12/29, 77	16/28, 79
Lepraric acid	27/40, 74	11/29, 77	21/28, 78
Stictic acid	32/42, 77	9/27, 77	18/28, 78
Alectoronic acid	33/43, 75	34/33, 76	17/31, 77
Physodic acid	25/42, 74	35/27, 76	18/29, 79
Gyrophoric acid	24/40, 74	42/30, 77	24/28, 79
Lecanoric acid	28/40, 75	44/30, 76	22/28, 79
Olivetoric acid	34/40, 76	37/28, 76	22/28, 78
Lividic acid	32/40, 74	35/29, 77	31/28, 78
Rangiformic acid	29/41, 75	38/29, 76	33/28, 80
Lobaric acid	30/40, 75	46/29, 76	38/29, 80
Psoromic acid	36/41, 73	41/27, 75	41/29, 76
Strepsilin	41/42, 78	21/27, 77	24/28, 78
Baeomycesic acid	39/42, 76	40/32, 79	42/28, 79
Anziaic acid	40/42, 76	59/29, 74	33/28, 76
Evernic acid	38/41, 76	61/32, 79	43/29, 77
Grayanic acid	38/40, 75	62/30, 77	44/28, 79
Polyporic acid	43/41, 73	28/27, 75	27/29, 76
α-Collatolic acid	40/41, 73	32/28, 75	35/28, 77
4-0-Demethylbarbatic acid	39/41, 77	60/32, 79	39/29, 78
Sekikiac acid	45/42, 78	57/27, 77	51/28, 79
Diffractaic acid	44/42, 76	64/32, 80	51/28, 78
Barbatic acid	44/42, 77	69/27, 75	52/29, 78
Divaricatic acid	39/39, 73	75/32, 80	51/29, 78
Didymic acid	44/42, 75	77/32, 80	52/28, 78
Perlatolic acid	44/43, 73	74/29, 73	52/28, 76
Norobaridone	50/40, 76	36/28, 76	21/29, 79
Zeorin	55/41, 76	42/30, 77	43/28, 80
Rhizocarpic acid	73/43, 76	41/29, 76	66/29, 78
Usnic acid	70/42, 76	57/28, 76	76/27, 76
Pinastric acid	72/41, 73	57/28, 76	76/27, 76
Vulpinic acid	76/42, 77	66/27, 77	82/29, 78

*The two numbers following the oblique (/) are the R_f of norstictic acid (N) and atranorin (A).

Partial data adapted from C.F. Culberson, 1972b. "Improved conditions and new data for the identification of lichen products by a standardized thin-layer chromatographic method." *J. Chromatogr.* 72:113–125.

High-Performance Liquid Chromatography. C.F. Culberson (1972a) first reported the use of HPLC to isolate and identify lichen compounds; since then, the number of lichen chemical investigations employing HPLC has increased dramatically. There are a number of advantages of this technique over other separation techniques.

1. Separation of compounds is much more rapid than with other techniques.

2. Results can be accurately quantified.

3. Mixtures that are difficult or impossible to separate by TLC can frequently be separated using HPLC.

HPLC makes use of small bore columns packed with variously treated spherical particles through which a solvent is pumped under pressure at relatively high speeds. Reverse phase columns are used most frequently for lichen extractions; these allow separations of a wide variety of sample types, polar, nonpolar, ionic and ionizable solutes, as well as compounds with varying molecular weights. Solutes in reverse phase liquid chromatography elute in a general order of decreasing polarity. Normal phase columns are used for compounds not easily separated using reverse phase LC.

As they leave the columns in the effluent, the separated compounds pass through a detector: a UV detector is used for aromatic substances and a refractive index detector is used for aliphatic substances. Detector signals are sent to a recorder that plots absorbance peaks against time. The peak height and retention time of the compound in the column enable an investigator to determine the identity of the compound and its concentration in the original extraction.

An example provided by Dr. C.F. Culberson will best illustrate the usefulness of HPLC for lichen chemical analyses. The foliose lichen *Cetraria pinastri* produces three secondary compounds: usnic (Usn), pinastric (Pin), and vulpinic (Vul) acids. Because Pin and Vul are so similar chemically, they are not easily separated using standard TLC approaches. Completely overlapping spots are seen when solvents A and C are used; solvent B separates the two, but because Usn sometimes leaves a long trail, especially when it is produced at relatively high concentration, it sometimes covers the Vul spot (Figure 8.10).

HPLC not only separates the two compounds, but provides information about their relative concentrations in the initial extraction (Figure 8.11). Unfortunately, Usn is not easily eluted in the reverse phase column used in this example. Its peak is a broad, flat, relatively indistinct one that probably

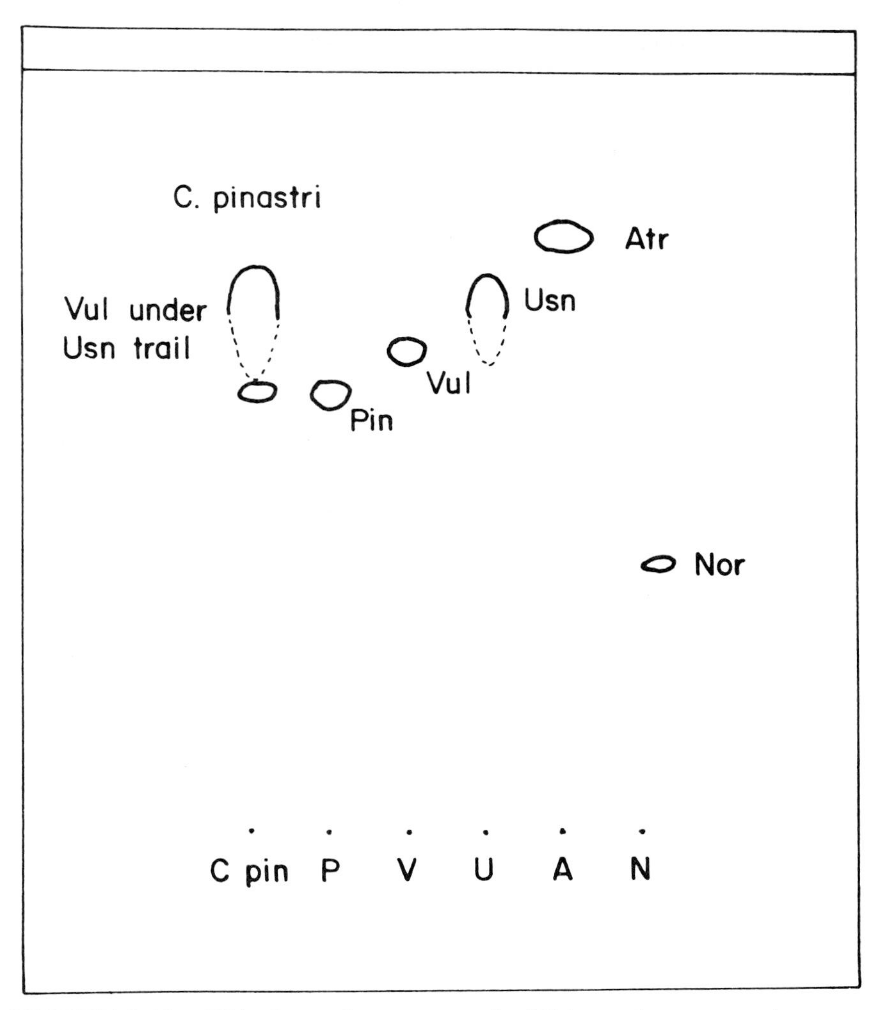

FIGURE 8.10. Thin-layer chromatograph of lichen substances and acetone extract of *Cetraria pinastri,* using solvent "B" of C.F. Culberson (1972b). Vul = vulpinic acid; Usn = usnic acid; Pin = pinastric acid; Atr = atranorin; Nor = norstictic acid.

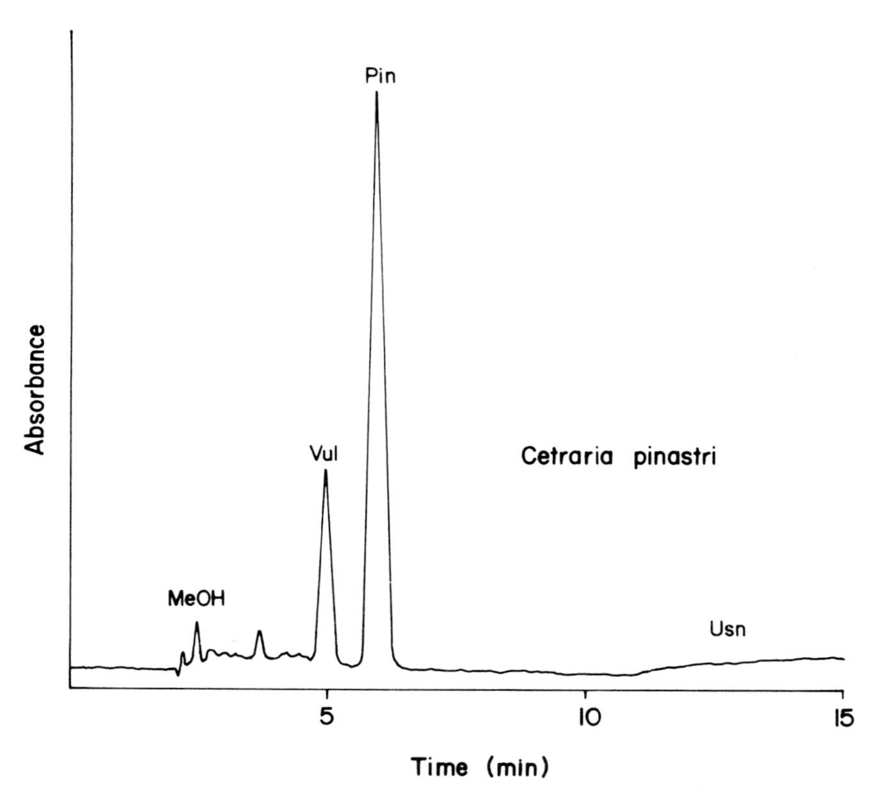

FIGURE 8.11. High performance liquid chromatography spectrum of MeOH extract of *Cetraria pinastri.* C_{18} column = ultrasphere ODS 5 μm 4.6 mm × 250 mm column. Solvent: MeOH-H_2O-HOAc (82:18:1.6 v/v/v); detector: 254 nm. Flow: 1 ml/min (1100 psi).

interferes with the analysis of other compounds. Use of a normal phase column is probably better for Usn and some other compounds as well (C.F. Culberson, 1972a).

Other Identification Techniques

In addition to the techniques already identified, Santesson (1973) mentions fluorescence and lichen mass spectrometry. Neither of these techniques requires isolation of compounds from lichen material.

Some lichen compounds fluoresce in long-wave (366 nm) ultraviolet light. Anthraquinones appear reddish, pulvinic acid derivatives, yellowish, and xanthones, bright orange to orange-red. Some depsides and depsidones fluoresce bright white to blueish (e.g., squamatic acid). These characteristics have been used to provide tentative identification of substances (Cěrnhorsky, 1950; Ozenda, 1951; Hale, 1956a; 1956b).

Lichen mass spectrometry (LMS) is used to indicate the presence of secondary compounds that sublime if heated at low pressure (Santesson, 1969). Mass spectra of the subliming compounds are recorded in the mass spectrometer and used for identification. This method is especially useful for the study of lichen pigments. Santesson (1973) has discussed the various applications that have been made.

Although all of the techniques discussed in this section are relatively reliable and of enormous value to the lichenologist, none gives a positive identification. As Santesson (1973) has mentioned, a positive identification requires that the compound be isolated so that its physical and chemical properties can be compared with authentic samples. Among the properties that are frequently checked are melting points and various spectral properties.

VARIATIONS IN CONCENTRATION

The extent to which secondary compound concentration varies in lichen thalli is only beginning to be appreciated; not surprisingly, little is known about what causes these variations. Hawksworth (1973; 1976) has reviewed much of the literature on concentration gradients in lichens and has discussed their taxonomic significance.

Some patterns have been recognized for many years. For example, the concentration of various cortical pigments is known to increase with light intensity (Hill and Woolhouse, 1966; Rundel, 1969), suggesting a light-screening role for these compounds (see Chapter 9).

Other concentration patterns also appear to be related to environmental gradients. In Japan, Hamada (1982b) found that salazinic acid in *Ramalina siliquosa* increased in concentration with habitat temperature; other environmental factors (precipitation and light intensity) seemed to be less important. No explanation for this very interesting relationship was given.

Other patterns appear to relate to thallus age. High concentrations of some lichen compounds are known to be produced only in young tissues. For example, in some *Cladonia* species, usnic acid (Fedoseev & Yakimov, 1960) and fumarprotocetraric acid (Mirando & Fahselt, 1978) concentrations are highest in podetial tips than in basal portions. Hamada (1982a) found a similar concentration gradient for salazinic acid in *Ramalina siliquosa*. Stephenson and Rundel (1979) found that vulpinic acid concentration decreased and atranorin increased as *Letharia vulpina* tissues grew older; they suggested that vulpinic acid was produced only by young *L. vulpina* tissues, whereas atranorin was produced continuously throughout the life of the thallus.

All of these studies beg a tantalizing array of questions. What do the compounds do? By what process and for how long are they formed? What environmental conditions control compound formation, and how responsive is the process to changing environments? As this area of research progresses and our ability to accurately measure compound concentrations increases, these and other questions undoubtedly will be answered.

LOCALIZATION OF LICHEN COMPOUNDS

Generally, lichen compounds are produced either in cortical or medullary regions of the thallus. This level of compound localization has long been known. The cortical component is usually one of two rather ubiquitous compounds, either atranorin or usnic acid (although lichexanthone, vulpinic acid, and parietin are also frequently produced in the cortex). The medullary compounds are usually much more diverse and include all the characteristic depsides and depsidones.

There are also a few cases of secondary compound production by reproductive tissues. Brightly colored pigments are sometimes found in apothecia. Examples include the red pigments rhodocladonic acid and haemoventosin, restricted to apothecia of *Cladonia* subgenus *Cocciferae* and *Haematomma ventosum*, respectively. Restriction of other compounds to reproductive tissues is less well known. W.L. Culberson (1969) discovered an interesting case of norstictic acid production in the hymenium of almost half of the fertile specimens of *Letharia california* (=*L. columbiana*) he examined. Even more interesting, norstictic acid, normally a medullary compound, was never detected in the medulla of these specimens.

Other types of localizations are beginning to be recognized as microchemical techniques improve and investigators spend more time looking for them. Restriction of depsidones to apical parts of *Cladina* podetia (Ahti, 1961) and *Alectoria tenuis* (Hawksworth, 1976) suggests that it may be a more widespread phenomenon than it appears at present. Nuno (1973) also discovered a possible localization of barbatic acid in podetia of *Cladonia strepsilis*.

It is apparent that our knowledge about the location and distribution of lichen substances within thalli is fairly limited. This is due mainly to technical difficulties. Most microchemical methods involve compound extraction, a procedure that obviously disrupts the original distribution of compounds in the thallus. A recently developed technique that avoids this problem involves the use of cathodoluminescence and X-ray microanalysis to visualize and identify fluorescent compounds within intact lichen thalli (Mathey & Hoder, 1978; Hoder & Mathey, 1980). Cathodoluminescence (CL, also color cathodoluminescence, CCL) is light emission stimulated by electron bombardment. In the column of a SEM, lichen material is exposed to an electron beam, causing fluorescent compounds to become luminescent. By comparing luminescence characteristics of these compounds to those of isolated pure compounds, an investigator can identify them and map their distributions in the thallus. Mathey and Hoder (1978) used CL and energy-dispersive, X-ray microanalysis (EDX) information to determine the distributions of several chlorinated xanthones in a number of different crustose species. In a related study (Hoder and Mathey, 1980), they used CCL to map the location of a xanthone and selected anthraquinones. In all species studied, they found that lichen compounds have distinct distribution patterns in the thallus.

9 Lichen Secondary Chemical Ecology

About 500 secondary chemicals are known to be produced by lichenized fungi; of these, over 100 are unique to lichens. Although there has been much interest in these compounds and their phylogenetic significance for the past century, serious experimental investigation of their biological role by lichenologists has not been done until relatively recently. This chapter reviews some of the recent attempts to explain the adaptive nature of lichen secondary compounds. For those unfamiliar with the distribution, chemistry, or biosynthesis of these compounds, Chapter 8 provides an introduction and suggests references for further reading. There is also a recent review of the ecological roles of lichen compounds by Rundel (1978).

Evidence for the adaptive nature of lichen secondary compounds is largely indirect. However, taken as a whole, the evidence accumulated to date suggests that, in certain ecological situations, production of secondary compounds by lichens confers on species that produce them an adaptive advantage over those that do not. If this is true, there are a number of evolutionary consequences involving, for example, lichen life history characteristics, distribution and abundance of lichen species, physiological ecology, and the nature of the lichen symbiosis itself.

Although experimental investigations into the ecological role of lichen secondary compounds have only begun appearing in the lichenological literature, they are surprisingly broad in scope, ranging from field studies of secondary compound production along environmental gradients to laboratory investigations of symbiont interactions that result in compound production. There have also been a number of papers on phytochemistry, especially theoretical works that attempt to generalize the ecological and evolutionary consequences of plant secondary compound production. Since there is already a wealth of information on the production and distribution of lichen compounds, and the methods used to study these compounds have been developed and perfected over many years, students interested in testing these theories as they relate to lichenized fungi will find data collection to be less difficult than with groups of plants for which less phytochemical information is available.

Rather than attempting an exhaustive discussion of the possible ecological roles of lichen compounds, the focus will be on the most recent results of investigations in this area. These results can be discussed in four separate categories: antibiotics, allelopathy, anti-herbivore compounds, and light-screening compounds. These are similar to the categories discussed by Rundel (1978) but do not include a discussion of rock mineralization. This presumed role of lichen secondary compounds is discussed in Chapter 6. Students will find more detailed historical information on lichen secondary compound production and discussions of additional research areas in C.F. Culberson (1969, including supplements by C.F. Culberson, 1970 and C.F. Culberson, W.L. Culberson, & A. Johnson, 1977a), Huneck (1973), Mosbach (1973), and Vartia (1973).

ANTIBIOTIC NATURE OF LICHEN COMPOUNDS

Early Work with Pathogenic Organisms

Burkholder et al. (1944) were the first to report that antibiotic substances were present in lichens. Burkholder and Evans (1945) tested extracts of 100 American lichen species for inhibitory activity against *Bacillus subtilis* and *Staphylococcus aureus*. Extracts of 52 species prevented the growth of either one or both of the bacteria used. These studies also suggested that lichen substances were most effective against gram-positive bacteria; gram-negative bacteria were, with a few exceptions, generally not affected.

These early studies stimulated further testing of lichen extracts against a variety of microorganisms, especially pathogenic bacteria and fungi, and numerous papers reporting antibiotic effects of lichen compounds appeared in the 1950s and 1960s (Vartia, 1973). Of particular interest was the apparent inhibitory effect of usnic acid on the growth and viability of the causative agent of tuberculosis, *Mycobacterium tuberculosis*. Other lichen compounds found to be especially effective were compounds related to lichesterinic acid and the orcinol-type depsides and depsidones. A number of reports appeared on this subject and, although the experimental results were quite variable, probably caused by differences in lichen extracts, culture methods and pathogen strains used, they demonstrated a potential use of lichen compounds as antibiotic substances.

Results of *in vitro* tests of lichen compound activity against pathogenic microorganisms were not nearly as exciting, owing to numerous problems associated with resorption, compound toxicity, allergic reactions, and so on (Vartia, 1973). The fact that lichen compounds have very poor water solubilities (Table 9.1) creates problems in clinically administering the compounds effectively. These problems have been alleviated to a certain extent, and a number of ointments containing lichen compounds have been produced and marketed in Europe. In Austria, the sodium salt of usnic acid is sold under the name "Usniakin." In Germany, "Evosin I" (containing usnic and evernic acids) and "Evosin II" (containing usnic, physodic, and physodalic acids) are sold; in Switzerland, an ointment called "Lichusnin" is sold. In Finland, after much work had been done to render usnic acid soluble, a compound called "Usno" was produced. This preparation gave good results in tests against skin diseases, and also against numerous yeasts (Capriotti, 1959; 1961), and in the treatment of *Trichophyton gallinae* (Virtanen & Kilpio, 1957).

In recent years, there has been little interest in lichen compounds as antibiotics; too many natural and laboratory-synthesized compounds are known that are either more effective or easier to administer than lichen compounds. This is perhaps unfortunate because some recent studies with fungal products very similar to lichen products have been yielding interesting results. Stark et al. (1978) isolated a number of compounds from the fungus,

TABLE 9.1. Solubility of Lichen Compounds in Water

Lichen Compound	*Amount Dissolved (mg/l)*
Depsides of the orcinol type	
Erythrin	57
Evernic acid	12
Depsides of the β-orcinol type	
4-*0*-demethylbarbatic acid	28
Atranorin	5
Depsidone of the orcinol type	
Lobaric acid	8
Depsidones of the β-orcinol type	
Fumarprotocetraric acid	47
Salazinic acid	27
Norstictic acid	23
Stictic acid	22
Psoromic acid	13

I.K. Iskandar & J.K. Syers. 1971. "Solubility of lichen compounds in water: Pedogenetic implications." *Lichenologist* 5:45–50. Reprinted with permission of Academic Press.

Chaetomium mollicellum that they called mollicellins. These compounds are depsidones and quite similar in structure to depsidones produced by lichen fungi. Eight of these mollicellins were assayed for mutagenicity and antibacterial activity in *Salmonella*/microsome tests. Two were mutagenic and bactericidal for *Salmonella typhimurum*. Two others, each of which contained chlorine, were bactericidal but not mutagenic. The authors pointed out that these compounds should be considered stable environmental mutagens that may be produced on foodstuffs during storage. It appears, in light of these findings, that similar testing of lichen depsidones is warranted.

Despite the decreased research effort on lichen antibiotics there is increasing interest in the antitumor activity of lichens (Shibata et al., 1968; Kupchan & Kopperman, 1975; Nishikawa et al., 1970; 1979; Takai, Uehara, & Beisler, 1979; Takeda et al., 1972). Lichens, particularly *Cetraria* species, have been used in the treatment of cancers for probably a thousand years (Hartwell, 1971); only recently, however, has the mechanism of tumor inhibition been investigated. From experiments done so far, it appears that antitumor activity is associated with the polysaccharide component of lichens; psoromic and usnic acids have also been shown to be effective against tumors.

Activity against Soil, Decay, and Mycorrhizal Fungi

Most studies of the antibiotic effects of lichen compounds have been done with pathogenic microorganisms in the laboratory. A few studies, however, have been designed to determine the extent to which chemical interactions between lichens and microorganisms occur in nature and the ecological consequences of such interactions.

Harder and Uebelmesser (1958) looked at changes in the frequency of occurrence of chytrids and other soil phycomycetes that apparently resulted from the presence of soil lichens. Only in very rare instances could fungi be baited from the soil collected under lichens; the reason for this could not be determined. However, the addition of ground-up lichens (*Parmelia caperata, P. scortea, P. dubia, P. physodes, P. sulcata*, and *Evernia prunastri*) to soil suspensions with known quantities of phycomycetes and an appropriate bait resulted in significantly less colonization than in control suspensions without lichens. In the Soviet Union, Evodokimova (1962) found that a ground cover of *Polytrichum commune, Cladonia* (= *Cladina*) *alpestris*, and *Stereocaulon paschale* was somehow highly toxic to *Azobacter* species in the underlying soil. Malicki (1965; 1967; 1970) has also demonstrated an inhibitory effect of *Cladonia* species and usnic acid on soil bacterial decomposition organisms. Since soil microorganisms have long been recognized as important components of soil ecosystems, particularly insofar as they participate in soil processes such as organic matter decomposition and essential element cycling, lichen compound inhibition of soil fungi may prove to be an important regulator of these processes.

Other studies have demonstrated inhibitory effects of water extracts of lichens on wood decay fungi (Henningsson & Lundström, 1970; Lundström & Henningsson, 1973). Lundström and Henningsson (1973) found that the fungistatic agents they extracted from *Hypogymnia physodes* were not acetone soluble, a result that suggested typical lichen phenolic compounds were not involved in this particular interaction.

Ectomycorrhizal fungi associated with tree roots are known to increase tree growth, often spectacularly, presumably by increasing essential element uptake by tree roots and inhibition of soil-borne pathogens around roots. Early experiments (Leibundgut, 1952) demonstrated that water extracts of lichen species inhibit mycorrhizae of trees and thereby retard their growth. Henningsson and Lundström (1970) also found an inhibitory effect of aqueous extracts of *Hypogymnia physodes* on mycorrhizal fungi. In a greenhouse experiment, Brown and Hooker (1971; reported in Brown & Mikola, 1974) moistened the soil around jack pine seedlings with a water extract of mixed reindeer lichens and compared the frequency of mycorrhizal infection of these seedlings to controls receiving only water. A significant infection reduction was observed in the number of mycorrhizae present on seedlings receiving the lichen extract; seedling growth was also reduced. All these observations suggested that lichen compounds can influence tree growth through the inhibition of mycorrhizae.

Brown and Mikola (1974) tested aqueous extracts of several lichen species against a wide range of mycorrhizal fungi and found that the extract from *Cladonia alpestris* was by far the most effective inhibitor of mycorrhizal fungi (Table 9.2). *Cladonia pleurota* (not shown in the table) was found to be almost as effective as *C. alpestris*, but, inasmuch as it was observed very rarely in the study area, it was not thought to be as influential under natural conditions.

To determine whether this obvious inhibition of mycorrhizal fungus activity by lichens has an effect on tree growth, Brown and Mikola (1974) measured labelled ^{32}P uptake by *Pinus sylvestris* seedlings infected and uninfected with various mycorrhizal fungi; aqueous lichen extracts were then tested for inhibition of ^{32}P uptake. Inoculation of seedlings did not always result in seedling infection; usually approximately 0.5 of the inoculated seedlings became infected. Except in the inoculated controls, infection resulted in increased ^{32}P uptake. Addition of *Cladonia alpestris* extract significantly reduced ^{32}P uptake (Table 9.3); extracts of *C. rangiferina* and *Cetraria islandica* had an inhibitory effect in uninfected seedlings and a stimulatory effect in infected seedlings. Thus, although it appears that *C. alpestris* is capable of influencing element uptake by tree seedlings through the inhibition of mycorrhizal fungi, the nature of the interaction and the specific role played by lichen compounds is still rather unclear.

TABLE 9.2. Effect of Aqueous Extracts of *Cladonia alpestris, C. arbuscula, C. rangiferina,* and *Cetraria islandica* on Selected Mycorrhizal Fungi Grown on Hagem Agar. X Indicates a Significant Inhibition at the 0.05 Level; + Indicates a Significant Stimulation at the 0.05 Level.

	Water Extracts			
Mycorrhizal Fungus Species	C. alpestris	C. arbuscula	C. rangiferina	C. islandica
Amanita muscaria	X	—	—	—
A. rubescens	X	—	—	X
Boletus bovinus	X	X	—	—
B. luteus	—	—	—	—
B. variegatus	X	—	+	—
Cenococcum graniforme	—	—	—	—
Corticium bicolor	X	X	X	X
Laccaria laccata	X	X	X	—
Paxillus involutus	X	X	X	X
Tricholoma flavobrunneum	X	—	—	—
T. imbricatum	X	—	—	X

Partial data adapted from R.T. Brown & P. Mikola. 1974. "The influence of fruticose soil lichens upon the mycorrhizae and seedling growth of forest trees." *Acta Forest Fenn.* 141:1–23.

TABLE 9.3. Effect of Aqueous Lichen Extracts on ^{32}P Uptake by *Pinus sylvestris* Seedlings Infected or Noninfected with Mycorrhizal Fungi

	^{32}P activity, mm^2	
Treatment	*Noninfected*	*Infected*
Inoculated control	171*	162
Inoculated + *C. alpestris* extract added	101	116
Inoculated + *C. rangiferina* extract added	77	201
Inoculated + *C. islandica* extract added	66	227

*Area of X-ray film exposed by ^{32}P in mm^2.

Partial data adapted from R.T. Brown & P. Mikola. 1974. "The influence of fruticose soil lichens upon the mycorrhizae and seedling growth of forest trees." *Acta Forest Fenn.* 141:1–23.

Nature of Antibiosis

The fact that most lichens grow very slowly, are perennial and persistent in their habitat, and are resistant to decay microorganisms that readily attack nonlichenized fungi suggests that lichen compounds have a primarily protective role. Despite the wealth of evidence demonstrating the antibiotic activity of lichen compounds, however, the mechanism of this activity remains totally unknown.

Lichen compounds are fungal in origin. Yet, they are only rarely produced by isolated mycobionts; at present, there is no known method to regularly induce formation of typical lichen compounds by isolated mycobionts (Ahmadjian, 1980a). This suggests that there is a regular participation by the phycobiont in the production of these compounds under natural conditions. The compounds may play some as yet undetermined role in initiating or maintaining the symbiotic state in lichens; however, not all lichens produce these compounds. Indeed, in some morphologically related groups of ascolichens, some species produce many compounds and others produce none. Lichen compounds are unknown in basidiolichens. It seems unlikely, therefore, that they are necessary for symbiosis.

Although isolated mycobionts rarely produce typical lichen compounds, they are known to produce antibiotics. Ahmadjian (1961) demonstrated antibiotic properties of isolated mycobiont extracts. Recent studies of lichens in the family Trypetheliaceae (Mathey, 1979) also suggested that isolated mycobionts produce antifungal compounds. The relationship between these mycobiont compounds and typical lichen compounds is not presently known.

How does one integrate these bits of information into a model that makes sense physiologically and ecologically? A survey of the literature on phenolic compound production by plants (Levin, 1971) classifies resistance to attack by microorganisms or animals as either constitutive or induced. Constitutive resistance is based on the presence of phenolic inhibitors prior to attack and is best suited to defense against animals. Induced resistance is based on the accumulation and modification of existing compounds in response to attack and is best suited to defense against bacteria, viruses, fungi, and nematodes. According to a current model of lichen secondary compound production (C.F. Culberson & V. Ahmadjian, 1980), lichen compounds may have roles that reflect both of these evolutionary approaches to defense.

Typical lichen compounds are produced from simple phenolic acid precursors that normally lead to distinctly different phenolic compounds in nonlichen fungi (C.F. Culberson & V. Ahmadjian, 1980). Production of typical lichen depsides, depsidones and dibenzofurans instead of these fungal phenolics appears to be linked to an inhibition of the phenolic acid decarboxylation step in lichens (see Figure 4.3 in Chapter 4). Mosbach and Ehrensvärd (1966) found that the lichen, *Lasallia pustulata*, but not the phycobiont of related *L. papulosa*, contained an orsellinate decarboxylase. This result suggested that only the fungus is capable of decarboxylating phenolic precursors. The reason the fungus does not do this under lichenized conditions is not completely known but may involve the alga in some way.

C.F. Culberson and V. Ahmadjian (1980) hypothesized that the lichen phycobiont produces an inhibitor of decarboxylases responsible for the production of typical lichen compounds. This is important for two reasons.

First, the phycobiont prevents the accumulation of biologically active, potentially phycotoxic quinones and phenolic products in the thallus. Second, the products of decarboxylation (typical lichen compounds) constitute a reservoir of defense compounds that could be hydrolyzed and released when the alga is disturbed; release of phenolics can also occur when the alga is disturbed and decarboxylase inhibitor production is interrupted. Since damage to the phycobiont results in the release of defense compounds, either through hydrolysis or decarboxylase inhibition, the whole process can be viewed as an induced defense system, where induction occurs through microorganism attack or herbivory.

There is some evidence to support these ideas (Mosbach & Ehrensvärd, 1966; Mosbach & Schultz, 1971; Schultz & Mosbach, 1971). Mosbach and Ehrensvärd (1966) discovered a depside-hydrolyzing esterase in cell-free extracts of the lichen fungus *Lasallia pustulata* and the phycobiont of *L. papulosa*. It appears, therefore, that both the alga and fungus have a depside-hydrolyzing capability and are capable of releasing phenolic precursors in response to attack. Since ingestion of lichen compounds by herbivores or contact by microorganisms may also result in hydrolysis, compound production may provide an additional direct constitutive protection.

According to Culberson and Ahmadjian's (1980) hypothesis, then, lichen compounds may best be viewed as storage forms of biologically active defense compounds. In a sense, they are constitutive defense compounds since their presence may provide a direct deterrent to attack, and hydrolysis to biologically active fungal phenolics can occur via a number of different mechanisms. Production of antibiotics may also be induced by damage to the thallus, especially the phycobiont. Further experimental work on this interesting subject is necessary before a detailed and unequivocal explanation of the process can be made.

ALLELOPATHY

Molisch (1937) coined the term "allelopathy" to refer to chemical interactions between all types of plants including microorganisms. It is derived from Greek words meaning mutual harm; hence, the current use of the term includes any direct or indirect harmful effect by one plant on another through the production of chemical compounds (Rice, 1974). It is of obvious importance in the study of inter- and intraspecific interactions, particularly insofar as mutual interference between plants of the same or different species influences the outcome of competition (Muller, 1969). It has also been implicated in phytoplankton and old field succession, nitrification inhibition, fire succession, and seed decay inhibition (Rice, 1974). Despite the increased interest in this subject in recent years, however, relatively little work involving lichens has been done. This section reviews the few studies done to date and suggests directions for future study.

Although the term allelopathy includes chemical inhibition of microorganisms, this aspect will not be treated here. Discussion of lichen chemical inhibition of microorganisms can be found in preceding sections. This section reviews work done on lichen chemical inhibition of vascular plants, bryophytes, and other lichens, and the effects of tree exudates, especially bark compounds, on corticolous lichen growth and viability.

Lichen Chemical Effects on Vascular Plants

Lichen compounds are known to inhibit germination and growth of vascular plant seedlings (Burzlaff, 1950; Follmann & Nakagava, 1963; Follmann & Peters, 1966; Rondon, 1966; Pyatt, 1967; Rathore & Mishra, 1971; Dalvi, Singh, & Salunkhe, 1972; Huneck & Schreiber, 1972; Reddy & Rao, 1978). Although these studies were done with commercial varieties of vascular plants, they demonstrated the allelopathic potential of lichens and suggested that chemical interactions of ecological importance involving lichens and vascular plants in nature are unquestionably possible. Rundel (1978) is correct to point out, however, that some of these studies do not control or consider the effects lichen compounds may have on mycorrhizal fungi. Since lichen compounds are known to inhibit mycorrhizal fungi, they may influence vascular plant growth through effects on mycorrhizal development and activity.

Huneck and Schreiber (1972) tested a number of lichen and liverwort secondary compounds for their effects on vascular plant growth using a number of bioassays. Potassium salts of lichen compounds were produced and tested at concentrations ranging from 10^{-3} to 10^{-7} M. A selected listing of data from one of these bioassays, a cress root growth assay (Table 9.4), demonstrates that many of the lichen compounds were quite inhibitory at 10^{-3} and 10^{-4} M; however, some of these same compounds were stimulatory at lower concentrations (e.g., atranorin-K, (-)usnic acid-K, pinastric acid-K). Planaic acid-K was stimulatory at all concentrations. Similar patterns were observed in other bioassays used (oat and pea seedling growth, oat coleoptile section extension, and lanolin paste method with bean). Buffard and Rondon (1977) also observed stimulatory effects of *Parmelia conspersa* extracts on the rooting of *Peperomia magnoliaefolia* cuttings. Liverwort compounds, though distinctly different in structure from lichen compounds, elicited similar vascular plant growth responses (Table 9.4).

There have been a few field studies done to determine the extent to which lichen chemical inhibition of vascular plant growth and seed germination actually occurs in nature. In a phytosociological study of east Canadian peat

TABLE 9.4. Growth of Cress Roots in the Presence of Different Concentrations of Lichen and Liverwort Secondary Compounds

	Concentration (M)				
	10^{-3}	*10^{-4}*	*10^{-5}*	*10^{-6}*	*10^{-7}*
Lichen compounds					
Aliphatic acids					
Caperatic-K[a]	13[b]	79	127	98	74
Roccellic-K	0	72	104	91	84
(+) protolichesterinic-K	75	72	69	81	97
Depsides					
Confluentic-K	87	118	97	127	111
Sekikaic-K	112	96	148	96	124
Atranorin-K	0	88	116	128	104
Evernic-K	7	91	88	94	81
Planaic-K	118	143	136	143	133
Depsidones					
Fumarprotocetraric-K	34	80	90	90	70
Salazinic-K	88	94	70	88	85
Psoromic-K	3	48	84	76	68
α-collatolic-K	30	76	73	112	73
Stictic-K	95	98	75	83	73
Lobaric-K	100	122	103	96	112
Virensic-K	11	70	112	108	100
Dibenzofurans					
(−) usnic-K	20	116	140	112	133
(−) isousnic-K	0	85	97	94	90
Pulvinic acid derivatives					
Vulpinic-K	0	43	74	69	92
Epanorin-K	63	100	100	123	100
Lepraric-K	0	17	94	98	67
Pinastric-K	32	104	132	108	120
Rhizocarpic-K	18	85	106	100	88
Liverwort Compounds					
Gymnocolin	0	55	115	80	94
Drimenol	0	20	50	115	110
Longiborneol	15	120	110	110	105
Longifolen	24	86	94	90	100
Lunalaric acid	62	90	68	97	76
Scapanin	0	40	100	125	90

[a]Potassium salts of lichen compounds were used throughout.

[b]All data are percentages of cress root length in presence of test compound as compared to controls.

Partial data adapted from S. Huneck & K. Schreiber. 1972. "Wachstumsregulatorische Eigenschaften von Flechten- und Moos-Inhaltstoffen." *Phytochem.* 11:2429–2434.

bog ecosystems, Fabiszewski (1975) suggested that lichens influence vascular plant community patterns through allelopathic effects. Ramaut and Corvisier (1975) observed vegetation patterns in the Belgian Ardennes that suggested an inhibitory effect of terricolous lichens. They tested this hypothesis by using aqueous extracts of *Cladonia impexa, C. gracilis*, and *Cornicularia muricata* in seed germination experiments with *Pinus sylvestris*. They found that extracts of *C. muricata* and *C. impexa* severely inhibited *Pinus* seed germination. Even when seeds germinated, seedling growth was much reduced. Pyatt (1967) made similar field observations of apparent lichen inhibition of vascular plants in a sand dune region of South Wales. Laboratory tests revealed an inhibitory effect of *Peltigera canina* extracts on several vascular plant seeds. Ott (1961) also demonstrated the inhibitory effect of pure lichen compounds on tree seed germination that suggested an ecological role, and Fisher (1979) has shown that jack pine and white spruce forests may be controlled to a certain extent by allelopathic substances produced by *Cladonia* species. Not all studies show an inhibitory effect of lichens, however. Gagnon (1966) reported that the presence of *Lecidea granulosa* improves the germination of *Picea maritima* seeds in Canadian forests. It is clear that further experimental work is needed before the ecological importance of lichen chemical inhibition of vascular plants is known.

Lichen Chemical Effects on Bryophytes

In his discussion of corticolous cryptogam community dynamics, Barkman (1958) mentioned that "chemical action" probably plays a role in influencing the outcome of species interactions, and he cited several descriptive papers that purported to show the existence of lichen chemical inhibition with other cryptogams. At the time Barkman was writing, however, so little information about chemical interference in cryptogam communities existed, it was difficult to discuss the subject with any authority.

Lichen-bryophyte interactions have been known since Zukal (1879) first observed the physical invasion of moss tissues by lichen hyphae. Bonnier (1888; 1899) was the first to demonstrate experimentally the ability of germinating lichen ascospores to parasitize and kill moss protonemata. He suggested that such microscopic competitive interactions occur widely in natural cryptogam communities. Keever (1957) noted that the presence of lichens on granite habitats caused protonemata of *Grimmia laevigata* to discolor but did not determine the cause of the inhibition. McWhorter (1921) also noticed the often highly destructive effects of terricolous *Cladonia* species on several species of mosses. He prepared stained thin sections of lichen-parasitized moss material obtained in the field and showed the presence of lichen hyphae in the meristematic tissues of moss leaves; both inter- and intracellular cell penetration were observed.

Heilman and Sharp (1963) published one of the first suggestions that lichen secondary compounds had an antibiotic effect on bryophytes. In the Great Smoky mountains, they observed instances of the lichen *Ocellularia subtilis* obviously overgrowing the liverwort *Frullania eboracensis*; even more dramatic was the ability of the slow-growing crustose lichen *Huilia* (= *Lecidea*) *albocaerulescens* to maintain itself in the presence of several species of mosses and liverworts. From these observations, the authors suggested that chemical action was influencing the patterns.

Lawrey (1977a; 1977b) was the first to determine experimentally the allelopathic effect of lichen secondary compounds on moss spore germination. In a preliminary experiment, Lawrey (1977a) found that spore germination of six moss species was significantly inhibited by acetone extracts of *Cladonia subcariosa, C. cristatella*, and *C. squamosa* (Table 9.5). A later experiment with pure lichen compounds (Lawrey, 1977b) demonstrated a significant inhibitory effect of the 4-*O*-methylated compounds, evernic and squamatic acids, on spore germination of *Funaria hygrometrica, Ceratodon purpureus*, and *Mnium cuspidatum*. However, no inhibitory effect was observed for norstictic acid, a hydroxylated compound, or *l*- or *d*-usnic acids. Lawrey (1977b) suggested, on the basis of this preliminary evidence, that lichen *O*-methylated compounds may be more effective allelopathic agents than hydroxylated compounds. This hypothesis was all the more appealing since Hale (1966) found a good correlation between production of *O*-methylated compounds by lichens and possession of advanced morphological traits that seemed to indicate that *O*-methylated compounds were evolutionarily advanced compounds.

A later study of the effects of lichen compounds on moss spore germination (Gardner & Mueller, 1981) also demonstrated the inhibitory effects of numerous lichen compounds on spore germination and sporeling growth of *Funaria hygrometrica*. Gardner and Mueller tested the effects of atranorin and usnic, lecanoric, evernic, vulpinic, stictic, fumarprotocetraric, and psoromic acids at four different concentrations on media buffered to pHs ranging from 5 to 8. They found that lichen compound toxicity was both pH- and concentration-dependent. All compounds tested were ineffective at concentrations of 2.7×10^{-5} M or less, and most compounds increased in toxicity as pH decreased (Tables 9.6 and 9.7). Toxicity patterns were found to be rather complex and very difficult to interpret ecologically. For example, some compounds did not inhibit germination but were effective in retarding sporeling growth. No relationship between toxicity and chemical structure (*O*-methylated vs. hydroxylated) was observed. Although two *O*-methylated lichen compounds (evernic and psoromic acids) were among the most

TABLE 9.5. Percent Spore Germination of Terricolous Moss Species on Filter Paper Treated with Acetone (Control) and Acetone Extracts of *Cladonia subcariosa, C. cristatella,* and *C. squamosa* (Treatments), Ovendried and Saturated with Various Concentrations of Hoagland's Solution. All Data Are Calculated on the Basis of a Minimum of 500 Counts in Each of Five Culture Plates for Three Concentrations of Hoagland's Solution (0 Percent, 25 Percent, 50 Percent) and for Each of Five Extracts of the Lichen Species Tested

	Time Required for Initial Germination	Percent Germination Treatments			
Moss Species	*(Days)*	*Control*	C. subcariosa	C. cristatella	C. squamosa
Funaria hygrometrica	2	95.6	0	0	0
Mnium cuspidatum	3	95.4	0	0	0
Amblystegium serpens	1	90.8	0	0	0
	3	91.2	71.4	15.1	21.4
Weissia controversa	10	90.0	0	0	0
Pohlia nutans	5	34.2	5.6	1.1	0.8
Physcomitrium pyriforme	7	26.2	0	0	0

J.D. Lawrey. 1977a. "Inhibition of moss spore germination by acetone extracts of terricolous *Cladonia* species." *Bull. Torrey Bot. Club.* 104:49–52. Reprinted with permission of the Torrey Botanical Club.

TABLE 9.6. Relative Toxicity of Several Lichen Acids to Spore Germination and Sporeling Growth of *Funaria hygrometrica*. Lichen Acid Concentration in All Cases Was 2.7 × 10^{-4}M and Agar pH Was 7.0

Lichen Compound	*Percent Germination*	*Sporeling Growth*
Vulpinic acid	37*	55
Usnic acid	62	49
Evernic acid	69	68
Psoromic acid	72	62
Lecanoric acid	94	84
Atranorin	97	85
Stictic acid	99	88
Fumarprotocetraric acid	100	80

*Data are percentages based on control spore germination and growth. At least six replicates were used.

Partial data adapted from C.R. Gardner and D.M.J. Mueller. 1981. "Factors affecting the toxicity of several lichen acids: Effect of pH and lichen acid concentration." *Am. J. Bot.* 68:87–95. Reprinted with permission of American Journal of Botany.

TABLE 9.7. pH Effect on the Relative Toxicity of Several Lichen Acids to Spore Germination of *Funaria hygrometrica*. Lichen Acid Concentration in All Cases Was 2.7 × 10^{-3}M

	Percent Spore Germination			
Lichen Compound	*pH 5*	*pH 6*	*pH 7*	*pH 8*
Vulpinic acid	0*	0	0	0
Psoromic acid	11	0	20	0
Fumarprotocetraric acid	41	97	97	97
Evernic acid	75	41	1.4	44
Lecanoric acid	94	86	90	96
Atranorin	97	98	97	99
Stictic acid	100	102	101	105
Usnic acid	102	98	25	0.9

*Data are percentages based on control spore germination. At least six replicates were used.

Partial data adapted from C.R. Gardner and D.M.J. Mueller. 1981. "Factors affecting the toxicity of several lichen acids: Effect of pH and lichen acid concentration." *Am. J. Bot.* 68:87–95. Reprinted with permission of American Journal of Botany.

effective in inhibiting spore germination and sporeling growth, stictic acid, another *O*-methylated compound tested, was one of the least inhibitory. Gardner and Mueller suggested that neither *O*-methylation nor any other single substitution on a lichen compound is likely to confer increased toxicity, regardless of the structure of the molecule or the organisms used in the assay. Thus, from these recent studies Lawrey's (1977b) hypothesis appears to require considerable modification.

Lichen Chemical Effects on Other Lichens

As Barkman (1958) and later Topham (1977) have mentioned, the level of competition between lichens in cryptogam communities is considered to be low compared with that observed in vascular plant communities. This is because lichens inhabit areas where other organisms cannot grow, making them "nature's pioneers, but also nature's poor relations" (Topham, 1977, p. 55). Or, as another eminent ecologist has put it, "... ecologically, lichens are hardly more than intricate scum on rock and tree surfaces that are too harsh for angiosperms" (Janzen, 1975, p. iv). Although most lichenologists would probably choose another way of phrasing it, the impression that lichens are well adapted to stressful environments but poorly adapted to competitive environments is fairly general (Grime, 1977).

Lichen-lichen interactions are observed frequently in nature, however, and competition is assumed to go on in lichen assemblages (Barkman, 1958; Topham, 1977). Furthermore, lichen secondary compounds are assumed to be capable of influencing interspecific competitive interactions. Unfortunately, little experimental work has been done to test this idea with lichens.

Occasionally, there are reports in the literature of spatial patterns in lichen communities that suggest allelopathic effects are reponsible. Barkman (1958) noted that lichen species growing below other species on a tree trunk will sometimes discolor and die, presumably through the action of lichen compounds excreted by the uppermost species. He gave other examples of lichen species pairs never found together despite the fact that they have exactly the same habitat requirements—patterns that suggest possible chemical interference between the species. Hilitzer (1925) also observed that when thalli of two lichen species are contiguous, the inferior competitor sometimes exhibits a dying marginal zone where it comes into contact with the superior competitor.

C.F. Culberson, W.L. Culberson, and A. Johnson (1977b) discovered an interesting nonrandom distribution of an epiphytic *Lepraria* species on chemically different species of *Parmelia* that suggested an allelopathic effect. Thalli of *P. loxodes*, which contained glomelliferic acid as the major medullary constituent and nine other substances, were found to have the *Lepraria* associated with them only very rarely (Table 9.8). In contrast, thalli of *P. verruculifera*, which contained divaricatic acid as the major constituent and only two other medullary compounds, were frequently observed to host the *Lepraria* epiphyte. The authors suggested that differences in secondary compound production may have been responsible for the patterns observed.

It is very difficult to design experiments in the field to determine the extent to which allelopathic interactions operate in lichen communities. Laboratory work with whole lichen thalli is hampered by the low viability of lichens in laboratory environments. However, lichen ascospore germination

TABLE 9.8. Distribution of a *Lepraria* Species Epiphyte on *Parmelia loxodes* and *P. verruculifera* Collected in Czechoslovakia

	Without *Lepraria*		With *Lepraria*	
	P. loxodes	P. verruculifera	P. loxodes	P. verruculifera
Number of specimens	26	20	4	55
Percentage by species	87	27	13	73
Weight (g)	5.96	1.00	0.23	4.52
Cover (cm^2)	129.0	39.4	6.0	139.0

Partial data adapted from C.F. Culberson, W.L. Culberson, & A. Johnson. 1977b. "Nonrandom distribution of an epiphytic *Lepraria* on two species of *Parmelia.*" *Bryologist* 80:201–203. Reprinted with permission from the American Bryological and Lichenological Society.

responses to lichen extracts and pure lichen compounds are relatively easy to measure and provide some information about the potential of lichen compounds to act as allelopathic agents in lichen communities.

Pyatt (1973) was the first to report effects of lichen compounds on lichen ascospore germination. He tested the effect of aqueous extracts of six lichen species on spore germination responses of *Xanthoria parietina* and *Pertusaria pertusa* and observed some inhibition. However, the effects were not particularly dramatic and it was impossible to determine from his results the effects of lichen compounds in causing the effect.

Whiton and Lawrey (1982) were the first to show lichen ascospore inhibition by pure lichen compounds. Using a technique similar to that of Gardner and Mueller (1981), they exposed ascospores of the lichenized fungus, *Cladonia cristatella*, and the nonlichenized fungus, *Sordaria fimicola*, to three lichen compounds (vulpinic, evernic, and stictic acid) on buffered agar ranging in pH from 4 to 7. The same concentration of lichen compounds (2.7×10^{-3} M) was used throughout the experiments; this concentration was found by Gardner and Mueller (1981) to inhibit *Funaria* spore germination. The test species were chosen because they were both ascomycetes that produce spores readily under laboratory conditions. After 24 hours' exposure to lichen compounds, spores of *C. cristatella* were slightly but significantly inhibited by vulpinic acid; there was no inhibition by evernic or stictic acids (Table 9.9). *Sordaria* spores, however, were severely inhibited by both vulpinic and evernic acids. These results demonstrated that lichen compounds are capable of functioning as allelopathic agents. Furthermore, they suggested that lichenized fungi may be more tolerant of the inhibitory effects of lichen compounds than nonlichenized fungi.

In an attempt to test this hypothesis further, Whiton and Lawrey (1984) did an additional set of spore germination experiments in which the effects of

TABLE 9.9. Effect of Vulpinic (VUL), Evernic (EVE), and Stictic (STC) Acids on Ascospore Germination of *Cladonia cristatella* and *Sordaria fimicola* on Buffered Media Ranging from pH 4 to pH 7

		% Germination			
Species	*pH*	*Control*	*VUL*	*EVE*	*STC*
C. cristatella	4	96.3* (1)	75.1 ± 7.0 (3)	96.1 ± 0.9 (4)	95.6 ± 1.0 (3)
	5	95.3 ± 1.2 (5)	72.6 ± 2.6 (5)	91.5 (1)	96.4 ± 1.9 (2)
	6	95.9 ± 0.2 (2)	71.5 ± 3.6 (10)	95.4 ± 1.5 (2)	97.1 ± 1.2 (3)
	7	2.5 ± 1.1 (2)	0 (3)	0.4 ± 0.4 (4)	2.7 ± 1.4 (3)
S. fimicola	4	97.7 ± 1.8 (4)	5.1 ± 4.2 (4)	1.9 ± 0.2 (4)	97.5 ± 0.7 (4)
	5	93.6 ± 0.8 (4)	9.4 ± 4.8 (4)	2.8 ± 1.1 (4)	98.6 ± 0.5 (4)
	6	96.6 ± 1.2 (4)	2.2 ± 0.6 (4)	4.6 ± 2.2 (4)	97.7 ± 1.2 (4)
	7	89.7 ± 7.1 (4)	2.8 ± 1.0 (4)	4.9 ± 2.7 (4)	98.6 ± 0.7 (4)

*Mean percent germination ± S.E. of mean. At least 200 ascospores were counted per plate; numbers in parentheses are number of plates counted per treatment. Variation about the mean thus represents plate to plate sample variation.

J.C. Whiton & J.D. Lawrey. 1982. "Inhibition of *Cladonia cristatella* and *Sordaria fimicola* ascospore germination by lichen acids." *Bryologist* 85:222–226. Reprinted with permission of the American Bryological and Lichenological Society.

vulpinic, evernic, and stictic acids, and an additional lichen compound, the cortical substance, atranorin, were tested using two crustose lichen species, *Graphis scripta* and *Caloplaca citrina*. Crustose species were chosen for these experiments because they were thought to depend more heavily on ascospores for reproduction and dispersal than species capable of asexual (vegetative) dispersal. The results of these experiments (Table 9.10) showed clearly that lichen ascospores are not any more tolerant of the inhibitory effects of lichen compounds than nonlichen ascospores. Lichen compound production appears, therefore, to be able to influence establishment of lichen ascospores in nature; however, the extent to which compounds might also influence whole lichen thallus interactions cannot be determined as yet.

Chemical Effects of Tree Exudates on Corticolous Lichens

Just as lichen species may influence the distribution of nearby cryptogams by virtue of their secondary compounds, lichens themselves may be limited in

TABLE 9.10. Effect of Vulpinic (VUL), Evernic (EVE), and Stictic (STC) Acids, and the Cortical Compound Atranorin (ATR) on Ascospore Germination of *Graphis scripta* and *Caloplaca citrina* on Buffered Media from pH 4 to 7

		% Germination				
Species	*pH*	*Control*	*VUL*	*EVE*	*STC*	*ATR*
G. scripta	4	98.54 ± 0.66*	0	0	99.56 ± 0.44	100.00
		(5)	(4)	(6)	(5)	(4)
	5	92.21 ± 5.13	0.14 ± 0.10	0.50 ± 0.20	77.13 ± 12.42	99.78 ± 0.22
		(11)	(3)	(4)	(6)	(3)
	6	96.67 ± 2.28	0	0	85.04 ± 11.76	95.52 ± 4.48
		(6)	(9)	(6)	(6)	(3)
	7	20.36 ± 12.20	0	0	13.85 ± 8.67	0
		(10)	(5)	(4)	(7)	(3)
C. citrina	4	0	0	0	0	0
		(4)	(5)	(4)	(2)	(1)
	5	0	0	0.18 ± 0.11	0	0
		(5)	(4)	(5)	(3)	(3)
	6	15.02 ± 3.02	0	0	9.75 ± 1.15	0
		(5)	(4)	(4)	(5)	(1)
	7	21.59 ± 5.63	0.32 ± 0.13	5.50 ± 1.75	11.02 ± 2.89	0
		(6)	(5)	(5)	(6)	(3)

*Mean percent germination ± standard error of the mean. At least 100 ascospores were counted per plate; numbers in parentheses are number of plates counted per treatment.

J.C. Whiton & J.D. Lawrey. 1984. "Inhibition of crustose lichen spore germination by lichen acids." *Bryologist* 87:42–43. Reprinted with permission of the American Bryological and Lichenological Society.

distribution by compounds produced by other plant species. For corticolous lichens, it is particularly important to consider the effects of the host phorophytes. There have been investigations of the nutrient contributions of host trees to corticolous cryptogams; for example, Tamm (1950) demonstrated that differential leaching of nutrients from tree canopies can affect the growth of epiphytic mosses; however, no direct effect of bark compounds was shown. The idea that lichen distribution may be controlled by tree bark chemistry has not been studied by lichenologists, although there have been numerous studies of host specificity or substrate specificity of epiphytic lichens, some of which also included bryophytes (Martin, 1938; Billings & Drew, 1938; W.L. Culberson, 1955a; 1955b; Hale, 1955; Barkman, 1958; Brodo, 1961b; Adams & Risser, 1971a; 1971b; Jesberger & Sheard, 1973; Stringer & Stringer, 1974; Gough, 1975; Slack, 1976). These studies established statistically significant associations between corticolous cryptogams and tree species but have not demonstrated a chemical or any other basis for these associations.

It is clear from the literature that numerous tree species produce antifungal compounds. Fawcett and Spencer (1969) discussed a number of naturally occurring compounds that are effective in controlling various fungal pathogens. Wood-decomposing fungi are also known to be inhibited by various volatile organic compounds that emanate from wood (Cobb et al., 1968; Hintikka, 1970; Rice, 1970; Fries, 1973; Väisälä, 1974). Despite the fact that these compounds may influence lichen distribution patterns in nature, nothing is known about their influence on lichen growth, spore germination, or dispersal ability.

A series of studies was done by Sivori and Jatimliansky (1965; 1970; Jatimliansky & Sivori, 1969) to explain the unusually sparse lichen cover they observed on the bark of *Laurus nobilis*. They initially tested bark extracts using a unicellular alga, *Scenedesmus obliquus* (Sivori & Jatimliansky, 1965), since they were unable to maintain whole lichen thalli very long under laboratory conditions. They observed a significant inhibitory effect and were ultimately able to isolate two inhibitory compounds (Jatimliansky & Sivori, 1969), actinodaphine and launobine, both aporphine alkaloids. A similar study was conducted with xylopine, an aporphine alkaloid produced by a different but related tree species, *Xylopa discreta* (Sivori & Jatimliansky, 1970). Xylopine was also observed to strongly inhibit growth of *Scenedesmus obliquus*, even at very low concentrations (Figure 9.1). The investigators suggested that tree bark alkaloids are potentially lichenocidal or at least lichenostatic compounds, especially insofar as they may severely inhibit lichen phycobionts.

Few investigations of the direct effects of tree bark compounds on lichen fungi have been done. Pyatt (1973) tested the effect of macerated tree bark on the ascospore germination responses of three lichen species (Table 9.11). He found a generally increased spore germination in the presence of tree bark

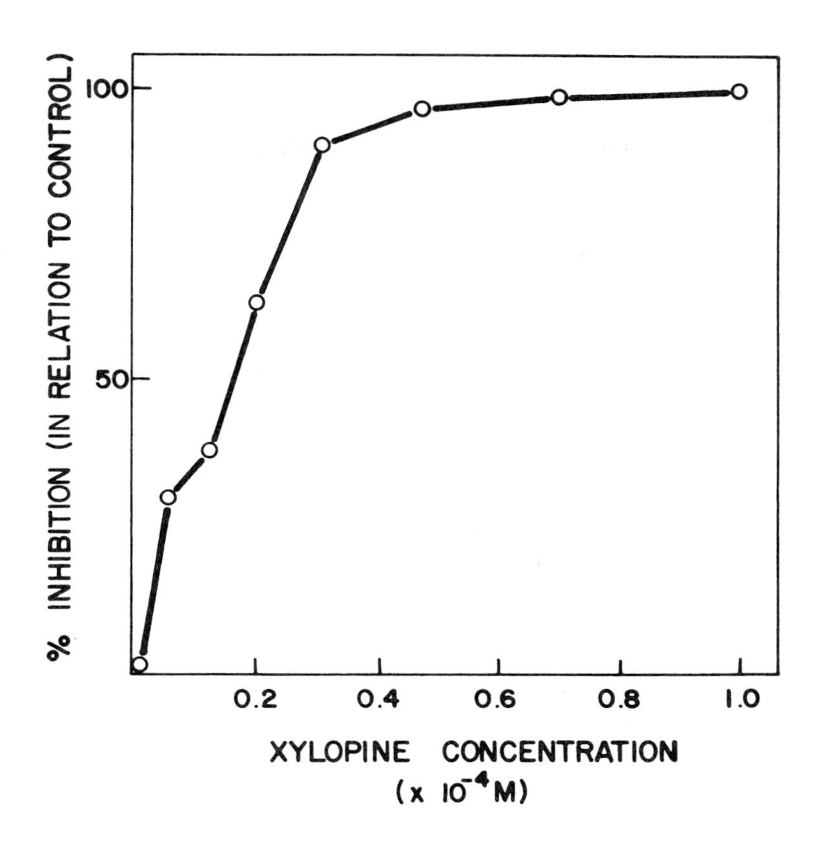

FIGURE 9.1. Xylopine inhibition of growth of *Scenedesmus obliquus.* Adapted from E.M. Sivori & J.R. Jatimliansky. 1970. "Effects of xylopine on physiological activities of *Scenedesmus obliquus.*" *Plant & Cell Physiol.* 11:921–926. Reprinted with permission of the Japanese Society of Plant Physiologists.

TABLE 9.11. Effect of Bark Extract on Lichen Ascospore Germination

	Percent Spore Germination	
Lichen Species	*Agar + Bark Extract*	*Agar*
Pertusaria pertusa	15	9
Thelotrema lepadinum	100	100
Lecanora subfusca	92	70

F.B. Pyatt. 1973. "Lichen propagules." In *The Lichens* (V. Ahmadjian & M.E. Hale, Jr., eds.), pp. 117–145. New York: Academic Press. Reprinted with permission of Academic Press and the author.

TABLE 9.12. Germination of *Xanthoria polycarpa* Ascospores Incubated for Seven Days at 15° C on Various Media

		Germinated Spores*	
Medium	*pH*	*Trial 1*	*Trial 2*
Bark-extract agar	5	135 a	148 a
Malt-yeast–extract agar	6	179 b	165 b
Lichen-extract agar	6	187 c	178 c
Water agar	6.5	188 c	180 c

*Each value is the mean obtained by counting 200 spores on each of five replicate plates. Means followed by the same letter are not significantly different ($p = 0.05$) according to LSD (for experimental details, see Ostrofsky & Denison, 1980).

A. Ostrofsky & W.C. Denison. 1980. "Ascospore discharge and germination in *Xanthoria polycarpa*." Reprinted by permission from *Mycologia* 72, pp. 1171–1179.

extracts, suggesting that corticolous lichens may require bark compounds for optimal growth and development. However, Ostrofsky and Denison (1980) found that ascospores of *Xanthoria polycarpa* exhibited lower germination responses on bark-extract agar than on other media, including water agar (Table 9.12), suggesting that bark compounds may play a role in inhibiting lichen ascospore germination. This area clearly requires further investigation.

ROLE OF LICHEN COMPOUNDS IN HERBIVORE DEFENSE

Interest in lichen-vertebrate and lichen-invertebrate associations has been growing in recent years, and some excellent reviews have appeared (Gerson, 1973; Gerson & Seaward, 1977 on lichen-invertebrate associations; Richardson & Young, 1977 on lichen-vertebrate associations). It is apparent from the information presently available that a number of vertebrate and invertebrate herbivores consume lichens. Compared to the level of herbivore damage observed for vascular plants, however, damage to lichens from grazing is generally very low. This observation has led some lichenologists to suggest that lichen secondary compounds are somehow involved in herbivore defense. In this section, some of the historical and recent studies of lichen compounds as defense compounds against both vertebrate and invertebrate grazers are reviewed. There is also a brief discussion of the various theories concerning the mechanism of compound toxicity in these interactions.

Invertebrate Herbivores

The idea that lichen compounds protect slow-growing lichen thalli from invertebrate grazers is not a recent one at all. Smith (1921) was the first to review what was, even then, a considerable body of observations and evidence on the subject of lichen chemical defense against invertebrates. She and later Coker (1967) summarized the contrasting views of Zukal (1895) and Zopf (1896) on the role of lichen compounds in protection of thalli from invertebrate grazers. Zukal (1895) argued that lichen compound production protected lichen thalli from herbivores, whereas Zopf (1896) held that lichen compounds afforded lichens very little protection. Zopf fed snails potato slices covered with various pure lichen compounds and found that vulpinic acid, a compound known at that time to be poisonous to vertebrates and invertebrates alike, was the only compound avoided. All other compounds were consumed "with great readiness" (Smith, 1921).

The Zukal-Zopf controversy remained unresolved after the work of Stahl (1904), who discovered that a snail (*Helix hortensis*) avoided thalli of *Parmelia caperata* and *Evernia prunastri* but consumed thalli that had been washed with a weak soda solution to remove lichen acids. Similar results were obtained with woodlice (*Oniscus murarius*) and earwigs (*Forficula auricularia*), suggesting that lichen compounds were responsible for a general herbivore avoidance behavior.

Until recently, these studies were the only experimental attempts to test Zukal's defense compound hypothesis. Mainly for this reason, the Zukal-Zopf controversy continued to cause controversy; lichenologists tended to take sides on the issue with little evidence to support either point of view. In their comprehensive review of lichen-invertebrate associations, Gerson and Seaward concluded "that the whole subject of 'protection from grazers' is still full of contradictions, and no general theory can as yet be formulated" (1977, p. 94). They compiled a list of invertebrate species known to graze lichens and the chemicals produced by the lichen species they were reported to consume. From this listing, it is difficult to discern patterns of ecological significance. Lichen compounds appear to afford lichens, at best, only partial protection from grazers.

Nevertheless, there have been a number of interesting observations published that suggest a defensive role for lichen compounds. Hale (1967) and C.F. Culberson (1969) commented that lichen herbarium specimens are seldom attacked by herbarium pests and seem to require less diligence insofar as protective measures (monitoring, poison strips, fumigation, etc.) are concerned; vascular plant specimens are frequently infested with numerous pests unless special precautions are taken. Both Hale and Culberson suggested that lichen compounds are responsible, at least in part, for this protection.

A number of interesting field observations of lichen-herbivore interactions have been made with barklice (psocids) grazers (Broadhead & Thornton, 1955; Broadhead, 1958; Laundon, 1971). Broadhead (1958) observed interesting food preferences of two lichenophilous psocid species, *Reuterella helvimacula* and *Eplipsocus mclachlani*, that he attributed to lichen compound production. Nymphs and adults of *R. helvimacula* consume both the algal layer and apothecia of *Lecanora conizaeoides*, whereas nymphs and adults of *E. mclachlani* graze only the apothecia. Since both species avoided medullary tissue of this lichen, Broadhead (1958) assumed that a medullary compound afforded protection against the grazers. As Gerson and Seaward (1977) later commented, however, this lichen species produces fumarprotocetraric acid in both the thallus and the apothecia, so the role of compound production in this interaction remains questionable.

Hale (1972) has reported similarly interesting feeding behavior patterns for a collembolan species, *Hypogastrura packardi*, that he believed were the result of lichen compound avoidance. The insect was observed to remove only the cortical and algal layers of *Pseudoparmelia baltimorensis*, leaving the medulla untouched (Figure 9.2). Production of the medullary compound protocetraric acid was assumed to be responsible for this feeding behavior.

Mites that consume lichens, mainly species from the suborder Oribatei, appear to exhibit few preferences that would suggest a protective role of lichen secondary compounds. Indeed, mites are frequently cultivated in the laboratory using lichens as food, either partially or entirely (Travé, 1963; 1969; Woodring & Cook, 1962). Gjelstrup and Søchting (1979) collected seventeen species of oribatids from thalli of the lichen *Ramalina siliquosa*, but found that only one species, *Phauloppia coineaui*, selected *R. siliquosa* as its exclusive food. The reason these generalist herbivores are able to consume such a wide range of lichen species, regardless of lichen compounds present, is perhaps because they are able to select only those portions of the lichen thallus that are not well defended. Lawrey (unpublished observations) has observed the feeding behavior of two common lichenophilous oribatid mites, *Eremaeus* sp. nr. *politus* and *Oribata quadripilis* on numerous species of lichens. The mites appear to graze the upper cortex and the underlying algal cells and leave medullary tissue untouched. Indeed, fecal pellets are often found to contain only algal cells, some still quite viable after egestion. This feeding behavior may be due to a preference for algal cells. However, since the medulla is the site in the thallus where the greatest diversity of secondary compounds is produced, another interpretation is that mites are avoiding contact with tissue containing compounds.

Oribatid avoidance of lichen compounds was tested (Lawrey, unpublished) using filter paper simulations of several lichen species. Filter paper strips were soaked with acetone extracts of several lichen species and placed in a fume hood to remove the acetone. After drying, the strips were treated with a bait consisting of a 5 percent aqueous yeast extract solution, and placed in a 10 gallon terrestrial microcosm containing soil, rocks, leaf litter, and plants collected from oak forests in northern Virginia. After 48 hours, the strips were removed and the oribatids found on each strip were counted; no attempt was made to identify them. More mites were found on control strips (acetone-soaked, dried, and baited) than on strips treated with lichen extracts (Table

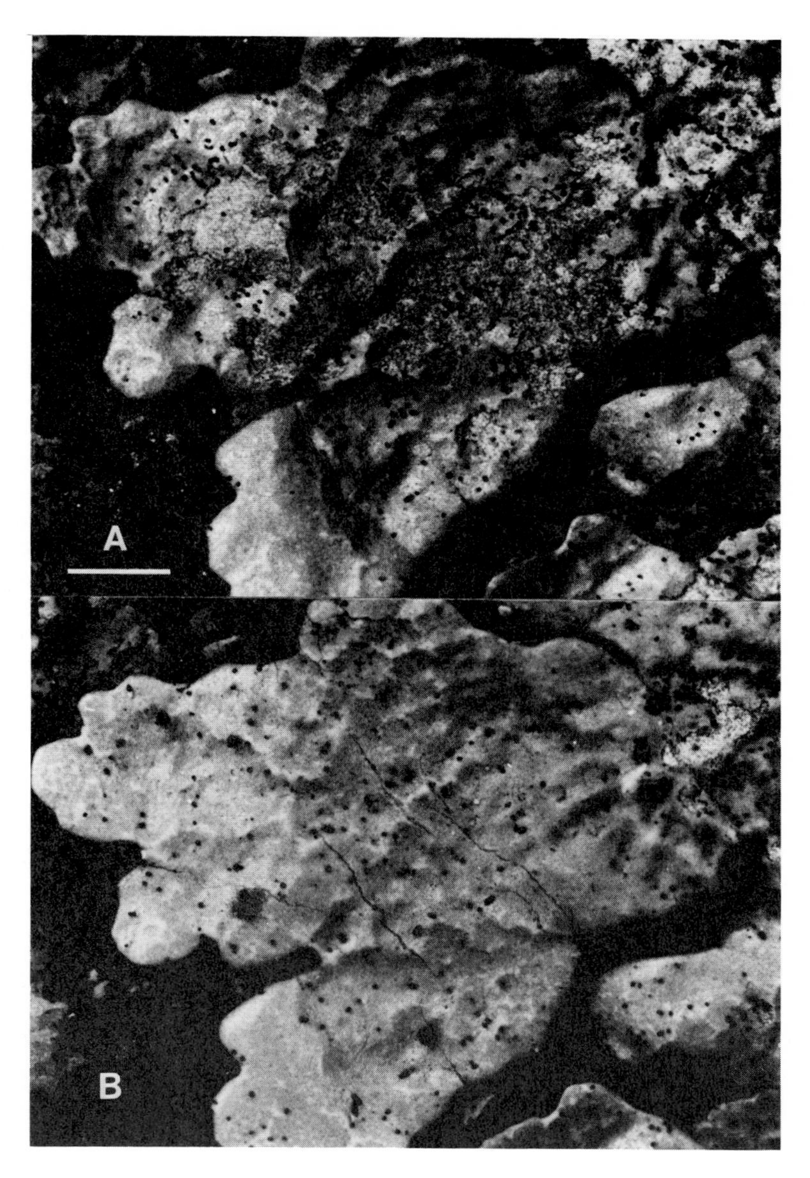

FIGURE 9.2. Single lobe of *Pseudoparmelia baltimorensis* photographed on Plummers Island, Maryland at different times showing different stages of infestation by the collembolan *Hypogastrura packardi,* bar = 1 mm. (a) October 4, 1970; (b) September 13, 1970. From M.E. Hale, Jr. 1972. "Natural history of Plummers Island, Maryland. XXI. Infestation of the lichen *Parmelia baltimorensis* Gyel. & For. by *Hypogastrura packardi* Folsom (Collembola)." *Proc. Biol. Soc. Washington* 85:287–296. Reprinted with permission of the Biological Society of Washington and the author.

TABLE 9.13. Number of Oribatid Mites Observed on Baited Filter Paper Strips Treated with Acetone Extracts of Three Lichen Species

Extracts	*Number of Oribatids per Strip*
Xanthoparmelia cumberlandia	93.41 ± 7.52*
Pseudoparmelia baltimorensis	115.21 ± 16.08
Lasallia papulosa	127.14 ± 18.17
Control	291.66 ± 24.98

*Data are mean numbers ± S.E. of mean. Sample size in all cases is four.

FIGURE 9.3. *Pallifera varia* from Stony Man Mountain, western Virginia, bar = 1 cm.

9.13). However, there appeared to be little difference in numbers of oribatids found on strips treated with different lichen extracts. These results suggested an antimite role for lichen compounds; however, so little evidence for mite avoidance of lichens is available in the literature, it is not possible to make a very strong argument for such a role at the present time.

Terrestrial gastropods are commonly observed feeding on lichens (Peake

TABLE 9.14. Occurrence of *Pallifera varia* on Ground Surface Substrates Located within Three 250 m² Quadrats at Three Sampling Times in Shenandoah National Park

	Sampling Times			
Substrate	*1*	*2*	*3*	*Total % Occurrence*
Leaf litter	7	11	3	16.3
Basidiomycetes				
Russula emetica	2	—	—	1.5
Lepista nuda	2	—	—	1.5
Lichens				
Aspicilia gibbosa	12	25	19	43.4
Aspicilia cinerea	4	14	7	19.4
Lasallia papulosa	2	3	5	7.7
Xanthoparmelia cumberlandia	1	3	1	3.8
Ochrolechia yasudae	1	1	2	3.1
Pseudoparmelia baltimorensis	2	—	1	2.3
Pertusaria sp.	—	1	—	0.7
TOTAL	33	58	38	—

J.D. Lawrey. 1980c. "Correlations between lichen secondary chemistry and grazing activity by *Pallifera varia.*" *Bryologist* 83:328–334. Reprinted with permission of the American Bryological and Lichenological Society.

& James, 1967). Coker (1967) reported feeding preferences of the slug, *Lehmannia marginata*, on the lichens, *Lobaria pulmonaria, Hypogymnia physodes*, and *Pertusaria pertusa*. The lichens were all grazed to some extent, but *P. pertusa* apothecia were preferred to thallus tissue. Also, the slug appeared to avoid thalli of *P. amara* growing in the same habitat. Since these lichens are all chemically distinct, secondary compound production may have been involved in eliciting the observed feeding patterns.

Yom-Tov and Galun (1971) reported on the feeding habits of two desert snail species in the Negev and Judean deserts. The snails fed on numerous plant species, including some lichens; however, they appeared to avoid lichen species that contained parietin, despite the fact that these were the most common lichen species in the habitat. The authors suggested that parietin functions as a protective agent for these lichens.

Lawrey (1980c) observed feeding behavior patterns of the slug, *Pallifera varia*, in western Virginia that strongly suggested a preference for certain lichen species. A specimen of *P. varia* is illustrated in Figure 9.3. On three separate sampling dates, slugs were observed on a number of surface substrates, including lichen thalli, in the study area (Table 9.14). During the first sample, some individuals were observed feeding on basidiocarps;

TABLE 9.15. Percent Coverage and Frequency of Saxicolous Lichen and Moss Species from 50 Randomly Located Rocks Sampled on a Stony Man Mountain Talus Slope

Species	*Percent Coverage*	*Percent Frequency*
Pseudoparmelia baltimorensis	12.98	72
Aspicilia gibbosa	11.46	52
Aspicilia cinerea	5.40	52
Xanthoparmelia cumberlandia	3.34	24
Huilia albocaerulescens	1.56	24
Lasallia papulosa	1.30	22
Parmelia omphalodes	1.22	24
Ochrolechia yasudae	0.84	22
Parmotrema crinitum	0.80	10
Parmelia rudecta	0.74	10
Grimmia apocarpa	0.72	22
Pertusaria sp.	0.34	10
Acarospora fuscata	0.22	4
Lepraria zonata	0.22	12
Usnea herrei	0.20	12
Cladonia furcata	0.20	4
Dicranum fulvum	0.18	10
Buellia disciformis	0.14	6
Cladonia subapodocarpa	0.08	8
Cetraria oakesiana	0.04	4
Hedwigia ciliata	0.04	4
Lecanora campestris	0.04	4
Ramalina intermedia	0.02	2

J.D. Lawrey. 1980. "Correlations between lichen secondary chemistry and grazing activity by *Pallifera varia.*" *Bryologist* 83:328–334. Reprinted with permission of the American Bryological and Lichenological Society.

thereafter, however, no basidiocarps were available. Slugs probably prefer basidiocarps to lichens when they are available, as the name of the family to which *Pallifera* belongs, Philomycidae, implies. However, lichens are readily consumed, especially thalli of *Aspicilia* species. To test the hypothesis that slug feeding involved definite preferences, the frequency and coverage of all saxicolous cryptogam species in the study area were determined (Table 9.15). From these field data, feeding preferences of *P. varia* were determined statistically using the following preference index:

$$P_i = \frac{\text{slug utilization frequency of lichen species i}}{\text{percent cover of lichen species i in community}}$$

Of the lichen species that exhibited relatively high coverage values, *Lasallia papulosa* and *Aspicilia* species were obviously preferred, and *Xanthoparmelia*

TABLE 9.16. Feeding Preference of *Pallifera varia* from Field Observations Made at Stony Man Mountain Study Area

Lichen Species	*Preference**
Lasallia papulosa	5.92
Aspicilia gibbosa	3.78
Ochrolechia yasudae	3.69
Aspicilia cinerea	3.59
Pertusaria sp.	2.06
Xanthoparmelia cumberlandia	1.13
Pseudoparmelia baltimorensis	0.17

*Preference was determined using the following index:

$$\text{Preference} = \frac{\text{Percent occurrence of slugs on resource (Table 9.14)}}{\text{Percent cover of resource in community (Table 9.15)}}$$

J.D. Lawrey. 1983. "Lichen herbivore preference: A test of two hypotheses." *Am. J. Bot.* 70:1188–1194. Reprinted with permission of the American Journal of Botany.

TABLE 9.17. Lichen Compounds Isolated from Fecal Material Produced 24 Hours after Sampling by 37 Individuals of *Pallifera varia*

Lichen Compound	*Number of Slugs Exhibiting Presence of Compound*	*Percent*
No chemicals	11	29.7
Aspicilin	11	29.7
Gyrophoric acid	9	24.3
Atranorin	5	13.5
Norstictic acid	3	8.1
Usnic acid	2	5.4
Unknown yellow pigment	1	2.7

J.D. Lawrey. 1980c. "Correlations between lichen secondary chemistry and grazing activity by *Pallifera varia.*" *Bryologist* 83:328–334. Reprinted with permission of the American Bryological and Lichenological Society.

cumberlandia and *Pseudoparmelia baltimorensis* were obviously avoided (Table 9.16). These results suggested that some species are clearly avoided by slugs and the likelihood of encountering species in the community is not solely responsible for the lichen choices made by slugs.

To test the hypothesis that secondary compounds are important in these food preferences, Lawrey (1980c) collected fecal material produced by slugs from the study area and produced thin layer chromatograms to determine whether or not lichens were consumed. The results (Table 9.17) showed quite definitely that lichen material had been consumed; numerous lichen compounds were identified on chromatograms. Also, the fact that a compound produced by *Aspicilia* species, aspicilin, was the most frequently observed lichen compound on these chromatograms, suggested that *Aspicilia* species were the lichens most frequently consumed by *Pallifera varia* individuals from field collections.

Laboratory food choice experiments with *Pallifera varia* further documented the avoidance of certain lichen species (Table 9.18). When presented with a choice between *Aspicilia* species or *Lasallia papulosa* and any of the other lichen species tested, *Aspicilia* species and *Lasallia* were always preferred. When presented with a choice between *Huilia albocaerulescens* and any of the other species, *Huilia* was always avoided. This was true even when *Huilia* was the only food choice available.

These lichen choice experiments with *Pallifera varia* suggested that secondary chemicals produced by certain lichen species influence the feeding behavior of slugs in nature. However, there are other factors besides secondary compounds that may be responsible for the observed feeding patterns. For example, lichen growth form, color, texture, and nutrient content may all influence slug food choice as much or more than lichen secondary chemistry. If these other factors could be experimentally eliminated, effects of lichen secondary compounds on slug activity could be more definitely determined. To do this, Lawrey (1983) used filter paper simulations of several lichen species from *Pallifera* habitats in western Virginia in slug food choice experiments under controlled laboratory conditions. These simulated lichen thalli were filter paper disks impregnated with acetone extracts of four lichen species, two apparently preferred by *Pallifera* (*Lasallia papulosa* and *Aspicilia gibbosa*) and two apparently avoided (*Xanthoparmelia cumberlandia* and *Pseudoparmelia baltimorensis*). After extracts were added to the disks, the acetone was allowed to evaporate in a fume hood and a 1 percent solution of aqueous yeast extract was added as a bait. Control disks (acetone added, evaporated, and yeast extract added) were also used. Disks were placed randomly in 10 gallon terrestrial microcosms, each containing at least 30 adult slugs. Slugs were observed to graze most heavily on control disks and disks treated with extracts of either *Lasallia* or *Aspicilia*, the preferred lichens; disks treated with extracts of *Pseudoparmelia* or *Xanthoparmelia*, the avoided lichens, were left almost totally untouched by slugs (Table 9.19). These results suggested that slug preference of certain lichen species and avoidance of others is due to lichen chemistry, and that the compounds responsible for this behavior are acetone soluble. Although pure lichen compounds were not used in this study, it is likely they are responsible, at least in part, for *Pallifera* avoidance behavior observed in the field and laboratory.

Pure lichen compounds have been shown in some recent experiments to deter feeding by assorted insect crop pests. Slansky (1979) has reported some results of feeding and growth experiments with larvae of the yellow-striped armyworm, *Spodoptera ornithogalli*, a polyphagous insect pest. He coated

TABLE 9.18. Results of Lichen Preference Experiments with *Pallifera varia*

Lichen Choices in Each Experiment	*Compounds Isolated from Lichens*	*Compounds Isolated from Slug Crops after One Week*
1. *Huilia albocaerulescens*	Stictic	Atranorin, aspicilin
Aspicilia gibbosa	Atranorin, aspicilin	—
2. *A. gibbosa*	Atranorin, aspicilin	Atranorin, aspicilin, norstictic
A. cinerea	Atranorin, aspicilin, norstictic	—
3. *A. gibbosa*	Atranorin, aspicilin	Atranorin, aspicilin
Pseudoparmelia baltimorensis	Atranorin, usnic, gyrophoric, caperatic, protocetraric	—
4. *P. baltimorensis*	Atranorin, usnic, gyrophoric, caperatic, protocetraric	Atranorin, usnic, gyrophoric, protocetraric
Xanthoparmelia cumberlandia	Norstictic, stictic, usnic, constictic, connorstictic	—
5. *Huilia albocaerulescens*	Stictic	—
6. *P. baltimorensis*	Atranorin, usnic, gyrophoric, caperatic, protocetraric	Gyrophoric, unknown
Lasallia papulosa	Gyrophoric, unknown	—
7. *L. papulosa*	Gyrophoric, unknown	Atranorin, aspicilin, gyrophoric, unknown
A. gibbosa	Atranorin, aspicilin	—

J.D. Lawrey. 1980c. "Correlations between lichen secondary chemistry and grazing activity by *Pallifera varia.*" *Bryologist* 83:328–334. Reprinted by permission of the American Bryological and Lichenological Society.

TABLE 9.19. Percent Area of Paper Disks Grazed by *Pallifera varia* after 48 Hours. Disks Were Treated either with Acetone (Control) or a Lichen Acetone Extract (Treatment) and, after the Acetone Was Allowed to Evaporate, the Disks Were All Baited with a 1 Percent Aqueous Yeast Extract Solution and Placed in *Pallifera* Culture Chambers

	Lichen Extracts			
	Lasallia papulosa	Aspicilia gibbosa	Xanthoparmelia cumberlandia	Pseudoparmelia baltimorensis
Control	18.15 ± 1.78*	30.31 ± 4.42	31.28 ± 2.08	17.61 ± 4.15
Treatment	14.27 ± 4.38	15.47 ± 2.74	1.69 ± 0.39	1.22 ± 0.78

*Mean percent area grazed ± S.E. of the mean. Sample size in all cases was four.

J.D. Lawrey. 1983. "Lichen herbivore preference: A test of two hypotheses." *Am. J. Bot.* 70:1188–1194. Reprinted with permission of the American Journal of Botany.

TABLE 9.20. Consumption of Broccoli Leaves Treated with Lichen Compounds by Larvae of *Spodoptera ornithogalli*

	Area Consumed (mm^2)		
Treatment	*Small Larvae (Probably 2nd Instar)*	*Intermediate Larvae (Probably 3rd Instar)*	*Large Larvae (Probably Penultimate or Ultimate Instar)*
Control (isopropanol)	41.3 ± 10.7*	275.7 ± 107.5	298.3 ± 19.2
Atranorin	41.3 ± 21.2	223.7 ± 91.1	270.3 ± 63.4
Vulpinic acid	13.3 ± 9.2	52.0 ± 27.4	112.7 ± 70.0
Atranorin and vulpinic acid	31.3 ± 15.2	71.0 ± 38.9	0

*Mean mm^2 leaf area consumed ± S.E. of the mean.

F. Slansky, Jr. 1979. "Effect of the lichen chemicals atranorin and vulpinic acid upon feeding and growth of larvae of the yellow-striped armyworm, *Spodoptera ornithogalli.*" *Environ. Entomol.* 8:865–868. Reprinted with permission of the Entomological Society of America and the author.

broccoli leaves with isopropanol solutions of atranorin, vulpinic acid, and a combination of the two, allowed the isopropanol to evaporate and compared larval feeding on these treated leaves with the level of feeding on controls to which only isopropanol was added and allowed to evaporate. The results (Table 9.20) demonstrated that vulpinic acid (at a concentration of 0.6 percent of leaf dry weight) significantly inhibited feeding by larvae of all sizes. Atranorin (at a concentration of 0.03 percent of leaf dry weight) had no significant effect. The inhibitory effect of vulpinic acid on larval feeding did not result in significant reductions in larval growth, however, as larvae forced to feed on vulpinic acid-coated leaves gained weight at the same rate as controls (Figure 9.4). Atranorin significantly reduced growth early in the experiment, but atranorin-fed larvae soon caught up with controls. Only larvae fed with leaves treated with both atranorin and vulpinic acid remained stunted at the end of the experiment. Mortality rates of larvae fed with lichen compounds were the same as for controls.

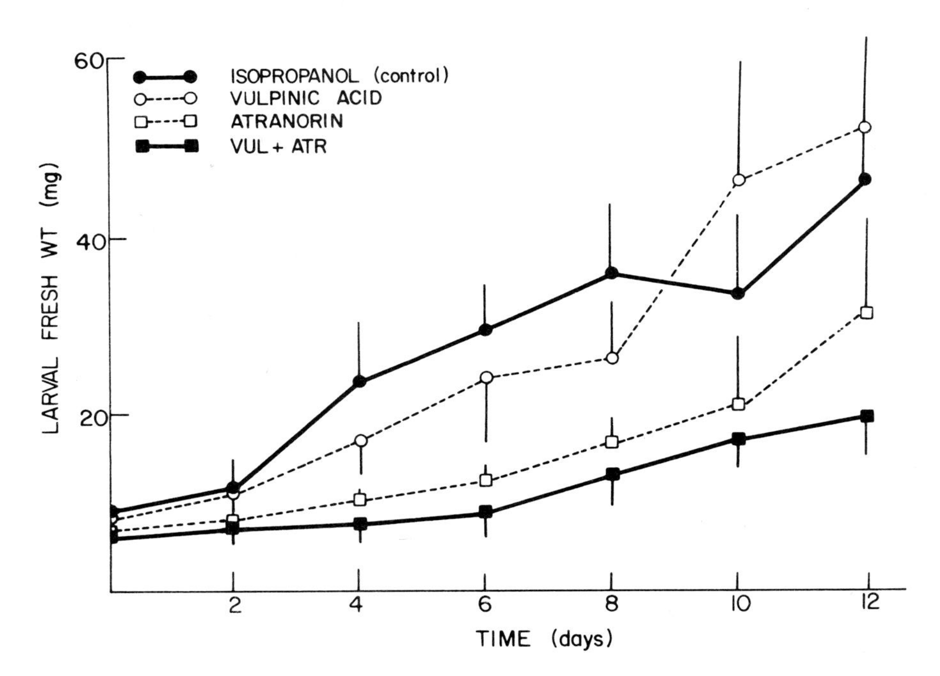

FIGURE 9.4. Mean fresh weight (± S.E.) of *Spodoptera ornithogalli* larvae fed broccoli leaves coated with lichen chemicals. From F. Slansky, Jr. 1979. "Effect of the lichen chemicals atranorin and vulpinic acid upon feeding and growth of larvae of the yellow-striped armyworm, *Spodoptera ornithogalli.*" *Environ. Entomol.* 8:865–868. Redrawn with permission of the Entomological Society of America and the author.

TABLE 9.21. Results of a Single Feeding Preference Experiment Involving Mid-Instar Larvae of *Peridroma saucia* Feeding on Cauliflower Leaves Treated with Lichen Compounds

Dish	*Control*	*Atranorin*	*Vulpinic*	*Atranorin & Vulpinic*
1	450*	411	235	9
2	515	262	4	9
3	377	578	81	342
4	255	437	91	168
Mean	399	422	103	132
Standard error of mean	56	65	59	79

*Data are mm^2 of leaf disks consumed by larvae.
Unpublished results provided by Frank Slansky, Jr.

Slansky concluded from these experiments and others involving herbivores cited in his 1979 paper (Table 9.21) that vulpinic acid can function as an antiherbivore compound, especially as a repellent, suppressant, or feeding deterrent.

Other investigators (Stephenson & Rundel, 1979) have also found that vulpinic acid can function as a feeding deterrent. Atranorin, on the other hand, appeared to function as an antibiotic, reducing growth in very young larvae. Since these results were obtained using concentrations of lichen compounds much lower than those found in many lichens in nature, Slansky suggested that lichen compounds may be quite potent antiherbivore defense compounds against generalist grazers, the kinds of herbivores most likely to consume lichens in nature.

Vertebrate Herbivores

Compared to the diversity of invertebrates that are known to feed on lichens, vertebrate grazers are very few in number. Richardson and Young (1977) reviewed the current literature on lichen-vertebrate associations and commented on the general paucity of information on these interactions, with the notable exception of lichen-reindeer/caribou interactions. They expressed the hope that further research in this area will reveal other relationships of ecological importance.

The fact that lichens are consumed in large quantities by caribou and reindeer has been recognized for many years. Most studies have focused either on the nutritional value of lichens as reindeer/caribou forage or on the bioaccumulation of radionuclides through the lichen-reindeer/caribou-man arctic and subarctic ecosystems (see Richardson, 1975 and Chapter 12 for discussions of these topics). Little interest has been shown recently in the role lichen compounds may play in regulating feeding by large vertebrate grazers. Considering the increasing interest in lichens as forage for introduced reindeer and caribou herds (Richardson & Young, 1977), however, this problem should be considered more closely.

Many studies of reindeer/caribou consumption of lichens lump all lichen species together without regard to chemistry. This is quite all right if Bergerud and Nolan (1970) and Bergerud (1972) are correct in their suggestion that caribou are opportunistic generalists that have few specific food preferences and feed on whatever is available. However, some studies (Ahti, 1959; Des Meules & Heyland, 1969; Holleman & Luick, 1977) indicate that animals are very selective with regard to the types of plants, including lichens, they consume. Soczava (1933, cited in Rundel, 1978) suggested that the low palatability of some lichens, particularly some *Parmelia* species and *Alectoria ochroleuca*, to caribou was the result of bitter chemical compounds produced by the lichens. des Abbayes (1939) reported that reindeer will not eat *Cladina arbuscula* or *C. rangiferina* (misnamed "reindeer moss"), both of which contain fumarprotocetraric acid, but will eat *C. alpestris* (= *C. stellaris*), which lacks this acid. Tobler's (1925) observation that penned reindeer refused to eat *C. arbuscula* but accepted *C. rangiferina*, which contains lower concentrations of fumarprotocetraric acid, contradicts des Abbayes' observations but supports the general hypothesis that fumarprotocetraric acid concentration levels in lichens influence caribou feeding behavior. Ahti (1959), although he recognized the fact that caribou exhibit differential feeding on lichens, doubted whether lichen acids can be responsible for this behavior.

The fact that lichens are generally poor quality forage for reindeer/caribou, especially with regard to protein (Rundel, 1978; Richardson, 1975) and fatty acids (Garton & Duncan, 1971), suggests that selective feeding on lichens by large vertebrate herbivores is controlled less by secondary chemistry than by other factors.

Antiherbivore Mechanisms

Whittaker and Feeny (1971) have classified interorganismic chemical effects based on whether the effect of the compound is beneficial or detrimental to the releasing or the receiving organism. Since, for numerous reasons cited in this chapter, lichen compounds appear to be able to function as antiherbivore deterrents, they are allomones according to Whittaker and Feeny's classification. Allomones generally give an adaptive advantage to the producing organism. They can do this one of several ways: as repellents, escape substances, suppressants, venoms, inductants, counteractants, and/or attractants.

Slansky (1979) has provided evidence that lichen compounds can function both as feeding deterrents (vulpinic acid) and as antibiotics (atranorin).

This suggests that the mechanism by which these compounds act in eliciting various feeding behavior responses in herbivores is multifaceted and, as yet, too complex to discuss with any degree of certainty. This does, however, leave us free to speculate on possible mechanisms of toxicity without fear of being contradicted by an overabundance of evidence.

The well-known antibiotic nature of lichen compounds (Vartia, 1973) suggests they may inhibit lichen herbivores through antibiosis of herbivore gut microflora, although no studies have been done to test this hypothesis. Gastrointestinal microorganisms have been reported for a number of animals (Alexander, 1971), and it is likely that microorganisms participate to varying degrees in normal digestive processes of lichen herbivores. Experiments need to be designed to test the hypothesis that lichen compounds are capable of such interference, possibly by observing herbivore gut walls directly using scanning electron microscopy, or by removing herbivore gut contents and assaying for microorganism activity.

Smith (1921) has commented that lichen compounds are rarely poisonous. However, she claimed that vulpinic acid produced by *Letharia vulpina* and pinastric acid produced by *Cetraria pinastri* are exceptions to this general their commercial use in the tanning industry was not very important. This observation, however, suggests that lichen compounds may have an antiherbivore function similar to that of many condensed and hydrolyzable tannins, namely, to tan the membranes of cells lining the gut, inactivate digestive enzymes, and render potential food proteins unavailable to the herbivore through tannin-protein complexing (Feeny, 1968; 1969; 1970; Levin, 1971; 1976; Van Sumere et al., 1975; Swain, 1977; 1979). This property, so common for other phenolic compounds, has never before been tested for lichen phenolics. A closely related subject for future research would appear to be lichen herbivore gut pH, since tannin-protein complexes are known to dissociate at high pH values (Boudet & Gadal, 1965; Goldstein & Swain, 1965; Williams, 1963) and positive correlations have been found between gut pH values and the tannin content of host plants consumed (Berenbaum, 1980).

In his review of lichen-arthropod associations, Gerson (1973) commented that lichen resistance to arthropod feeders may be due, in part, to the chelating ability of lichen compounds. Lichen compounds are capable of chelating metals (Syers, 1969; Iskandar & Syers, 1971; 1972; Williams & Rudolph, 1974; Ascaso & Galvan, 1976b). Recognition of this property of lichens has led to the study of their importance in biochemical weathering and pedogenetic processes. Chelating agents are also known to inhibit growth of herbivores by complexing essential elements (Sell & Schmidt, 1968). Therefore, lichen compounds may be capable of the same antiherbivore role.

Smith (1921) has commented that lichen compounds are rarely poisonous. However, she claimed that vulpinic acid produced by *Letharia vulpina* and pinastric acid produced by *Cetraria pinastri* are exceptions to this general rule. She mentioned that vulpinic acid has been used to poison wolves. Animal carcasses, stuffed with a mixture of ground glass and vulpinic acid, are offered to wolves. Whether the ground glass by itself is capable of killing a wolf unfortunate enough to eat the adulterated carcass has never been made clear. It is clear, however, that some lichen compounds are quite toxic. Toxicities of some compounds have been determined using laboratory animals; the LD_{50} for vulpinic acid has been found to be 150 milligrams/kilogram body weight for mice (Rundel, 1978). Other toxicities can be found in Vartia's review.

LIGHT SCREENING AGENTS

Despite the fact that most lichens are photophilous and exhibit numerous similarities to sun leaves of vascular plants (Barkman, 1958), lichens also have adaptations thought to provide protection against strong light. Barkman (1958) discussed some of the early experimental work done to demonstrate this. Butin (1954, cited in Barkman, 1958) found that individuals of the same lichen species growing in the sun had thicker upper cortical layers than those growing in shade, and that there was a corresponding increase in the light compensation point for these sun lichens. In a study of several species in the family Teloschistaceae, Schulz (1931, cited in Barkman, 1958) found that the pruinosity and parietin content of the upper cortex frequently increased in strong light, suggesting that lichen compounds may serve a protective role against high light intensities. Ertl (1951) also suggested that cortical lichen pigments reduced light intensities that may severely harm light-sensitive lichen phycobionts.

Rundel (1978) has discussed the potential role of lichen compounds as light screening agents. He briefly mentioned his own earlier work with *Cladonia subtenuis* in North Carolina (Rundel, 1969), in which he discovered an interesting correlation between usnic acid concentration in the cortex of individuals of *C. subtenuis* and the light intensity of the habitat. According to Figure 9.5, usnic acid concentrations varied from 0.13 percent in individuals from low light habitats (3 percent full sunlight) to 2.82 percent in open habitats (51 percent full sunlight), strongly suggesting that usnic acid production may be a response, under environmental control, to potentially damaging light intensities.

Other studies, especially those done with lichens containing the anthraquinone parietin (Scott, 1964; Hill & Woolhouse, 1966; Richardson, 1967) have demonstrated changes in lichen pigment concentration along light intensity gradients that indicate a phenotypic control over lichen pigment production. Hill and Woolhouse (1966) collected thalli of *Xanthoria parietina*, a common parietin-producing lichen in Britain, from habitats that differed in light intensity. Material was collected from tree trunks, roofs of farm build-

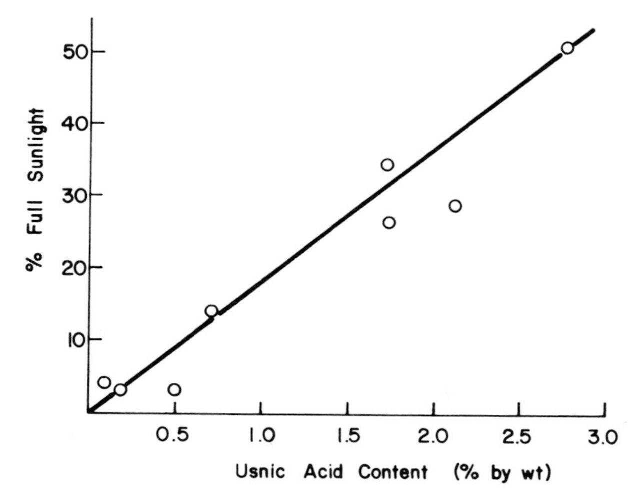

FIGURE 9.5. Percentage of usnic acid by weight in the terminal 5 mm of the podetia of *Cladina subtenuis* from habitats with various light intensities. From P.W. Rundel. 1969. "Clinal variation in the production of usnic acid in *Cladonia subtenuis.*" *Bryologist* 72:40–44. Redrawn with permission of the American Bryological and Lichenological Society.

ings, and seashore rocks immediately above the high water mark; measurements of gross anatomy, particularly thickness of the cortex and medulla, parietin concentration, and chlorophyll concentration were taken. Thalli from the three sites were significantly different in thallus thickness and parietin and chlorophyll concentration (Table 9.22). Thalli collected from open habitats (roofs and rocks) consistently produced thicker upper cortices and medullae and had higher concentrations of parietin; thalli from the more shaded tree habitats had thicker algal layers and produced more chlorophyll, on both a dry weight and a surface area basis. These results were interpreted as evidence for the environmental control over variations in anatomy and chemical composition of *Xanthoria* from different habitats.

Richardson (1967) collected further evidence for the environmental control over morphology and chemistry in *Xanthoria* species by transplanting thalli from a number of habitats to habitats that differed significantly in environmental conditions, including light intensity. He found that transplanted thalli survived very well and frequently showed changes in morphology and parietin content that appeared to be responses to a changed microenvironment. This technique has rarely been applied in studies of lichen compound production (see Fahselt, 1979; 1981 for recent discussions) but is of obvious value in documenting the extent to which lichen compound production correlates with physical environmental conditions. Although such corre-

TABLE 9.22. Anatomical and Chemical Measurements of Thalli of *Xanthoria parietina* Collected from Different Habitats

	Trees	*Roofs*	*Rocks*
Thallus thickness (μm)			
Upper cortex	17.6 ± 1.3	18.1 ± 0.9	25.4 ± 1.2
Algal layer	46.4 ± 6.0	36.9 ± 3.1	34.2 ± 3.2
Medulla	52.1 ± 3.2	72.3 ± 6.4	111.9 ± 7.1
Lower cortex	21.5 ± 1.7	20.7 ± 0.9	27.3 ± 1.3
Total	137.6	147.8	198.6
Dry weight (mg/cm^2)	7.24 ± 0.68	8.22 ± 0.33	10.99 ± 0.45
Chlorophyll			
mg/g dry wt.	4.1 ± 0.2	3.5 ± 0.4	1.7 ± 0.2
μm/cm^2	31.5 ± 2.2	28.0 ± 1.0	18.1 ± 1.4
Parietin			
mg/g dry wt.	11.2 ± 3.3	47.9 ± 6.9	37.5 ± 2.9
μm/cm^2	76.6 ± 21.5	391.5 ± 59.6	408.1 ± 24.6

Partial data adapted from D.J. Hill & H.W. Woolhouse. 1966. "Aspects of the autecology of *Xanthoria parietina* agg." *Lichenologist* 3:207–214. Reprinted with permission of Academic Press.

lations do not prove a cause-and-effect relation between the two variables, they show the capacity of lichens to respond to changing environmental conditions and indicate the multifaceted nature of the response, a response that appears to include but is not restricted to production of high concentration of pigments.

Rao and LeBlanc (1965) proposed a biological role for the cortical compound atranorin that is quite different from those discussed thus far. Recognizing that cortical pigments and thick cortical tissues are able to reduce the light intensities considerably before they reach the algal layer, they investigated the possibility that some lichens may have a mechanism of increasing light available for photosynthesis. They found that the fluorescence spectrum of atranorin corresponds to the blue peak of the absorption spectrum for chlorophyll. This suggested to them that atranorin could function as an accessory photosynthetic pigment by absorbing shorter wavelengths of light unavailable for photosynthesis and reemitting light at longer wavelengths able to be absorbed by chlorophyll. They contrasted this light absorbing role of atranorin with the light quenching role of other cortical compounds, notably usnic acid.

Recent papers have questioned the accessory pigment hypothesis of Rao and LeBlanc (Klee & Steubing, 1977; Rundel, 1978). Rundel (1978) noted that atranorin is not restricted to species from shaded habitats, nor are there any correlations between atranorin concentration and light intensity. Indeed, it appears that atranorin is produced at fairly constant rates throughout thallus development (Stephenson & Rundel, 1979). Finally, Rundel argued that the frequent joint occurrence of atranorin and usnic acid in lichen species calls into question the opposite roles proposed for the two compounds. It is difficult to conceive of a long-term selective environment that favors both usnic acid for light screening and atranorin for light capture; as Rundel said, "the maintenance of cortical compounds of opposing purposes does not seem energetically realistic in an evolutionary sense, although extremely low concentrations of atranorin in usnic-containing species may be vestigial" (1978, p. 161).

SOME TESTABLE HYPOTHESES REGARDING LICHEN CHEMICAL PRODUCTION

Completely integrating lichen secondary compound production into an evolutionarily meaningful framework will require a good deal more experimental work, both at the whole organism level and at the cellular and molecular levels. As this information slowly accumulates, however, it is perhaps useful to consider in a general way the evolutionary consequences of lichen compound production; that is, the patterns of lichen chemical investment that would be expected in certain ecological situations and the energetic trade-offs that would be necessary to accommodate such patterns. Attempts to formulate concepts that relate plant secondary chemical investment to other aspects of a plant's life history characteristics have proven valuable in the development of a growing body of theory for vascular plants and animals. If there are "rules" that determine the ways organisms allocate energy, and if lichens are more than mere "intricate scum" (Janzen, 1975) governed by the same rules, there is a need to consider what the rules are and how they might apply to lichens.

In this section, some tentative predictions are developed. Some evidence to support a few of the predictions is also considered. However, despite their intellectual attractiveness, many of the predictions are unsupported by any data and should therefore be treated with a certain degree of reader skepticism until they can be verified by experimental testing. It is important that they be considered now, however, because the progression of any scientific endeavor from a purely descriptive phase to a more experimental stage of development requires the formulation of a theoretical framework within which experiments make sense. A consideration of the material in this chapter suggests that lichen chemical ecology has reached this stage of development.

In this area of lichen chemical ecology, as in vascular plant chemical ecology, a conceptual framework is just beginning to emerge, and the result has been an increased interest in conducting experiments and testing the various theories. Indeed, this should continue to be the case as far as lichen investigations are concerned, because in many ways lichens are biologically unique organisms that exhibit numerous characteristics quite different from vascular plants. Development of a body of theory that accounts for the origin and maintenance of various patterns of secondary compounds in lichen groups may therefore begin with a consideration of theories that relate to vascular plant secondary compound production. Ultimately, however, it will be necessary to consider how the unique aspects of lichen associations influence the way lichens solve ecological and physiological problems over evolutionary time.

Prediction 1: Lichen Compounds Are Energetically Expensive; The Cost of Production Will Be Proportional to Their Adaptive Value

Basic to all the current theoretical discussions concerning various patterns of plant chemical investment is the idea that each individual plant has a finite and limited amount of energy available to use in solving all of its ecological and physiological problems. Allocations of energy to secondary compounds represent only one avenue of energetic investment; plants must also invest energy in growth, reproductive structure, competitive structures, and maintenance. There are theoretically limitless patterns of energy allocation; however, it should be possible to predict, at least in a general way, the pattern a given individual will exhibit if one knows something about that individual's selective environment. Selection should favor production of secondary compounds only in environments where secondary compound production is adaptive. Otherwise, energy that could have been used for something else has instead been squandered on compounds, and is essentially wasted. As Rhoades and Cates have stated,

> ...in the total milieu of grazing and other pressures upon the plant, it is postulated that secondary substances must have a net positive effect on plant fitness, since if they did not, metabolic costs associated with their production and sequestration should cause the elimination of genotypes producing these substances from the plant population. (1976, p. 170)

Over evolutionary time, then, selection will tend to remove individuals who waste energy on useless secondary compounds.

If this idea is correct, then lichens will be especially conservative in their investment strategies. Lichens are organisms with notoriously low growth rates and productivities. With so little total photosynthetic production available to invest, it is unlikely that lichens will waste any on compounds with little or no adaptive value.

It has been suggested that lichen compounds are metabolic waste products and have no real biological role. The fact that some lichen groups invest heavily in a wide diversity of compounds, and others exhibit zero chemical investment or very low investments, however, suggests that compound production is not universal but optional. Lichen compounds are also relatively complex molecules that are expensive to produce, especially in the large quantities frequently measured for various lichen groups (Hale, 1974; C.F. Culberson, 1969). It is difficult to imagine how such expenditures of energy could have no adaptive value beyond the mere disposal of metabolic waste products.

It is true that the adaptive nature of lichen compounds is far from completely understood. This chapter illustrates the relative paucity of reliable evidence for any of the proposed biological roles of lichen compounds. However, it is becoming increasingly apparent as more experimental work is done that compound production does influence lichen success in nature and may therefore increase lichen fitness in certain selective environments.

Prediction 2: Advanced Groups of Lichen Compounds Will Contribute More to an Individual's Total Fitness and Production Will Be More Expensive Than Less Advanced Groups; The Increased Expense Will Be Proportional to the Increased Adaptive Advantage

This idea is intuitively appealing but, unfortunately, is very difficult to test in any group of plants, including lichens. Hale (1966) and W.L. Culberson and C.F. Culberson (1970) have discussed the idea of advanced versus primitive lichen compounds. Hale (1966) has suggested that *O*-methylated compounds tend to be associated with morphologically advanced lichens. *O*-methylation of hydroxylated lichen compounds represents a metabolic substitution that probably requires the mediation of a single enzyme, *O*-methyl transferase. Vascular plants that have the ability to carry out this substitution appear to be morphologically more advanced than those that do not (Swain, 1963). Hale found that this pattern was also generally true in the lichen genus *Parmelia (sens. lat).* Morphologically advanced groups tend to produce *O*-methylated compounds; morphologically less advanced groups tend to produce hydroxylated compounds. The patterns are not absolutely clear-cut, but the tendencies are generally apparent.

W.L. Culberson and C.F. Culberson (1970) discussed a number of important phylogenetic trends in the evolution of the Lecanorales, the lichen order that exhibits the greatest chemical diversity. As in Hale's (1966) study, these trends are based on the assumption that the more advanced compounds are those that require more steps in their synthesis. *O*-methylation represents one such biosynthetic process that can be used to compare the phylogenetic status of groups from the standpoint of chemical evolution. Other chemical processes that can be used, in addition to *O*-methylation, include:

1. Depsidone formation via oxidative cyclization of *para*-depsides.
2. Substitution at the 3- and 5-carbon positions rather than the 1-carbon position.
3. Instead of short side chains, long side chains are constructed from condensation of additional malonate units.
4. Chlorination of unchlorinated compounds.

As both Hale (1966) and W.L. Culberson and C.F. Culberson (1970) have accurately pointed out, it is inappropriate to begin ranking lichen species solely on chemical criteria. There are too many gaps in our knowledge and the phylogenetic pathways responsible for the apparent patterns in chemical diversity we are able to measure are far too complex to be understood at present. However, the underlying assumption in lichen chemosystematics, that a lichen compound's biosynthetic pattern relates in a meaningful way to the phylogenetic status of a lichen or lichen group capable of producing that compound, is a logically compelling assumption and will undoubtedly stimulate further valuable research.

If it is true that advanced lichen compounds are those that require many steps for their synthesis, these compounds should also require a greater energetic investment than similar compounds requiring fewer biosynthetic steps. Furthermore, this increased cost should require energetic trade-offs that will be reflected in other characteristics of the lichen's life history.

Theoretically, in order to be cost-effective, a lichen that produces an expensive, advanced compound rather than a less expensive, precursor compound will derive an advantage of some sort as a result. Otherwise the increased expense to produce the advanced compound would be strongly selected against. More advanced lichen compounds should therefore be more adaptive than less advanced compounds, if all other factors remain the same.

Practically no evidence is available to test these ideas. For example, there is no evidence that advanced lichen compounds are really more costly to produce than less advanced compounds; it is assumed that the addition of a step to a biosynthetic process is costly, but no measurements have been made. Furthermore, no direct evidence is available to suggest a greater adaptive advantage for so-called advanced compounds. Lawrey (1977b), on the basis of very little evidence, suggested that *O*-methylated compounds are more effective antibiotics and better able to inhibit moss spore germination than hydroxylated compounds. However, as Gardner and Mueller (1981) have pointed out, this is apparently an oversimplification. Their data suggest that lichen compound toxicity is not associated with a single chemical characteristic but is instead the result of complex interactions between a number of different factors.

This lack of data notwithstanding, the search for evolutionarily meaningful patterns of lichen secondary compound production will have to be based in the future on a better understanding of the biological role of these compounds and the energetic costs associated with their production.

Prediction 3: Protection from Lichen Herbivores Will Be Most Important in Juvenile Thalli and in Reproductive Tissue

If lichen compounds function as allelopathic or antiherbivore agents in nature, and this chapter provides some evidence that they do, then one would expect that young, metabolically active lichen tissues would contain the highest concentrations of compounds. These tissues are capable of the highest photosynthetic rates and are responsible for a large portion of the thallus growth. Damage to these tissues would therefore be more costly to overcome than damage to older senescent tissues.

This argument would also apply to juvenile thalli and reproductive tissues. Investment in secondary compounds that have a defensive or competitive function would have the greatest pay-off in individuals most susceptible to removal by herbivores or exclusion by competitors. Individuals most likely to be defended, therefore, are young reproductive units, since herbivore damage to or competitive interactions with juveniles, either ramets (vegetatively produced individuals) or genets (sexually produced individuals), results in elimination of entire individuals from the lichen population.

Although these ideas have not been tested directly, there is a study available that strongly suggests that lichen chemical production is frequently an age-specific process. Stephenson and Rundel (1979) were able to detect different gradients of atranorin and vulpinic acid production in *Letharia* species that correlated with tissue age. They found that vulpinic acid was highest in concentration in youngest tissue, whereas the concentration of atranorin increased steadily with tissue age (Table 9.23). Other studies have shown that lichen compounds are frequently more concentrated in thallus tips than in older tissues (Nourish & Oliver, 1976; Fedoseev & Yakimov, 1960, cited in Stephenson & Rundel, 1979); no study prior to that of Stephenson and Rundel had ever demonstrated a positive correlation between compound concentration and tissue age.

These patterns of compound concentration in *Letharia* species suggest different biological roles for atranorin and vulpinic acid. Vulpinic acid is probably only produced in young tissues; production apparently stops as tissues get older, resulting in a reduced concentration of vulpinic acid as the lichen biomass of older regions increases and compounds are removed by leaching. This would support the generally held supposition that vulpinic acid is an antiherbivore compound. As Stephenson and Rundel say in their paper, "branch tips (of *Letharia*) are probably the most metabolically active portion of the lichen and are susceptible to predation because of their exposure to herbivores at the outer surface of lichen mats" (1979, p. 265).

The biological role of the cortical compound atranorin has not been determined, although there has been some speculation that it can act as an accessory photopigment by absorbing energy at wavelengths not usable in photosynthesis and reemitting energy at photosynthetically active wavelengths (Rao & LeBlanc, 1965). This would not explain the gradients of atranorin concentration in *Letharia*, however. Whatever its biological role, if there is a role, atranorin production in *Letharia* is clearly different from that of vulpinic acid. It appears to be produced at the same rate in tissue of different ages, resulting in a steady increase in concentration as tissue ages.

Stephenson and Rundel's (1979) study demonstrated that lichen com-

TABLE 9.23. Variation of Atranorin and Vulpinic Acid Concentrations with Tissue Age in a *Letharia columbiana* Thallus

Tissue Age	*Atranorin*	*Vulpinic Acid*
Youngest (thallus tips)	0.006*	3.3
Intermediate	0.16	2.1
Oldest (basal)	0.59	1.0

*Concentrations expressed as percent dry weight of lichen tissue.

Partial data adapted from N.L. Stephenson and P.W. Rundel. 1979. Reprinted with permission from *Biochem. Syst. Ecol.*, 7, N.L. Stephenson & P.W. Rundel, "Quantitative variation and the ecological role of vulpinic acid and atranorin in the thallus of *Letharia vulpina* (Lichenes)." Copyright © 1979, Pergamon Press, Ltd.

pounds are not produced at the same rates in lichen tissues of different ages. It also suggested that younger tissues may contain higher concentrations of defense compounds than older tissues. Further studies are needed to determine if similar patterns of compound production can be found in:

1. thalli of different ages,
2. reproductive versus nonreproductive tissues,
3. vegetative thalli versus thalli bearing reproductive structures.

Prediction 4: Total Lichen Investment in Secondary Compounds Will Reflect the Cost of Replacing Resources Lost to Herbivores; The Cost of Replacing Resources Will Be a Complex Function of a Lichen's Growth Rate, Reproductive Effort, Dispersal Ability, and Nutrient Element Content

Several theoretical studies have appeared recently that attempt to relate a plant's investment in defense compounds to other energetic costs incurred by the plant in its particular physical and biotic environment. Janzen (1974) predicted that a plant's investment in defense compounds will reflect the cost of replacing resources lost to herbivores. The greater the cost to replace lost tissues, the more likely a plant will invest in defense compounds. A lichen's ability to replace resources lost to herbivores should depend on its growth rate and dispersal ability. A greater growth rate should reflect a lower cost to replace lost tissues, and a higher dispersal ability should allow a palatable species to escape severe herbivore damage by dispersing over a larger area of space than herbivores can efficiently search. If Janzen's (1974) hypothesis applies to lichens, then those species that invest heavily in defense compounds will be found only in environments subject to herbivory; furthermore, the environments will be those in which the lichen is somehow limited in its ability to recover from herbivory.

Mooney and Gulmon (1982) have also discussed various trade-offs between photosynthetic capacity, nitrogen content, and longevity in vascular plant leaves that should be favored in environments that vary in the level of herbivore susceptibility. They argue that a plant's carbon gain per leaf is a function of photosynthetic capacity and longevity minus maintenance costs and losses to herbivores. Leaf photosynthetic capacity and nitrogen content are positively correlated, and both of these factors confer increased susceptibility to herbivores.

Similar correlations for lichen production, nutrient content, and herbivore susceptibility cannot be generated at this time because of a lack of information. However, there are several generalizations that can be made.

1. Lichens maintain much lower concentrations of essential elements than vascular plants.

2. Lichens moistened with solutions of essential elements grow faster than moistened but unfertilized thalli (Hakulinen, 1966; Jones & Platt, 1969).

3. There are apparently a number of lichen adaptations designed to facilitate element uptake and reduce element loss (Farrar, 1976c; 1976d).

Coupling these rather widely held generalizations with Mooney and Gulmon's (1982) and Janzen's (1974) predictions would lead to the following specific predictions for lichens.

1. Lichen growth rate is a function of photosynthetic capacity (a complex interaction of thallus morphology, algal biomass, and environmental fluctuations), longevity and essential element content, minus maintenance costs (including investment in defense compounds) and losses to herbivores.

2. In environments without herbivores present, lichens that maintain relatively high levels of essential elements will have higher growth rates than morphologically similar species with low element contents.

3. In environments with herbivores present, undefended lichens that produce high concentrations of essential elements will be qualitatively more valuable to herbivores and therefore be subjected to greater herbivore damage than morphologically similar undefended lichens with low element contents.

4. In environments with herbivores present, tissues of lichens with high element contents lost to herbivores will be more costly to replace than tissues of morphologically similar species with low element contents, since nutrient acquisition and storage by lichens is assumed to be energetically costly.

5. From prediction (4) and Janzen's (1974) hypothesis, then, lichens subjected to herbivore damage will have a greater tendency to invest in defense compounds if essential element content is high than if element content is low.

There are some data available to test these ideas. Rundel (1978) suggested that lichens will be most likely to produce defense compounds when they contain high concentrations of nitrogen. He argued that nitrogen is in very short supply for lichens and probably lichen herbivores as well. Given a choice between unprotected thalli high in nitrogen and low in nitrogen, herbivores would be expected to consume thalli presenting the greatest concentration of nitrogen if all other factors remained the same. Protection of high concentrations of nitrogen through the production of defense compounds should there-

TABLE 9.24. Total Nitrogen Content of Selected Lichen Species

	Percent N
Nostoc symbiont	
Lichina confinis	3.6–3.9
Peltigera canina	3.24–4.05
P. polydactyla	3.6–4.5
P. praetextata	4.7
Pseudocyphellaria anthrapsis	3.91
P. intricata	2.98
Sticta sylvatica	4.0
Nostoc in cephalodia	
Lobaria laetivirens	2.2
L. pulmonaria	2.7
Peltigera aphthosa	1.9–3.4
Placopsis gelida	0.9–1.3
Stereocaulon alpinum	0.96
S. vesuvianum	0.73–0.77
Pulvinic acid derivatives	
Letharia columbiana	1.79
L. vulpina	2.10
Pseudocyphellaria flavicans	2.07
Green algal symbionts	
Anaptychia fusca	0.90
Cetraria islandica	0.51
C. nivalis	0.40
Cladonia foliacea	0.65
C. impexa	0.33
Cornicularia aculeata	0.38
Evernia prunastri	0.84
Lecanora atra	0.69
Ochrolechia parella	0.61–0.70
Physcia adscendens	1.0–1.3
Ramalina siliquosa	0.9–1.28
Usnea subfloridans	0.58

fore be favored in lichens subject to herbivory. He provided some data from the lichen literature to support his hypothesis (Table 9.24). Not too surprisingly, lichens with N_2-fixing, blue-green algal symbionts (either as phycobionts or cephalodia) generally maintained the highest concentrations of nitrogen. However, of those lichens without blue-green symbionts, those with the highest nitrogen concentrations also produced pulvinic acid derivatives, compounds that have been shown to be defense compounds (Rundel, 1978;

Slansky, 1979; Stephenson & Rundel, 1979). Lichen protection of high nitrogen concentrations through production of defense compounds thus appears to be a general characteristic of lichens.

Lawrey (1983) also observed correlations (presumably chemically induced) between lichen preference by the slug, *Pallifera varia*, and essential element content that supported Rundel's (1978) hypothesis. Individuals of *P. varia* were found to exhibit relatively high preferences for *Lasallia papulosa* and *Aspicilia gibbosa* and lower preferences for *Xanthoparmelia cumberlandia* and *Pseudoparmelia baltimorensis* in a species-rich lichen community located in the Shenandoah National Park in western Virginia (Table 9.16). One of the initial hypotheses developed to explain these preferences was that preferred lichens had high concentrations of essential elements and avoided lichens were low in element concentration. Lichens are known to contain low levels of essential elements and herbivores were expected to maximize essential element concentrations; therefore, lichens found to sequester high concentrations of essential elements were expected to be preferred by herbivores if all other factors remained the same. However, actual measurements of element concentration in the preferred and avoided lichens (Table 9.25) revealed that lichen species preferred by herbivores contained significantly lower concentrations of essential elements (particularly N and P) than lichen species avoided by herbivores (Figure 9.6).

These results demonstrated that lichen herbivore avoidance (escape) mechanisms can be quite complex. They apparently represent a balance between high growth rates that depend on nutrients and maintenance that depends on defense compounds. The trade-offs between growth, nutrients, and defense compound production are difficult to analyze, however. As Mooney and Gulmon (1982) accurately point out, high nutrient contents contribute equally to increased growth and herbivore damage.

Herbivore escape mechanisms in lichens could therefore favor any one or more of the following:

1. high growth rates to replace tissues lost to herbivores,

2. high dispersal rates to allow a palatable species to escape severe herbivore damage by outdispersing herbivores,

3. low essential element concentration to discourage herbivore activity by reducing resource quality,

4. production of defense compounds to inhibit herbivore grazing.

So few studies have been done to relate growth rate, nutrient content, and secondary compound production in lichens that these hypotheses cannot be rigorously tested at the present time. Lichen investments in growth and reproduction (including vegetative dispersal), nutrient acquisition, and

TABLE 9.25. Concentration of Elements Found in Selected Lichen Species from Stony Man Mountain Study Area, Shenandoah National Park

	Lichen Species			
Element	Aspicilia gibbosa *(3)*[a]	Lasallia papulosa *(4)*	Pseudoparmelia baltimorensis *(4)*	Xanthoparmelia cumberlandia *(3)*
P	1052.57 ± 22.74[b]	1056.50 ± 46.28	1383.25 ± 65.93	1299.33 ± 101.08
K	2679.67 ± 250.23	3998.25 ± 270.17	3528.25 ± 208.47	3474.00 ± 409.18
Ca	371.93 ± 19.47	225.48 ± 36.30	5364.25 ± 916.39	409.80 ± 9.76
Mg	786.73 ± 25.25	388.53 ± 29.51	492.08 ± 39.54	519.07 ± 54.73
Mn	33.42 ± 0.79	17.56 ± 1.70	53.49 ± 2.83	31.11 ± 3.09
Fe	1729.33 ± 56.00	328.90 ± 62.98	906.25 ± 81.06	923.53 ± 147.64
B	1.19 ± 0.20	2.04 ± 0.13	3.05 ± 0.35	1.29 ± 0.21
Cu	6.39 ± 0.48	24.64 ± 1.45	30.51 ± 5.69	25.56 ± 4.38
Zn	30.72 ± 0.68	161.60 ± 7.15	85.52 ± 5.37	81.67 ± 3.10
Al	1447.67 ± 17.47	347.85 ± 54.95	1415.00 ± 178.10	993.03 ± 49.06
Na	58.10 ± 1.77	32.43 ± 2.91	57.36 ± 6.48	58.97 ± 3.35
N	1.33 ± 0.04[c]	1.26 ± 0.04	1.49 ± 0.02	1.52 ± 0.02

[a]Numbers in parentheses are sample sizes for each species.

[b]Mean $\mu g/g \pm$ S.D. for all elements except N.

[c]Mean % ± S.D.

J.D. Lawrey. 1983. "Lichen herbivore preference: A test of two hypotheses." *Am. J. Bot.* 70:1188–1194. Reprinted with permission of the American Journal of Botany.

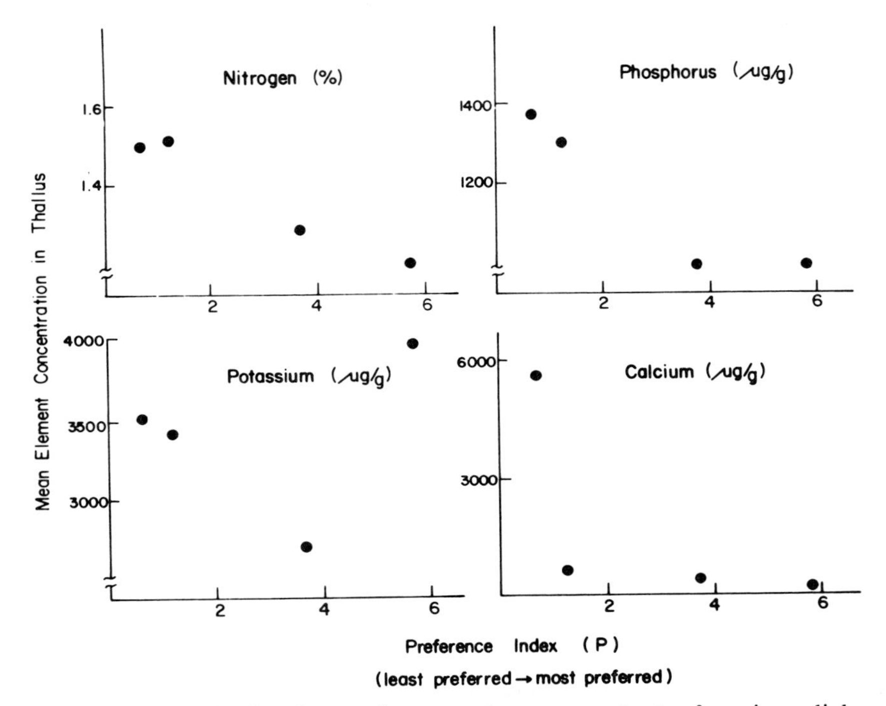

FIGURE 9.6. Ordinations of mean element content of various lichen species on preference to the lichen herbivore *Pallifera varia.* From J.D. Lawrey. 1983. "Lichen herbivore preference: A test of two hypotheses." *Am. J. Bot.* 70:1188–1194. Reprinted with permission of the American Journal of Botany.

secondary compound production appear to be very closely coupled, however. This suggests that future experimental investigations of lichen life history strategies that consider these factors will go a long way toward unravelling the relative energetic investments lichen species make in environments that differ significantly in the level of herbivory.

Prediction 5: Investment in Lichen Secondary Compounds Will Relate to the Apparency of Lichen Thalli to Herbivores

According to Feeny (1976), a plant's susceptibility to discovery by herbivores should influence not only the level of energetic investment in defense compounds but also the kind of defense compounds produced. Feeny called a plant's susceptibility to herbivore damage "plant apparency" and proposed several predictions that relate it to chemical defense. Following is a summary of these predictions.

1. Plants that exhibit characteristics that make them bound to be found (e.g., large size, persistence over successional time, clumped distributions, and high frequencies) will generally exhibit a greater total investment in chemical defense compounds than species that are relatively unapparent to herbivores (e.g., small, ephemeral, and rare).

2. Unapparent plants will exhibit a greater tendency to produce toxins that are effective at low concentrations against specific enemies (qualitative defense compounds), whereas apparent plants will tend to produce broad spectrum inhibitor compounds that are most effective at high concentrations (quantitative defense compounds).

3. Allocation of chemical defenses to component parts of plants will vary through time as the apparency of the parts and the importance to plant fitness change.

4. Interspecific chemical diversity will be lower in communities dominated by apparent plants than in those dominated by unapparent plants.

Feeny (1976) provided a limited amount of evidence from vascular plant-insect investigations to support these predictions; however, so little evidence is presently available that it is difficult to consider any of them particularly robust. Feeny expressed optimism that the predictions will generally stand the test of time, in spite of refinements that will undoubtedly have to be made to account for specific patterns of investment in specific plant-animal associations.

These predictions have not been tested in lichen communities, nor is there information presently available from lichen-herbivore studies that could be used to test the predictions. However, there are a number of important generalizations concerning lichen chemical patterns that can be logically derived from Feeny's predictions. The following is a partial listing of these generalizations.

1. Because lichens are slow-growing, long-lived, and persistent in most habitats, they should be relatively apparent to herbivores.

2. Lichen thalli exhibit patchy spatial distributions and should therefore tend to be consumed by generalist rather than specialist herbivores.

3. If lichens invest in defense compounds at all, they should invest in quantitative inhibitors that are effective against a wide range of generalist herbivores.

Lichen compounds share many characteristics of tannins, compounds Feeny used to illustrate properties of quantitative inhibitors. Both groups of compounds are phenolics, they are produced in relatively high concentrations, and they are frequently produced by a wide array of often unrelated groups of species. Tannins have been shown to be digestibility-reducing compounds that are able to complex both starch and proteins (Feeny, 1970). It is possible that lichen compounds have similar characteristics; however, no research on this has yet been done.

If lichen compounds are quantitative inhibitors, then investigators must accurately document patterns of lichen compound production in quantitative as well as qualitative terms. In future ecological studies of lichen chemical production, therefore, it will probably be insufficient merely to demonstrate the presence of specific compounds; rather, the concentration of each compound will have to be measured to indicate the total investment level that a lichen makes to chemical defense. If there are correlations between total phenolic compound production by lichen species and their apparency to herbivores, such patterns will undoubtedly emerge. However, until rigorous empirical studies are done, we are left with a number of interesting predictions for which there are no data, a situation we hope will soon be changed.

10 Lichen Community Ecology

Lichen populations in nature coexist and interact with other populations, including vascular plants, animals, and other cryptogams that live in the same habitat. We call these *communities,* recognizable and functionally integrated assemblages of interacting populations that inhabit a given space and time.

Lichen community investigations have helped to explain how lichen distribution patterns reflect underlying environmental factors. Although such studies were mainly descriptive in the early part of this century, recent investigators have taken a much more experimental approach. This has resulted in part from an increased interest in testing newly developed theories in the areas of community patterning, equilibrium, and dynamics, theories that have not been tested in lichen communities until relatively recently.

In this chapter, aspects of lichen communities most frequently investigated, especially development (succession), structure, and diversity, are discussed in light of recent theoretical predictions. Also, recent developments in the study of lichen competition, island biogeography, and habitat selection are reviewed and discussed.

LICHEN COMMUNITY DEVELOPMENT (SUCCESSION)

Community development or succession is thought to be a more or less directional, predictable, and cumulative change in species composition with time. As a concept, succession is relatively old, and it has undergone a number of important changes as ecologists, particularly plant ecologists, have modified their views about community organization. Early concepts

(Clements, 1916; 1936) assumed community development proceeded in much the same way as organismal development. Clements regarded climax (mature) communities as idealized assemblages that progressed naturally through immature, mature, and senescent developmental stages. This reflected the view held by all ecologists of that period. Modern approaches that most closely approximate these early discrete-unit concepts (e.g., Braun-Blanquet school of phytosociology) consider succession in much the same way. In general, these approaches assume patterns of community change to be directional and predictable sequences of stages caused by internal and external adjustments between species and the abiotic environment. This deterministic view of succession also characterizes recent concepts derived from general system theory (Margalef, 1968; Odum, 1969).

In opposition to these deterministic views of succession, several probabalistic views have emerged (MacArthur, 1958; Anderson, 1966; Leak, 1970; Waggoner & Stephens, 1970; Botkin, Janak & Wallis, 1972; Horn, 1975). These views assume that individual longevity and the replacement of one individual by another, of the same or a different species, determines the course of succession. According to this view, changes in species composition in a community occur because of the unique biological abilities of invading species to use resources not utilized, or utilized less efficiently, by species presently occupying the community. This idea has been tested preliminarily in vascular plant communities (Horn, 1975); lichen ecologists have not considered it.

Studies of lichen community development are quite numerous and varied in the lichen literature. A number of reviews are available, including summaries of studies (Barkman, 1958; Topham, 1977). Some of the better known of these studies will be briefly discussed in the next section. Most of the lichen succession literature is descriptive, reflecting the deterministic view of succession held by most investigators. It is difficult to draw ecologically significant generalizations from many of these studies because they are not designed to test hypotheses; instead, the objective appears to be a detailed identification of temporal changes in lichen communities for a specific region. A greater emphasis on hypothesis testing is expected in future investigations, however, because theoretical treatments of succession as a general process are appearing with increasing frequency in the ecological literature (MacArthur & Connell, 1966; Drury & Nisbet, 1973; Horn, 1974; Bazzaz, 1979). In an effort to show how lichen succession studies fit into this emerging theoretical framework, lichen succession will be considered in terms of trends predicted for general life history characteristics, an area that has been the subject of much recent interest.

Corticolous Lichen Community Development

Lichen community development occurs on all substrates, although the rates of development may vary considerably from one substrate type to another.

Perhaps the greatest amount of effort has been spent studying epiphytic lichen species composition on tree bark. By their nature, these substrates are relatively unstable, leading some investigators to question whether successional change in the sense it is used in vascular plant communities even occurs in epiphytic communities (Kershaw, 1964; Yarranton, 1972; Topham, 1977). Corticolous lichen successions vary with tree species and at different locations on the trees. The complexity of the developing community is also obviously limited by the longevity of the host tree.

These problems notwithstanding, many of the more interesting lichen succession studies have involved corticolous communities. Degelius (1964; 1978) has studied the development of epiphytic vegetation on twigs of numerous tree species throughout Scandinavia. His study of epiphyte colonization of *Fraxinus excelsior* twigs (Degelius, 1964) represents the most detailed investigation of pioneer corticolous lichen and moss community dynamics done to date, and a number of important trends were observed. He found that crustose species were seldom the first colonizers of *Fraxinus* twigs, except when the bark was very smooth. This contradicts a generalization frequently mentioned in the older literature—that succession proceeds from crustose to foliose to fruticose species (Topham, 1977). The first colonizers on rough bark were foliose species; several years passed before smooth bark colonizers appeared, and these were frequently crustose species (Table 10.1). In his notes on the competitive abilities of various lichen colonists of *Fraxinus* twigs,

TABLE 10.1. Typical Pioneer Species on *Fraxinus excelsior* Twigs in Scandinavia

Pioneer Species	*Age of* Fraxinus *Twigs on Which Lichens Were First Observed (Yrs)*
On rough bark	
Xanthoria polycarpa	3–5
Physcia tenella	3–5
Parmelia sulcata	3–6
P. subaurifera	3–5
P. exasperatula	3–5
P. stellaris	4–5
On smooth bark	
Arthopyrenia punctiformis	7–10
Lecanora carpinia	8–9
Lecidea elaechroma	8–9
Rinodina sophodes	8–9

Modified from G. Degelius. 1964. "Biological studies of the epiphytic vegetation on twigs of *Fraxinus excelsior*." *Acta Horti Gotob.* 27:11–55. Reprinted with permission of Acta Universitatis Gothoburgensis and the author.

Degelius (1964) indicated that some crustose species were far better competitors than others, although foliose species tended to be superior competitors generally. Hale (1974) also found that foliose species frequently colonized young *Fraxinus americana* twigs in Minnesota. These results suggested that lichen growth form is not a universal predictor of ecological success in epiphytic environments.

Other studies of corticolous lichen community development generally have made use of more indirect approaches than Degelius used. Harris (1971) investigated vertical and successional distribution patterns of corticolous lichens on oak and birch trees in South Devon. Although there was considerable tree-to-tree variation in the patterns, he found a number of interesting distributions that correlated well with microclimatological factors and individual species physiology. The mean percentage cover of the three most abundant species, *Pseudoparmelia caperata, Hypogymnia physodes*, and *Parmelia sulcata*, plotted against tree age at various sample heights (Figure 10.1), showed *P. caperata* to have the highest coverage values at lower, older tree heights of both oak and birch. *Hypogymnia physodes* and *P. sulcata* were dominant in oak and birch canopies and exhibited reduced coverages on lower, older surfaces. These results are strong, indirect evidence for a shift of corticolous lichen community structure as trees grow in height and the microclimate of tree bark surfaces changes with age.

Terricolous Community Development

Development of lichen communities on soil appears to be closely coupled to environmental conditions created by the developing vascular plant community. Robinson (1959) studied *Cladonia* succession in old fields of various ages in North Carolina and found that most species frequented soils of intermediate age (Figure 10.2). The exception was *Cladonia capitata*, the species found earliest and latest in the sequence. This species was apparently able to tolerate a wide range of environmental conditions, although it was not found in abundance in either the two-year-old stand or the climax stand. The explanation for the observed increase in *Cladonia* abundance at intermediate stages of succession has not been determined experimentally but apparently has to do with two factors: a requirement for organic matter for *Cladonia* squamule colonization early in succession, and an inability to tolerate shade in later stages of succession. The requirement for humus and an open environment is a common feature of soil lichen successions (Topham, 1977).

A number of soil lichen successions have been studied on sand dune environments of known age. Alvin (1960) studied a system of sand dunes on South Haven Peninsula, Studland Heath, Dorset, that are known to support vascular plant communities of various ages. He found that a definite succession of lichen species, primarily *Cladonia* species, occurred as the vascular

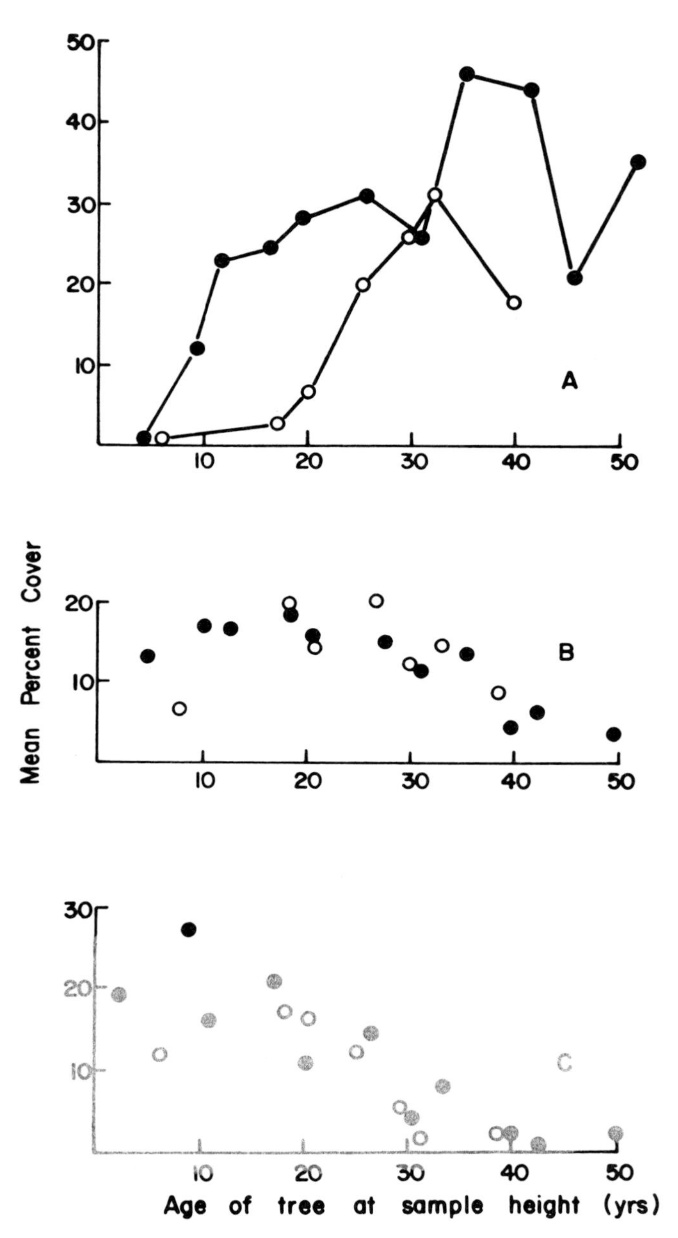

FIGURE 10.1. Mean percentage cover of three abundant corticolous lichens in South Devon, U.K., plotted against tree age at sample height; (a) *Pseudoparmelia caperata;* (b) *Hypogymnia physodes;* (c) *Parmelia sulcata;* ● = oak; ○ = birch. From G.P. Harris. 1971. "The ecology of corticolous lichens. II. The relation between physiology and the environment." *J. Ecol.* 59: 441–452. Redrawn with permission of Blackwell Scientific Publications.

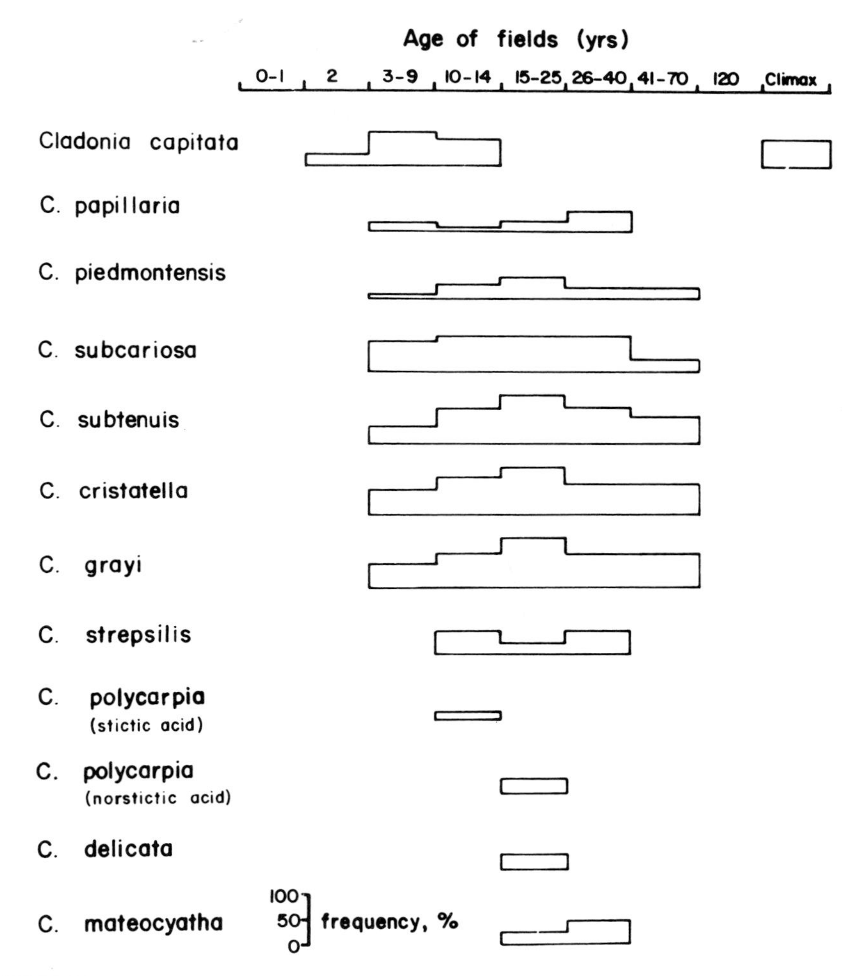

FIGURE 10.2. Frequency of lichens in abandoned fields and reforested areas of known age in the Piedmont region of North Carolina. From H.R. Robinson, 1959. "Lichen succession in abandoned fields in the Piedmont of North Carolina." *Bryologist* 62: 254–259. Reprinted with permission of the American Bryological and Lichenological Society.

plant community changed. The pioneer species were almost always associated with mosses or organic matter; later colonists gradually replaced these pioneers as *Calluna*-dominated vegetation replaced dune grass vegetation. Soil factors, particularly decreased soil pH, were cited as possible causes of these replacements. Other studies of dune systems (Watt, 1947; Gimingham, 1964) have also included data on lichen successions; however, the edaphic and/or biotic factors responsible for the observed changes have not been investigated experimentally.

Kershaw and colleagues have done a number of ecological investigations in a raised beach ridge system in the Hudson Bay coastal tundra in Canada. The system is made up of a number of ridges of different age that support lichen-heath vegetation in various stages of development. Larson and Kershaw (1974) examined ridges of about 250 to 800 years in age and were able to document a developmental sequence coupled with soil development. In a similar study, Pierce and Kershaw (1976) investigated the colonization of young ridges; they found correlations between environmental and edaphic factors and the development sequence. Especially important were soil peat thickness and organic matter content; definite trends in soil pH and essential element concentrations were not observed.

Other successional studies have been done in subarctic areas where fires are relatively frequent. Maikawa and Kershaw (1976) described the postfire recovery sequence of black spruce-lichen woodland in a large drumlin field southeast of Great Slave Lake, Northwest Territories, where sites representing various stages of development could be located. The investigators recognized four distinct phases in the sequence:

1. an initial 20-year period during which the vegetation was dominated by *Polytrichum* species, notably *P. piliferum*, with *Lecidea granulosa* and *L. uliginosa* as associates;

2. a subsequent 40-year period of *Cladonia*-dominated vegetation, involving especially *C. stellaris* and *C. uncialis*;

3. a 20-year period in which spruce-*Stereocaulon* vegetation dominated with *S. paschale* forming an almost pure ground cover;

4. a period after about 130 years characterized by canopy closure and replacement of lichen cover by mosses.

These results suggested that cyclic natural fires in the past have maintained the spruce-lichen woodland vegetation type in many parts of the subarctic.

An interesting cyclic succession between terricolous mosses and *Baeomyces rufus* has been reported by Jahns (1982). Colonies of the lichen and numerous moss species coexist in moist, shady habitats, especially along roadbanks. Jahns and Ott (1982) found that thalli of *B. rufus* live only about three years in a single location. They must therefore constantly recolonize a habitat that supports numerous competitors. The question is: Is *B. rufus* restricted to bare habitats, or can it recolonize habitats dominated by mosses?

Jahns found that moss-dominated portions of the habitat were thickly powdered with *Baeomyces rufus* soredia. These soredia stick to moss leaves, lodge between leaflets, and as they grow, they coalesce to produce clearly visible lichen thalli. The moss cushion dies and quickly rots beneath the growing lichen colony. During the first two years of *B. rufus* colonization, apothecia develop and large, scalelike vegetative diaspores (schizidia) are produced rather than soredia. Schizidia are apparently distributed over relatively short distances and serve to consolidate the developing lichen colony. In the third year of development, few apothecia are produced and the thalli begin to decompose. During this interval, moss species are able to colonize *B. rufus* habitats. However, heavy soredia production by *B. rufus* during thallus decomposition enables the lichens to disperse over the largest possible habitat area.

This interesting study appears to be an example of an equilibrium situation between a lichen species and mosses, with neither apparently able to cover a large enough area to exclude the other completely. It is also one of the very few studies to show the adaptive nature of age-specific changes in lichen life history characteristics, particularly reproductive (or disperal) characteristics, a subject which will be mentioned again in a later section.

Saxicolous Community Development and Pedogenesis

Saxicolous lichen community development will be considered from two quite different points of view. The first is the presumed importance of lichens in rock degradation, humus accumulation, and pedogenesis during primary succession on rock. The second and less prevalent view is that of rocks as island habitats on which lichen community dynamics important in island biogeography have recently been studied (Armesto & Contreras, 1981). Only the first view will be considered here, and that very briefly. The second will be taken up in a later section.

Saxicolous lichens have been assumed to contribute to soil formation ever since Linnaeus' time (Smith, 1921; Winterringer & Vestal, 1956). The role of lichen secondary compounds in this process has been particularly interesting (Syers, 1969; Iskandar & Syers, 1971; 1972; Williams & Rudolph, 1974; Ascaso & Galvan, 1976b). Another recent development includes the study of lichen element accumulation through the use of various microelement probe devices (Noeske et al., 1970; Hallbauer & Jahns, 1977; Wilson, Jones, & Russell, 1980; Jones, Wilson, & McHardy, 1981; Vidrich et al., 1982). Some fairly good reviews of the role of lichens in pedogenesis are available (Syers & Iskandar, 1973; Tansey, 1977) and should be consulted by those interested.

Lichens are thought to facilitate rock weathering by both physical and chemical processes (Tansey, 1977; Syers & Iskandar, 1973). The physical processes include rhizine penetration into rock substrates and thallus expansion and contraction on rock surfaces. Fry (1927) described in great detail the disintegration of rocks through the expansion and contraction of saxicolous lichen thalli.

Chemical weathering effects include mineral chelation by lichen secondary compounds and production of CO_2 and oxalic acid (Tansey, 1977), compounds known to take part in various mineralization processes. Lichen secondary compounds are definitely capable of forming metal complexes with rock minerals (Rundel, 1978; Syers & Iskandar, 1973; Tansey, 1977), although the importance of this process to rock weathering in nature is probably overstated. Respiratory CO_2, dissolved in water, may be more important in promoting rock decomposition, although no studies have been done (Tansey, 1977). Other investigators have suggested that oxalic acid production by lichens may promote rock decomposition by removing calcium and forming an insoluble calcium oxalate deposit in the thallus (Mitchell, Birnie, & Syers, 1966; Syers, Birnie, & Mitchell, 1967).

Regardless of the process, it is apparent that the presence of lichens on rocks influences surface weathering. Jackson and Keller (1970) found that *Stereocaulon vulcani* accelerated the rate of chemical weathering of Hawaiian lava flows. The weathering crust of lichen-covered rock was found to be much thicker than that of bare rock, and the lichen-covered crust contained higher amounts of Fe and lower amounts of Si, Ti, and Ca. Other investigators have also documented lichen-induced chemical changes in rocks, although it is still not clear precisely what mechanisms are involved (Ascaso & Galvan, 1976b; Galvan, Rodriguez, & Ascasco, 1981).

The role of lichens in rock weathering has been the subject of heated debates in the past. Lichens and mosses were thought to be extremely important in terrestrial primary succession, initiating processes that resulted ultimately in the destruction of rocks and the formation of soils suitable for vascular plant colonization. Many early ecology texts (e.g., Weaver & Clements, 1938; Clements & Shelford, 1939; Emerson, 1947) encouraged this view. The fact that lichens and mosses are frequently the first rock colonizers (e.g., primary colonization of Surtsey studied by Brock, 1973) and contribute to rock weathering would tend to support this classical view of lichens as rock disintegrators, humus collectors, and soil formers. However, Cooper and Rudolph (1953) argued convincingly against this idea, suggesting instead that lichens make relatively insignificant contributions to pedogenesis. Topham (1977) discussed other more recent studies that generally support Cooper and Rudolph's argument. Although it is not nearly as controversial now as it was a few decades ago, the question of whether or not lichens play an important role in pedogenesis has not been fully resolved.

Changes in Reproductive and Competitive Investments—*r*- and *K*-Selection

According to the theory of *r*- and *K*-selection developed by MacArthur and Wilson (1967), life history characteristics of organisms will vary at different stages of succession, reflecting changes in the selective environment as communities develop. Early in succession, selection is supposed to favor heavy investment in net production and reproductive effort, thereby increasing colonizing ability and the exploitation of new habitats. As a consequence, however, investments in maintenance and competitive ability are reduced. Later in succession, as the level of competition for resources increases, selection favors a greater investment in competitive ability and lower reproductive effort. Other attributes of the two strategies are listed in Table 10.2.

Recent studies of seral vascular plant communities (Newell & Tramer, 1978; Werner, 1976) have demonstrated that early successional plants are good colonists, grow rapidly, and produce numerous small seeds; later successional species tend to devote less energy to reproduction, produce fewer larger seeds, and invest in maintenance and competitive ability, including secondary chemicals (Coley, 1980). Although it is probably not as simple as it is sometimes considered to be, *r*- and *K*-selection appears to represent trade-offs between investments in productivity, reproduction, and colonizing ability on the one hand, and investments in maintenance and competitive ability on the

TABLE 10.2. Correlates of *r*- and *K*-Selection

	r-*Selection*	K-*Selection*
Stage in succession	Early	Late
Climate	Variable and/or unpredictable; uncertain	Fairly constant and/or predictable; more certain
Mortality	Often catastrophic, nondirectional, density independent	More directed, density dependent
Population size	Variable in time, usually below carrying capacity	At or near carrying capacity
Intra- and interspecific competition	Variable, often lax	Usually intense
Selection favors	1. Rapid development	1. Slower development
	2. High maximal rate of increase, r_{max}	2. Greater competitive ability
	3. Early reproduction	3. Late reproduction
	4. Small body size	4. Large body size
	5. Single reproduction	5. Repeated reproduction
	6. Many small offspring	6. Fewer larger offspring
	7. Short length of life	7. Longer length of life

Source: E.R. Pianka, *American Naturalist,* 104:592–97. Copyright 1970 by the University of Chicago Press.

other. No organism is a perfect *r*- or *K*-strategist; rather, a continuum is thought to exist between these two extreme strategies (Pianka, 1970).

Although there are no lichen studies that test these theories directly, there are indications that reproductive characteristics of pioneer lichen species are different from those of later colonists. Topham (1977) provided some examples in her review of lichen colonization, succession, and competition. In general, she believes that as lichen community development proceeds, species are more likely to exhibit increased diaspore size, reduced diaspore number, and longer maturation times prior to reproduction. All of these trends support *r*- and *K*-selection theory.

Degelius (1964) observed a number of interesting patterns that relate to this discussion of *r*- and *K*-selection. He found that the species diversity of epiphytes on *Fraxinus* twigs increased continuously with twig age. However, he observed the greatest number of species on 10-year-old twigs; older twigs had fewer epiphyte species (Figure 10.3). This demonstrated the effect of competitive exclusion on community composition with time. Degelius mentioned throughout his paper instances of competitively weak pioneer species

FIGURE 10.3. Number of lichen species on Scandinavian *Fraxinus* shoots of various ages. From G. Degelius. 1964. "Biological studies of the epiphytic vegetation on twigs of *Fraxinus excelsior*." *Acta Horti Gotob.* 27: 11–55. Redrawn with permission of Acta Universitatis Gothoburgensis and the author.

being replaced by stronger competitors that enter the community later in succession. Not all pioneers were found to be weak competitors, however; thalli of *Parmelia sulcata* sometimes colonized relatively young twigs and dominated in competitive interactions with other species. The crustose species, *Lecanora carpinea* and *Lecidea elaeochroma*, were also good competitors early in succession. In general, however, ephemeral species poor in competition tended to be very early colonists that exhibited good reproductive and dispersal characteristics.

Degelius (1964) also observed that pioneer lichen species exhibited differences in the maturation time required before first reproduction. Apothecial species tended to reproduce earlier than species bearing soredia or isidia (Table 10.3), although there was apparently no difference in the colonization time of the various species. This is an interesting observation because, as Topham (1977) has mentioned, spores are generally smaller than soredia and

TABLE 10.3. Age of Pioneer Lichen Colonizers of *Fraxinus* Twigs when Apothecia and Vegetative Diaspores (Soredia and Isidia) First Appeared

Species	*Age of Twigs (Yrs) Bearing Lichen*	*Age of Lichen (Yrs) Exhibiting Apothecia or Diaspores (Approx.)*
Sexual		
Bacidia chlorococca	6	4
Lecanora carpinea	6	2
Lecidea elaeochroma	7	3
Parmelia aspera	8	5
Physcia stellaris	6	3
P. tenella	7	5
Rinodina sophodes	7	2
Xanthoria lobulata	5	2
X. parietina	7	5
X. polycarpa	4	2
Sorediate		
Parmelia subaurifera	8	6
P. sulcata	10	8
Physcia adscendens	7	4
P. dubia	9	7
P. orbicularis	7	4
P. tenella	6	4
Isidiate		
Parmelia exasperatula	6	4

G. Degelius. 1964. "Biological studies of the epiphytic vegetation on twigs of *Fraxinus excelsior*." *Acta Horti Gotob.* 27:11–55. Reprinted with permission of Acta Universitatis Gothoburgensis and the author.

isidia and may represent a much lower energetic investment per reproductive (or dispersal) unit. Production of spores would therefore tend to be associated more closely with early reproduction (in the broad sense, including asexual reproduction) than production of soredia and/or isidia.

Unfortunately, no studies of lichen succession have been specifically designed to study life history characteristics. However, it is beginning to become apparent that there are definite and predictable life history patterns in lichen successions. If *r*- and *K*-selection operate in lichen populations, there should be a number of important trends during succession, not only in diaspore size and competitive ability, but also investment in secondary compounds, longevity, physiological plasticity, niche specialization, and other characteristics. Experimental studies in these areas are obviously needed.

SPECIES DIVERSITY

Two important and easily measured aspects of community composition are the number of species in a community (species richness) and the distribution of individuals among the species (equitability). These aspects are frequently quantified using various indices of species diversity (Peet, 1974). Species diversity estimates reflect the underlying availability of resources and habitats in the community and the degree to which these are apportioned among the species; therefore, they can be used as a first approximation of community composition in relation to resource availability.

The study of lichen communities has included some interesting investigations of species diversity in relation to complex environmental gradients. These will be discussed briefly. There has also been interest in lichen community dynamics in terms of island biogeography theory, which attempts to relate community diversity to colonization and competitive characteristics of species on habitat islands. This has not been actively investigated by lichenologists, but recent studies suggest that island biogeography concepts can be profitably tested in lichen community studies.

Species Diversity Changes along Environmental Gradients

Hoffman and Kazmierski (1969) studied the epiphytic vegetation on *Pseudotsuga menziesii* on the Olympic Peninsula, Washington and discovered a number of interesting distributional patterns that appeared to relate to the habitat moisture availability. Bryophyte assemblages were found in mesic habitats, and lichen-dominated assemblages were in more xeric habitats. Hoffman (1971) later analyzed his original data using various species diversity indices. The simplest estimate, species number or species richness, changed dramatically as the habitat became drier, with lichens consistently outnumber-

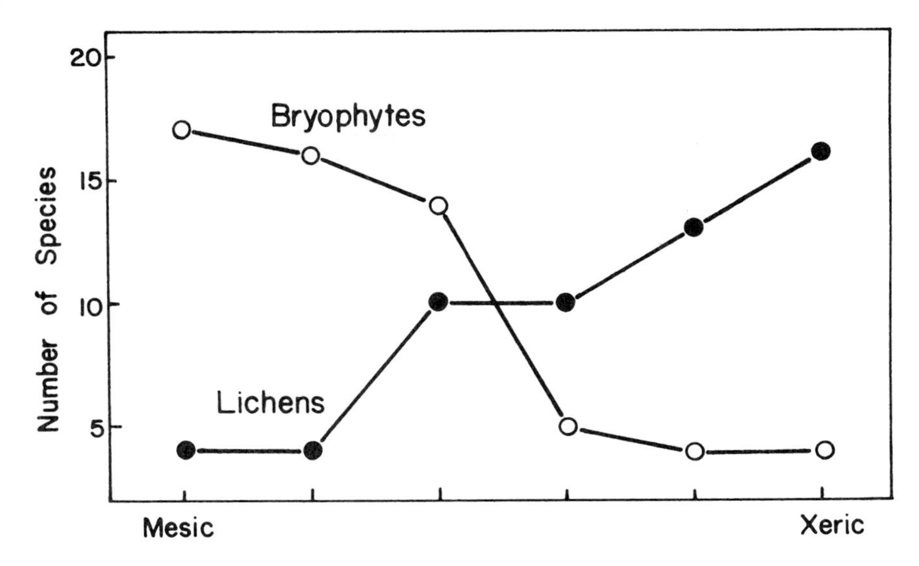

FIGURE 10.4. Total number of bryophyte and lichen species encountered in sampling six vascular plant vegetation types on the Olympic Peninsula, Washington, arranged from mesic to xeric. From G.R. Hoffman. 1971. "An ecologic study of epiphytic bryophytes and lichens on *Pseudotsuga menziesii* on the Olympic Peninsula, Washington. II. Diversity of the vegetation." *Bryologist* 74: 413–427. Reprinted with permission of the American Bryological and Lichenological Society.

ing mosses on dry sites (Figure 10.4). Other more complicated estimates (e.g., Simpson's Index), several based on information content, demonstrated that the total diversity increased as the habitat became drier. Whereas mosses tended to dominate the mesic sites, a greater equitability of species, including numerous lichen species, was observed at xeric sites.

Nash et al. (1979) studied changes in lichen and vascular plant communities in central and south-central Baja California at different distances from the Pacific coast. They found that lichen species richness declined from the coast inland, although no particular pattern was found for vascular plants (Table 10.4). Distinct shifts in the distribution of certain lichens along the gradients were also observed, suggesting that some species have fairly specific microhabitat requirements. The authors were able to show a clear correlation between lichen species richness and atmospheric moisture. Coastal areas were influenced by frequent fogs formed where cold ocean currents meet the warm desert land; inland from the coast, the climate is more continental. These differences in atmospheric moisture were also correlated with the physiologi-

TABLE 10.4. Species Richness and Mean Cover of Lichen and Vascular Plant Communities along Two Transects away from the Pacific Ocean. Species Lists and Individual Cover Values Are Available in the Original Paper

	Transect 1 (El Arco, 28° N)			Transect 2 (San Carlos, 25° N)			
Distance Inland from Coast (km)	*0*	*31*	*45*	*0*	*30*	*50*	*70*
Lichens							
Number of species	15	16	3	25	21	9	2
Total cover (%)	15.0	5.7	0.2	15.2	6.3	0.9	0.1
Vascular plants							
Number of species	7	15	13	12	9	17	16
Total cover (%)	29.6	26.4	24.0	38.9	17.9	45.4	45.1

T.H. Nash et al. 1979. "Lichen vegetational gradients in relation to the Pacific coast of Baja, California: The maritime influence." *Madrõno* 26:149–163. Reprinted with permission of the California Botanical Society.

cal responses of the dominant lichen species from each habitat. The coastal dominant, *Roccella babingtonii*, maintained high rates of photosynthesis along a broad range of thallus saturations (40-100 percent); the inland dominant, *Ramalina complanata*, had a narrow photosynthetic optimum at 30–40 percent saturation. It appears from these experiments that coastal species are adapted to higher atmospheric moisture conditions.

Saxicolous Lichen Communities and Island Biogeography

Island biogeography theory was originally formulated by MacArthur and Wilson (1963; 1967) to account for the relative impoverishment of island communities compared to mainland communities. Generally, the theory hypothesizes that the number of species of a given group of organisms on an oceanic island will be the result of an equilibrium between immigration and extinction rates, and that these rates will be functions of the island area and the distance away from the mainland.

Although a number of interesting tests of the theory have involved oceanic islands, there has been some recent interest in testing the theory in continental situations where the islands are different-sized habitat patches. Isolated lakes, mountaintops, and other similarly distributed habitat islands exhibit many of the same characteristics of oceanic islands as far as colonists are concerned and should therefore represent valuable natural laboratories for island biogeographic research.

If rocks colonized by lichens are islands in the sense of MacArthur and Wilson (1967), saxicolous lichen species diversity should relate in a predictable fashion to island area and distance between islands. Armesto and Contre-

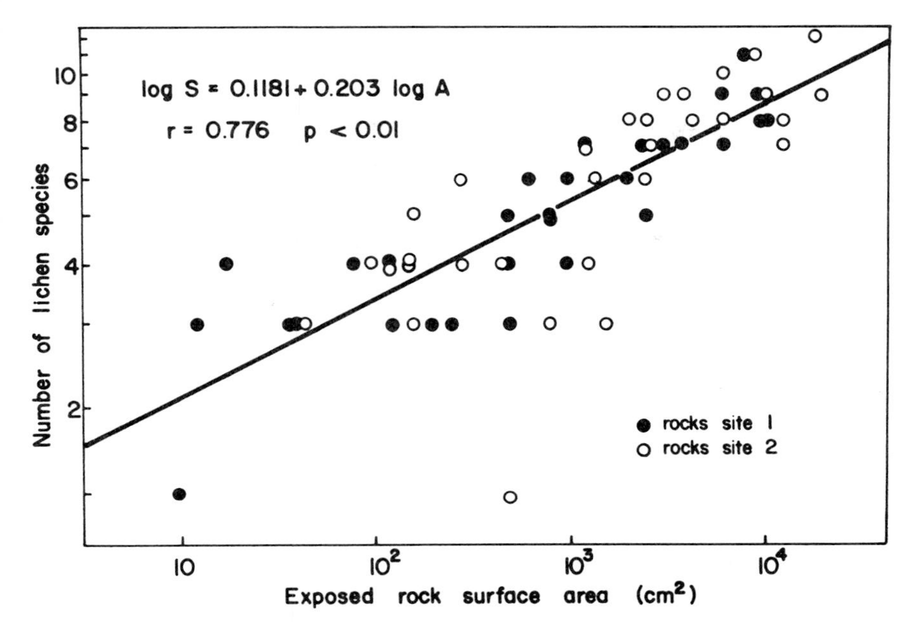

FIGURE 10.5. Relation between the number of lichen species on rocks and the exposed rock surface area for two sites in central Chile. Redrawn and reprinted from *American Naturalist*, Vol. 118: 597–604, J.J. Armesto & L.C. Contreras and the University of Chicago Press. Copyright 1981 by the University of Chicago.

ras (1981) tested this idea at two different sites in the Estación Experimental Agronómica of the Universidad de Chile. One site was an open area with numerous granitic rocks of different sizes; the other was covered by *Acacia caven* shrubs with granitic rocks located mainly in open spaces between the shrubs. For randomly located rocks at each site, rock surface area, distance from the nearest rocks, and lichen species richness were measured. Although distance between rocks was not found to be very important, a good correlation was observed between island area and lichen species richness for both sites (Figure 10.5). The following equation has been used to describe the relationship between species richness and island area:

$$S = CA^z$$

where S is species number, A is island area, C is a constant of undetermined biological meaning, and z is a coefficient that appears to relate to immigration ability (MacArthur & Wilson, 1967). Armesto and Contreras calculated a z

TABLE 10.5. Values of *z* Taken from the Literature for Patchily Distributed Habitats on Continents, Nonisolated Continental Areas and Oceanic Islands

Fauna or Flora	*Type of Island*	z	*Source*
Terrestrial invertebrates	Caves	0.72	Culver et al. (1973)
Acarines	Cushion plants	0.42–0.69	Tepedino & Stanton (1976)
Mammals	Mountaintops	0.43	Brown (1971)
Carabids	West Indies[a]	0.34	Darlington (1943)
Land plants	Galapagos[a]	0.33	Preston (1962)
Lichens	Rock surfaces	0.20	Armesto & Contreras (1981)
Ants	New Guinea[b]	0.17	Wilson (1961)
Mammals	Sierra Nevada[b]	0.12	Brown (1971)

[a]Oceanic archipelagoes.
[b]Nonisolated continental sample areas.

value of 0.203 (Figure 10.5). This low *z* value is similar to values obtained in continental studies (Table 10.5), supporting the hypothesis that there were few barriers to lichen dispersal in this particular system, and that rock islands are saturated with lichen species, probably the result of extremely low extinction rates. This further suggested that a lichen species is not easily displaced once it becomes established on a given rock surface. There may be species replacements resulting from competitive interactions or environmental disturbance, but these take place so slowly they are practically undetectable. The extreme stability of saxicolous lichen communities in nature (Frey, 1959; Hawksworth & Chater, 1979; Pentecost, 1980) would tend to support this view.

COMPETITION

Competition occurs when two or more individuals interfere with or inhibit each other, frequently when some common resource is in short supply (Pianka, 1976). Since it is advantageous for competitors to avoid each other if possible, competition tends to promote divergence of species resource utilization patterns over ecological and evolutionary time, thereby generating ecological diversity.

When one studies animal communities, it is relatively easy to view competitive interactions in terms of resource utilizations. However, all plant species use the same resources: light, water, nutrients, and space. For plant communities, then, a better definition of competition is that of Grime: "the tendency of neighboring plants to utilize the same quantum of light, ion of a mineral nutrient, molecule of water or volume of space" (1977, p. 1170).

Competition has not been frequently investigated in lichen communities,

TABLE 10.6. Factors Expected to Influence Competition between Lichens

Factors	*Weak Competitor*	*Strong Competitor*
Growth rate	Low	High
Growth form	Crustose	Foliose, fruticose
Chemistry	No allelochemics produced	Allelochemics produced
Size	Small	Large
Herbivore palatability	High	Low

probably because lichens are not thought to be particularly good competitors (Berner, 1970; Janzen, 1975; Topham, 1977; Grime, 1977). Direct or indirect evidence for competition in lichen communities is available, however, and there have been a number of interesting recent studies that demonstrate the dynamic nature of equilibrium lichen communities, due in part to interspecific competitive interactions. These will be discussed briefly.

There is a good deal more indirect than direct evidence for lichen competition. Malinowski (1911; 1912) observed a number of instances of thallus overgrowth in saxicolous crustose species and commented on the importance of such phenomena in lichen community dynamics. Photographic analyses (Frey, 1959; Hawksworth & Chater, 1979) have also documented the dynamic nature of lichen species assemblages.

Barkman (1958) suggested that epiphytic lichens and bryophytes might be subjected to a number of different competitive interactions, especially the following;

1. mechanical destruction,
2. suffocation,
3. competition for light,
4. chemical interference.

He illustrated each of these forms of competition with examples from the available literature and suggested that competition for light was the most important. Other factors thought to influence the outcome of lichen competition are listed in Table 10.6. An additional listing is available in the review by Topham (1977).

Pentecost (1980) has recently investigated the means by which saxicolous lichens compete for space on rock substrates. He suggested that four factors are important:

1. colonization rates and densities of species,
2. radial growth rates,
3. types of contacts between species,
4. rates of thallus senescence.

Using techniques developed by Stebbing (1973) for epiphytic vegetation, he characterized the competitive interactions between various species observed at two sites in Wales. He found that interactions between crustose species resulted in one of the following outcomes:

1. One species overgrows the other.
2. A truce condition results when neither species grows at the point of contact.
3. One species grows epiphytically on the other.

Interactions between foliose species rarely resulted in a truce; one species usually overgrew the other, and the winner was determined by the size and growth rate of each species. Interactions involving crustose and fruticose species also generally resulted in overgrowth of one species by the other.

One aspect of Pentecost's (1980) analysis was the consideration of the effects of lichen senescence and "window" regeneration. Many lichen species exhibit a form of senescence—not true physiological senescence since the entire organism is not changed physiologically—in which the central portion of the thallus dies and falls away, leaving a window of bare substrate available for colonization by either the same or a different species. Depending on the rate of senescence and the colonizing rate of each species in the community, changes in community composition could occur through overgrowth of individuals from the inside out. The outcome of competition might also be different depending on whether species interact at outer margins or window margins. Pentecost observed an interaction between *Aspicila calcarea* and *Caloplaca heppiana* that demonstrated this possibility. One species (*A. calcarea*) was the stronger competitor when outer margins came into contact and the other (*C. heppiana*) was stronger when *A. calcarea* windows were colonized by *C. heppiana*.

Few considerations of lichen competition have involved the measurement of resource utilization patterns. Lawrey (1981) investigated saxicolous lichen competition in terms of the utilization of light. Communities located on two Potomac River islands, Plummers Island and Bear Island, Maryland, were chosen for study because they varied significantly in species diversity. The Plummers Island community was much less diverse, a result of air pollution fallout from a nearby highway. Two foliose lichen species, *Pseudoparmelia baltimorensis* and *Xanthoparmelia conspersa*, dominated the communities on both islands. However, the species showed obvious differences in light preference; *P. baltimorensis* preferred shaded rocks and *X. conspersa* sunny rocks (Figure 10.6).

A comparison of these light utilization patterns on Plummers Island with those on Bear Island revealed an interesting difference for *Xanthoparmelia conspersa* that appeared to relate to its competitive ability. Whereas the shape

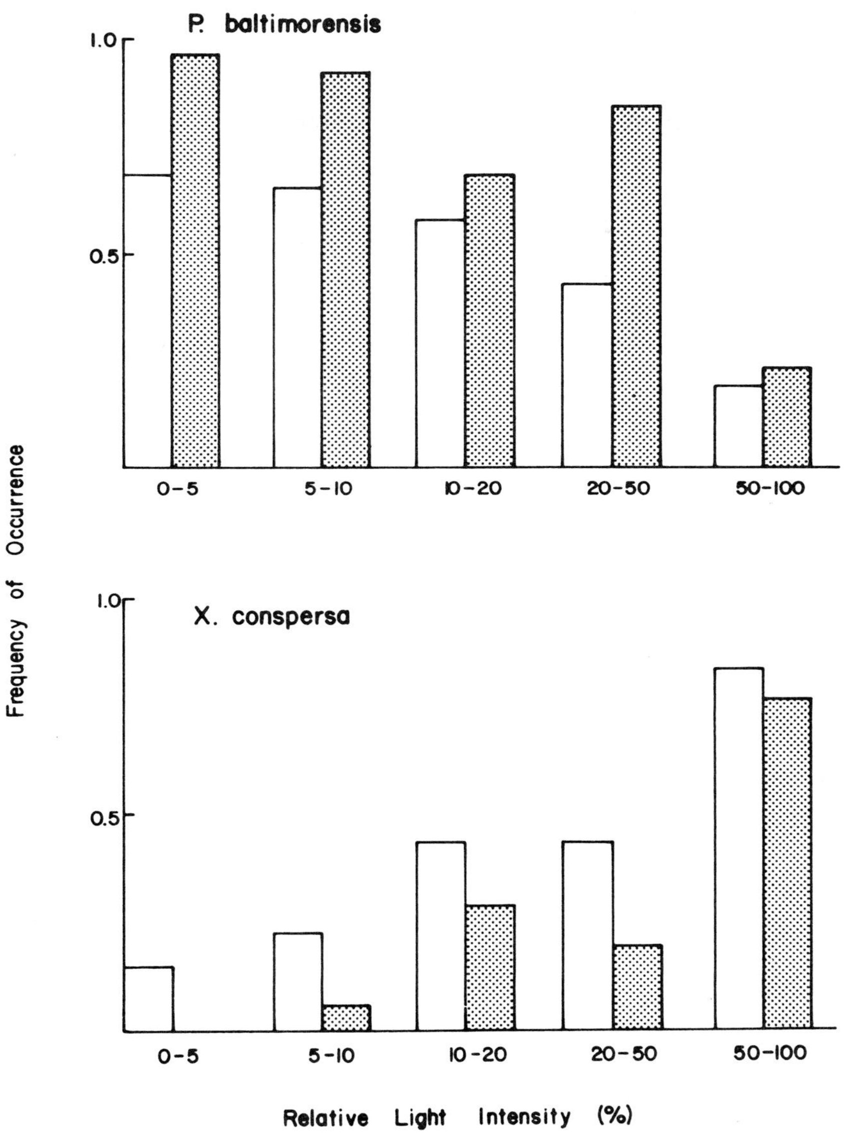

FIGURE 10.6. Frequency of occurrence of (a) *Pseudoparmelia baltimorensis* and (b) *Xanthoparmelia conspersa* on rocks receiving various light intensities on Plummers Island (open histograms) and Bear Island (stippled histograms), Maryland. From J.D. Lawrey, 1981. "Evidence for competitive release in simplified saxicolous lichen communities." *Am. J. Bot.* 68: 1066–1073. Reprinted with permission of the American Journal of Botany.

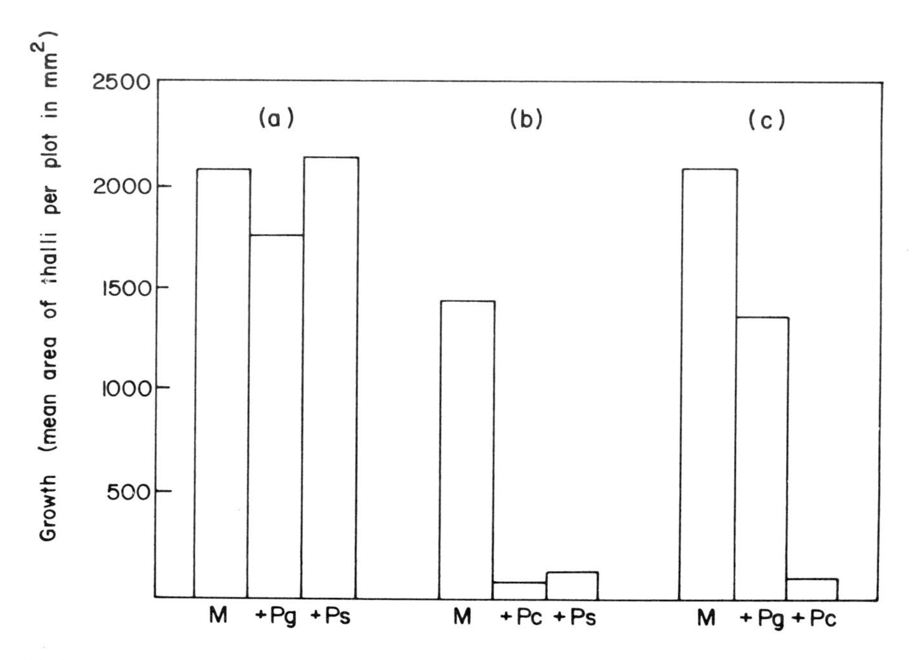

FIGURE 10.7. Growth of thalli. (a) *Parmelia* (=*Xanthoparmelia*) *conspersa* (PC); (b) *Parmelia* (=*Melanelia*) *glabratula* (PG); (c) *Parmelia saxatilis* (PS) in monoculture (M) and various mixtures (+). From R.A. Armstrong, 1982. "Competition between three saxicolous species of *Parmelia* (lichens)." *New Phytol.* 90: 67–72. Redrawn with permission of Academic Press and the author.

of the *Pseudoparmelia baltimorensis* light curve was the same on both islands, the curve for *X. conspersa* was much broader on Plummers Island, with higher frequencies in low and intermediate light intensities than on Bear Island (Figure 10.6). Lawrey (1981) attributed this shift in light utilization by *X. conspersa* to a reduction in competition on Plummers Island that resulted from community simplification. A shift in the resource utilization pattern of a species in response to removal of a competitor (or competitors) is called ecological or competitive release (MacArthur & Wilson, 1967; Pianka, 1978) and is considered strong indirect evidence for competition.

Direct evidence for competition in lichen communities is less abundant in the literature, and experimental evidence is practically nonexistent. Frey (1959) used transplant experiments of various *Cladonia* species in Switzerland to measure competitive abilities. He found that transplanted mat-forming *Cladonia* species frequently overgrew other types of soil *Cladonias*.

Armstrong (1982) has also used lichen transplants to investigate competition. He glued thallus fragments of *Parmelia saxatilis, Xanthoparmelia conspersa*, and *Melanelia glabratula* ssp. *fuliginosa* to pieces of slate in various pure and mixed patterns and measured the relative growth of each species after three years. The glueing was found not to influence lichen growth. He found that, although no species eliminated another in mixtures, they apparently varied considerably in their ability to tolerate competition. When compared to their growth in monoculture, *P. saxatilis* and *M. glabratula* exhibited reduced growth in the presence of *X. conspersa* (Figure 10.7), suggesting that *X. conspersa* is a superior competitor when the three coexist in nature. *Parmelia saxatilis* had an intermediate competitive ability and *M. glabratula* was apparently always inferior, probably because its growth rate is much slower than the others. Although the mechanisms responsible for this growth interference were not studied, the technique developed by Armstrong lends itself to these sorts of direct investigations of lichen competition. It will undoubtedly contribute greatly to our understanding of lichen competition as interest in experimental lichen ecology grows.

HABITAT SELECTION

Despite their apparent ability to grow on a wide variety of substrates, most lichen species are found on only one of the three most commonly used substrate types: trees, rocks, or soil. This is habitat selection in its broadest sense. However, it is becoming increasingly apparent that lichen species exhibit habitat selection within broad substrate types; for example, corticolous (tree bark-inhabiting) species tend to be found either on hardwoods or conifers, saxicolous (rock-inhabiting) species tend to be found on certain types of rock or in certain light intensities, and terricolous (soil-inhabiting) species tend to be found on certain types of soil. These trends are ecologically important because they suggest that interacting lichen species in species-rich communities are not all using the same resources in the same way.

A number of lichen habitat-selection studies have been done, of which only a small sample will be discussed here. Reviews are provided by Brodo (1973) and Slack (1976) for natural substrates and Brightman and Seaward (1977) for man-made substrates.

Habitat Specificity Exhibited by Chemically Distinct Populations

Several recent studies have demonstrated that lichen populations are not chemically or ecologically homogeneous. Some morphologically uniform lichens have different chemical races with distinct distributions and ecological requirements. Since chemical differences in lichens are assumed to have a genetic basis, studies of ecological differentiation into chemical groups (races, subpopulations, and demes) can provide valuable evidence for evolutionary patterns in lichens.

The first suggestion that lichen chemical races differed ecologically was made by W.L. and C.F. Culberson (1967), who found that each of the six chemical races of *Ramalina siliquosa* colonized a distinctly different zone in the rocky intertidal region of North Wales. This study demonstrated that chemical differentiation within lichen groups was not simply an odd but essentially meaningless characteristic of lichens but an important indicator of the underlying physiological and ecological differentiation that influences the rate and direction of evolution in all plant and animal groups.

The Culbersons have been able to demonstrate this ecological differentiation by chemical races in several different lichen groups. For example, evidence was given for the parallel evolution of asexual chemical morphs lumped by conventional taxonomy under the name *Parmotrema hypotropum* from chemically similar sexual morphs of (again by convention) *P. perforatum* (W.L. Culberson, 1973; W.L. Culberson & C.F. Culberson, 1973). This was the first evidence that chemically variable asexual morphs are polyphyletic and not the product of random and essentially meaningless chemical differentiation. Poelt (1972) first proposed this theory, and it has since been supported by a number of chemical studies.

Other studies have also demonstrated ecological differentiation by chemically distinct lichen groups (Wetherbee, 1969; Nash & Zavada, 1977; Bowler & Rundel, 1978). Nash and Zavada (1977) found that closely related *Xanthoparmelia* chemotypes have several habitat preference patterns in the Sonoran desert. The patterns of species establishment on five different rock types were nonrandom for several—an example of three species is provided in Table 10.7—which suggested to the authors that significant ecological differentiation among closely related but chemically distinct sympatric populations had occurred in the past.

Recently, W.L. and C.F. Culberson (1982) demonstrated that the foliose *Pseudoparmelia caperata*, a lichen species that is common worldwide on rocks and trees, is primarily corticolous in most of eastern North America, its place on rock being taken by the closely related, exclusively American *P. baltimorensis*. The two species are similar morphologically but different chemically and appear to exhibit definite habitat preferences, possibly reflecting past interspecific competition between the two. This study, and all studies of this sort, emphasize the value of lichen compounds as genetic markers in population and community studies. It is expected that further study of chemical variability in terms of ecological and distributional patterns will yield evolutionarily significant information.

TABLE 10.7. Average Frequency of Occurrence of Three *Xanthoparmelia* Species on Different Substrates. Numbers Not Connected by an Underline Are Statistically Different ($\alpha = 0.05$)

X. lineola					
% occurrence	1.25	3.00	9.17	10.78	17.52
Substrate*	An	Da	Gr	Tf	Ba
X. mexicana					
% occurrence	0.42	7.44	7.47	16.84	33.23
Substrate	Gr	Tf	Da	Ba	An
X. plittii					
% occurrence	0.00	0.33	2.51	6.70	17.33
Substrate	Da	Gr	Ba	An	Tf

*Abbreviations: Gr = granite, Da = dacite, Ba = basalt, An = andesite-basalt complex, Tf = rhyolite tuff.

Partial data adapted from T.H. Nash & M. Zavada. 1977. "Population studies among Sonoran Desert species of *Parmelia* subg. *Xanthoparmelia* (Parmeliaceae)." *Am. J. Bot.* 64:664–669. Reprinted with permission of the American Journal of Botany.

Habitat Specificity in the Intertidal

Zonation of lichens in seashore environments is a well-known phenomenon (e.g., Kärenlampi, 1966; Sheard, 1968; Fletcher, 1973a; 1973b; 1976). Seawater is thought to be the most critical factor influencing the distribution of species, although pH, substrate chemistry, and frequency of wetting and drying are also known to affect species from the various zones differently.

Fletcher (1976) presented evidence from a number of physiological experiments suggesting that salinity is perhaps secondary in importance to other factors. Species collected from the littoral zone (seashore) were found to withstand continuous immersion better than species from the supralittoral zone (maritime regions some distance from the seashore), regardless of salinity. Species from the supralittoral zone were better able to tolerate repeated drying and wetting; once again, salinity was secondary in importance. These physiological response patterns correlated well with the habitat conditions that characterize the littoral and supralittoral zones.

Salinity has been found to be important in the establishment phase of intertidal lichens. Ramkaer (1978) collected fertile lichen specimens from zones on the west coast of Sweden variously influenced by seawater inundation and measured their spore germination responses to different salinities. Species closest to the littoral fringe (*Verrucaria maura* and *Caloplaca marina*) showed a preference for salinities close to that of seawater (approximately 3 percent); these species also had the greatest tolerance of higher salinities (Figure 10.8). Species from zones farther away from the littoral fringe (*Xanthoria parietina* and *Caloplace scopularis*) had fairly high tolerances of salinities up to about 3 percent; *Lecanora muralis,* most removed from the shore, was very intolerant of high salinities. Similar response patterns were observed when hyphal growth rates of the lichens were compared. These results suggested that true littoral zone lichens have different responses to salinity in the establishment phase, which may explain the remarkably distinct lichen zonation patterns observed in intertidal regions.

Habitat Specificity on Arid Zone Soils

The fact that arid zone lichens are important in binding soil and preventing erosion has stimulated research into the mechanisms of substrate specificity. Terricolous lichens from the arid zone of southeastern Australia have been extensively studied by Rogers and colleagues (Rogers & Lange, 1972; Rogers, 1972a; 1972b; 1974). In this region, lichen distributions are apparently related to numerous soil factors, especially surface hardness, pH, and calcium content, although it is difficult to separate these factors from prevailing climatic factors (Rogers, 1972b); soil texture has also been found to be an important limiting factor (Rogers, 1974).

Terricolous lichens from arid regions in Israel are also known to prefer certain soil types, although the soil factors responsible for these preferences were not known until fairly recently. Garty, Gal, and Galun (1974) found that climate and the shrinkage rate of these desert soils, due in part to the clay content, type of clay present, and lime content, influenced the patterns of terricolous lichen colonization. On soils with relatively high shrinkage rates, *Squamaria* species were unable to establish themselves; instead, they were frequently found on mosses and rocks, and the community was dominated by *Psora decipiens* and *Dermatocarpon hepaticum*, species able to tolerate high soil shrinkage.

The investigators suggested that the different responses of lichens to soil shrinkage rates may be related to thallus morphology. They reasoned that continuous expansion and contraction of the substrate would tend to tear *Squamaria* thalli, whereas the squamulous construction of *D. hepaticum* and *P. decipiens* would not be as severely disturbed.

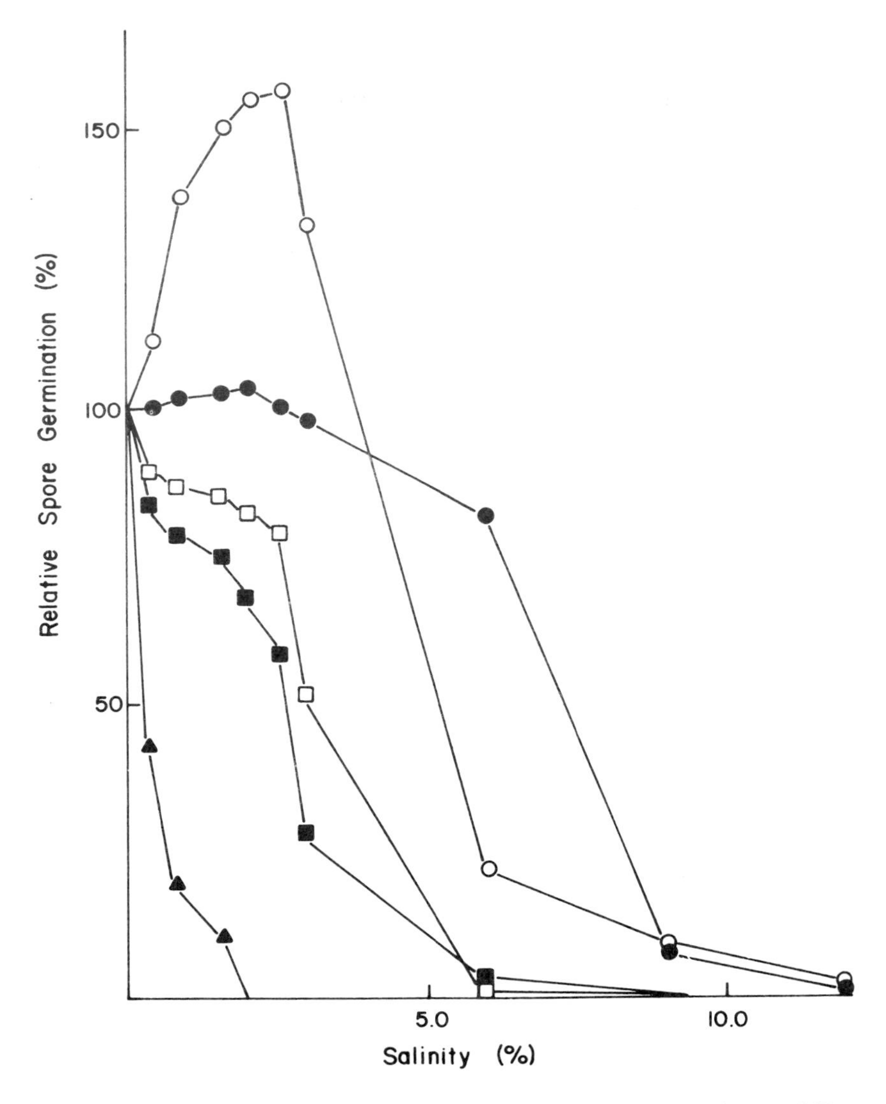

FIGURE 10.8. Relative spore germination of lichens from different Swedish intertidal zones in different salinity media. (●) *Verrucaria maura*; (○) *Caloplaca marina*; (■) *Xanthoria parietina*; (□) *Caloplaca scopularis*; (▲) *Lecanora muralis*. Adapted from K. Ramkaer, *Bot. Tidsskrift* 72:119–123 (1978), *Nordic Journal of Botany*, with permission.

Habitat Specificity on Rocks

Saxicolous lichens exhibit different types of habitat specificity. They may frequent certain kinds of rock, establishing themselves most commonly on calcareous rocks, for example (Asta & Lachet, 1978). They may also exhibit preferences for certain rock microhabitats (Foote, 1966; Yarranton & Green, 1966; Bates, 1975; West & Stotler, 1977). Slope and aspect are frequently found to influence distribution patterns on rocks (Wirth, 1972; Armstrong, 1975; Pentecost, 1979; Larson, 1980b).

Pentecost (1979) has developed a unique ordination technique that allows an investigator to plot species distributions in relation to slope and aspect simultaneously. Using this technique, he found that some lichens growing on boulders in North Wales had cover patterns that varied with slope and aspect. *Lecidea tumida* and *Fuscidea tenebrica* were distributed evenly over the rock surfaces, with *F. tenebrica* the dominant of the two; however, *F. cyathoides* and *Rhizocarpon geographicum* appeared to avoid steep north-facing slopes. Use of this technique by other investigators might help to simplify and standardize microhabitat preference data collected in saxicolous communities.

Habitat Selection by Foliicolous Lichens

Leaf surfaces are frequently colonized by lichens and mosses in tropical regions, and a number of interesting investigations of these communities have

TABLE 10.8. Relative Frequencies and Confidence Intervals (in percent) for Some Foliicolous Lichens on Fern and Dicotyledonous Leaves from Chocó, Colombia

	Relative Frequencies, %	
Species	*Fern Leaves*	*Dicot Leaves*
Gyalectidium filicinum	0–10.45*	18.84–44.12
Mazosia phyllosema	13.15–43.99	3.82–22.10
Phyllobathelium epiphyllum	18.07–50.49	5.17–24.45
Porina epiphylla	55.99–86.84	30.91–57.97
Stirtonia sprucei	10.79–40.63	0–8.84
Strigula elegans	0–8.53	10.90–33.54
S. nemathora	25.98–59.72	3.82–22.10
Tricharia albostrigosa	0–8.53	12.43–35.71
Trichothelium epiphyllum	52.72–84.42	5.17–24.45

*Confidence intervals of the relative frequencies of occurrence on each leaf type at the 95 percent significance level (see Nowak & Winkler, 1975 for details on data analysis).

Partial data adapted from R. Nowak & S. Winkler. 1975. "Foliicolous lichens of Chocó, Colombia, and their substrate abundances." *Lichenologist* 7:53–58. Reprinted with permission of Academic Press.

appeared (Santesson, 1952; Nowak & Winkler, 1970; 1972; Vězda, 1975). Many foliicolous lichens exhibit preferences for certain kinds of leaves. An analysis of the foliicolous lichen flora of Chocó, Colombia (Nowak & Winkler, 1975) revealed that some species were restricted to dicotyledonous leaves and others to fern leaves. Of those that are found on both leaf types, some had higher frequencies on one or the other (Table 10.8), although there were generally wide variations for all species.

Santesson (1952) has discussed the peculiar environmental constraints common to all foliicolous lichens. Of these, the most important is leaf persistence. Obligately foliicolous species are most often found on leaves with long durability, although there are some species reported from deciduous (annual) leaves. Generally, lichens found on leaves of very long durability are not obligate foliicoles but also belong to the corticolous flora, indicating that the obligately foliicolous lichens are perhaps restricted to their habitats because they are relatively poor competitors in other habitats. This hypothesis needs to be tested experimentally.

11 Endolithic Lichens

Lichens occupy some of the most inhospitable terrestrial environments on earth, from hot, arid and semiarid regions (Rogers, 1977) to cold, polar deserts (Lindsay, 1977). This impressive ability to tolerate extreme environments is partly physiological (e.g., the well-known freezing tolerance of many polar lichens) and partly behavioral (e.g., some lichens avoid extreme conditions by occupying protected microhabitats). The physiological tolerance of lichens has been actively studied (Kappen, 1973); only recently has attention been paid to lichen behavioral responses to extreme environments.

This chapter will discuss one of these behavioral responses, endolithic growth. Endolithic lichens live inside rocks in hot and cold desert environments too extreme to support normal epilithic growth. Lichens are not the only organisms capable of endolithic growth, nor are they the most successful. Recent investigations have revealed entire communities of microorganisms (mainly algae and microfungi) that occupy this unusual ecological niche in a variety of habitats.

Most of what we know about the biology of endolithic organisms comes from the work of E.I. Friedmann and various colleagues. His review with M. Galun of desert algae, lichens, and fungi (Friedmann & Galun, 1974) is the best introduction to the field and identifies a number of important references.

Although the free-living algae and fungi are numerous and ecologically important, the discussion here will be limited to endolithic lichens and will concentrate mainly on the recent work of Friedmann and coworkers on the biology of Antarctic cryptoendolithic lichens. The recent work of David Porter on marine endolithic lichens will also be mentioned briefly.

TABLE 11.1. Lithobiontic Terminology

Epiliths	Colonize external surfaces of rocks
Endoliths	Colonize the interior of rocks
Chasmoendoliths	Colonize fissures and cracks in the rock (*chasm:* cleft)
Cryptoendoliths	Colonize structural cavities within porous rocks, including spaces produced by euendoliths (*crypto:* hidden)
Euendoliths	Penetrate actively into the interior of rocks forming tunnels that conform with the shapes of their bodies; rock-boring organisms (*eu:* good, true)

S. Golubic, E.I. Friedmann, & J. Schneider. 1981. "The lithobiontic ecological niche, with special reference to microorganisms." *J. Sedimentary Petrology* 51:475–478. Reprinted with permission of the Society of Economic Paleontologists and Mineralogists and E.I. Friedmann.

TERMINOLOGY

Golubic, Friedmann, and Schneider (1981) recently proposed a standardized terminology to be used in discussions of lithobionts (organisms that live both in and on hard rock substrates). It combines information about location of organisms on or in the substrate with biological information. According to this scheme (Table 11.1), *epiliths* live on the substrate surface, and *endoliths* live within the substrate to one degree or another. *Endoliths* are of three types: *chasmoendoliths* occupy fissures and cracks within rocks and may be partially exposed; *cryptoendoliths* occupy pores and other preexisting structural cavities within the rock; *euendoliths* are boring microorganisms that penetrate into relatively soluble rock substrates such as carbonates and phosphates.

As Golubic, Friedmann, and Schneider (1981) have pointed out, not all microorganisms can be clearly classified by this scheme. For lithobiontic lichens, however, few growth forms intermediate between epilithic and the various endolithic types are known to occur. Therefore, the terminology proposed by Golubic and coworkers will be used throughout this discussion.

CRYPTOENDOLITHIC LICHENS

Hot Desert Cryptoendoliths

Cryptoendolithic lichens are not frequently found in hot deserts, although Hale (personal communication) mentions the occurrence of *Lecidea* species within porous sandstones in western North American deserts. Those endolithic lichen species known from desert regions, primarily species of the Verrucariaceae, have their thallus partly or entirely sunk in the substrate (Friedmann & Galun, 1974) and are therefore not truly cryptoendolithic.

The dominant cryptoendolithic organisms in hot deserts are prokaryotes, including both blue-green algae and bacteria (Friedmann, 1980). These organisms have been discussed in a number of articles (Friedmann, Lipkin, & Ocampo-Paus, 1967; Friedmann, 1971; Friedman & Galun, 1974). They generally occur as monospecific populations in light-colored, porous rocks of various types. Their main source of water is thought to be early morning dewfall. Friedmann (1980) has suggested that the combination of high temperatures and extreme aridity imposes environmental stresses so severe that eukaryotic organisms are unable to survive.

Arctic-Alpine Cryptoendoliths

Endolithic lichens of arctic-alpine areas are generally modified stages of species that normally exhibit epilithic growth. True cryptoendolithic lichen growth in these regions is not known, but these organisms have not been adequately studied. Friedmann (1971) cites Bachmann (1915) and Ercegović (1925) as the basic sources of information about these lichens.

Antarctic Cryptoendoliths

The dry valleys are not lifeless, however. Friedmann and colleagues have found that numerous microorganisms are able to live here inside certain types of rock. In 1974, cryptoendolithic cyanobacteria were discovered in the radiation. The area presents a conspicuously lifeless aspect (Figure 11.1), rocky, ice-free, and barren.

The dry valleys are not lifeless, however. Friedmann and colleagues have found that numerous microorganisms are able to live here inside certain types of rock. In 1974, cryptoendolithic cyanobacteria were discovered in the Beacon sandstone formation of the dry valleys (Friedmann & Ocampo, 1976). Later investigations (Friedmann, 1977; 1982) indicated that the cyanobacterial communities were relatively rare; the microflora appeared to be dominated instead by various chasmoendolithic and cryptoendolithic lichens, mainly *Buellia, Lecanora, Lecidea*, and *Acarospora* species.

Cryptoendolithic lichens characteristically inhabit a narrow subsurface zone up to 10 millimeters thick in the sandstone and form colonies from a few centimeters to a meter in diameter, depending on the availability of substrate (Friedmann, 1982). The thallus is organized in a fashion similar to epilithic lichens, but true mycobiont tissues are rarely formed; instead, fungal hyphae fill the available pore space within the rock. A scanning electron micrograph

FIGURE 11.1. Landscape in the high-desert areas of the dry valleys, University Valley, Southern Victoria Land, Antarctica. From E.I. Friedmann. 1982. Reprinted from *Science*, Vol. 215, pp.1045–1053. Copyright 1982 by the American Association for the Advancement of Science.

of these pore spaces (Figure 11.2) shows the rather loose association of fungal and algal cells within the sandstone.

In cross section, a typical cryptoendolithic lichen thallus exhibits a number of distinct bands (Figure 11.3). A black zone (Figure 11.3a) is observed just below the rock surface and usually contains numerous cells of a *Trebouxia* phycobiont; below this is a white zone (Figure 11.3b) of mycobiont medullary tissue and then a green layer (Figure 11.3c) of non-lichenized green algal cells (one population has recently been identified as *Hemichloris antarctica* Tschermak-Woess & Friedmann (Tschermak-Woess & Friedmann, 1984). There is also sometimes a blue-green layer of free-living unicellular cyanobacteria (Figure 11.3d). Thus, a single cryptoendolithic lichen may in reality be a complex community of interacting microorganisms.

Cryptoendolithic lichens only occasionally form sexual reproductive structures. These are always produced on the rock surface and consist of

FIGURE 11.2. Phycobiont cells and mycobiont hyphae in airspaces within Beacon sandstone (×1000). From E.I. Friedmann. 1982. Reprinted from *Science*, Vol. 215, pp.1045–1053. Copyright 1982 by the American Association for the Advancement of Science.

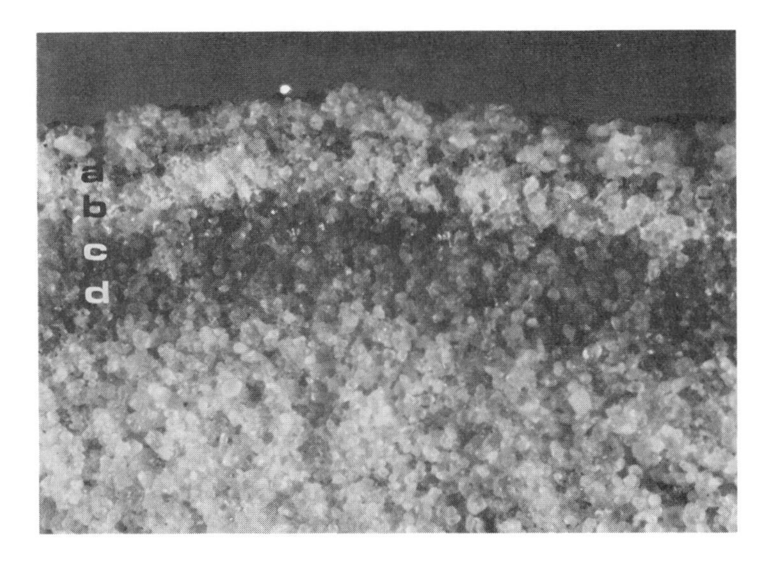

FIGURE 11.3. Cryptoendolithic microorganisms in vertically fractured Beacon sandstone. (a) Black lichen zone; (b) White lichen zone; (c) Green zone of free-living *Hemichloris antarctica* Tschermak-Woess & Friedmann; (d) Blue-green zone of free-living unicellular cyanobacteria. From E.I. Friedmann. 1977. *Antarctic J. U.S.* 12: 26–30. Reprinted with permission from the National Science Foundation and the author.

algae-containing areoles (Friedmann, 1982). Pycnidia are immersed in the cortex and prothalloid layers of the areoles; they may also be found in the pigmented cryptoendolithic layer (Friedmann, 1982).

Lichen compounds are sometimes produced by cryptoendolithic lichens (Hale, personal communication). These compounds are always produced in the subsurface layers, and their production enables investigators to tentatively identify sterile cryptoendolithic material. It is obvious, however, that identifications based on chemistry alone are tentative at best.

Chasmoendolithic lichens frequently colonize cracks that form in numerous dry valley rocks, particularly granite, granodiorite, and dolerite. These lichens exhibit a more typical areolate appearance and frequently produce reproductive structures (Figure 11.4). They may become epilithic under certain circumstances (Figure 11.5).

Colonization Patterns, Weathering, and Mineral Mobilization

In the dry valleys, Beacon sandstone surfaces with northern exposures are more frequently colonized by cryptoendolithic lichens than those facing south

FIGURE 11.4. Silicified Beacon sandstone fractured vertically along an existing fissure showing chasmoendolithic lichens (×3.3). From E.I. Friedmann. 1977. *Antarctic J. U.S.* 12: 26–30. Reprinted with permission from the National Science Foundation and the author.

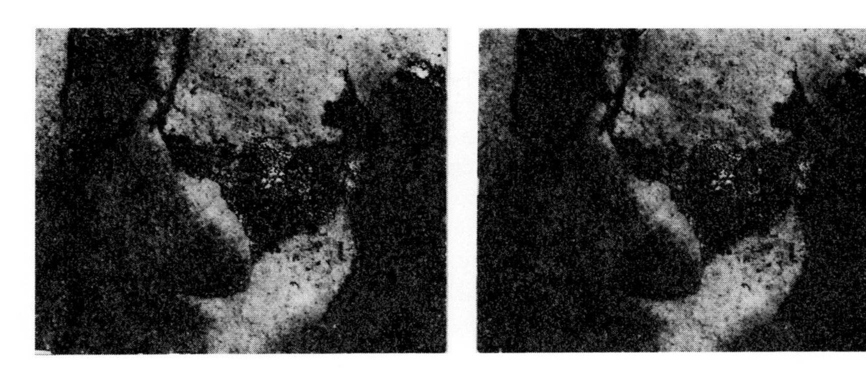

FIGURE 11.5. Stereophotograph of epilithic areolate stage of cryptoendolithic *Buellia* sp. where surface crust was lost as a result of exfoliation. Natural size. From E.I. Friedmann. 1982. Reprinted from *Science*, Vol. 215, pp. 1045–1053. Copyright 1982 by the American Association for the Advancement of Science.

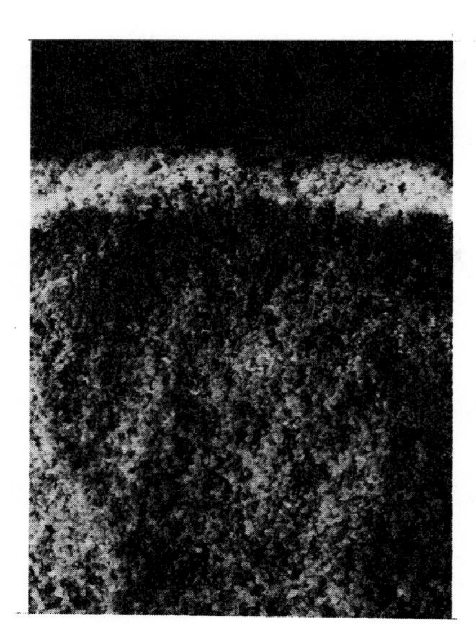

FIGURE 11.6. Cryptoendolithic lichen in iron-rich sandstone, University Valley, Antarctica. The lichen zone is bleached white, whereas zones above and below are darker. Results of chemical analysis of each layer are in Table 11.2. From E.I. Friedmann. 1982. Reprinted from *Science,* Vol. 215, pp. 1045–1053. Copyright 1982 by the American Association for the Advancement of Science.

because they receive more solar radiation and are less exposed to katabatic (downslope) winds descending from the ice plateau (Friedmann, 1982). Apparently, subsurface sandstone layers are solubilized by lichen hyphae, and the upper surface peels away to expose the lichen tissues. Hyphae then penetrate the rock again, and additional layers are sliced off. This biogenous weathering results in a characteristic pockmarked surface on northern sandstone surfaces.

Cryptoendolithic lichens from Beacon sandstones are also apparently able to mobilize metals, particularly iron compounds, which gives the lichen zone a bleached appearance (Figure 11.6) and probably also contributes to the exfoliative pattern. Mineral analysis of the different rock layers (Table 11.2) shows that metal compounds mobilized by the lichen tissues are being carried away by water in two directions: downward with snow melt and upward by capillary movement (Friedmann, 1982). The lichen zone contains very low concentrations of metals.

TABLE 11.2. Concentrations of Metals at Different Levels in a Sample of Beacon Sandstone Colonized by Cryptoendolithic Lichens. The Data Are Given as Percentages of the Weight of the Sample (See Friedmann (1982) for Experimental Details).

Zone	Na_20	K_20	*Mg0*	*Ca0*	*Mn0*	*Fe0*	*Zn0*	*Pb0*
Upper rock crust	0.03	0.09	0.02	0.06	0.005	0.60	0	0.0003
Leached lichen zone	0	0	0	0	0	0.01	0	0
Dark brown zone below leached lichen zone	0	0.01	0	0	0	0.44	0	0.0010
Bedrock	0	0.05	0.01	0	0.001	0.18	0	0.0003

ECOPHYSIOLOGY OF ANTARCTIC CRYPTOENDOLITHIC LICHENS

Water and temperature appear to be the most important limiting factors for cryptoendolithic lichens in the dry valleys. Investigations carried out so far indicate that temperature and moisture conditions below the rock surface are significantly more favorable for lichen growth than conditions at the surface (Kappen, Friedmann, & Garty, 1981; Kappen & Friedmann, 1983). These observations combined with results of ecophysiological studies of the lichens themselves suggest that cryptoendolithic lichens are not better adapted to colder or drier conditions than are crustose lichens from more favorable maritime Antarctic regions (Friedmann, Friedmann, & McKay, 1981; Kappen & Friedmann, 1983). Rather, "the special adaptive achievement of cryptoendolithic lichens is their ability to occupy [their] niche by changing their pattern of growth from plectenchymatous organization to a filamentous growth form" (Friedmann, Friedmann, & McKay, 1981, p. 68).

Water

Melting snow is thought to be an important source of water (Friedmann, 1978). Meltwater entering the porous sandstone is retained many days after a snowfall. Just below the surface, therefore, relative humidities are consistently higher than at the surface where evaporative water loss is high (Kappen, Friedmann, & Garty, 1981). The requirement for meltwater probably also explains the absence of lichens on steep or vertical rock faces where snow cannot accumulate (Kappen, Friedmann, & Garty, 1981).

Temperature

The subsurface lichen zone is not only more moist; it is also generally warmer and subjected to smaller temperature fluctuations than the rock surface. One series of measurements made on a sandstone boulder at Linnaeus Terrace, Asgard Range on December 25, 1980 (Friedmann, Friedmann, & McKay, 1981) showed that during a period of 42 minutes, surface temperatures fluctuated between -1.8 ° C and +5.9 ° C and crossed the 0 ° C line 14 times. During the same period, however, temperatures in the lichen zone never went below 0 ° C and ranged from 1.7 ° to 6.1 ° C.

Gas Exchange

Gas exchange experiments indicate that cryptoendolithic lichens are not physiologically different from other polar lichens, although results of these experiments are frequently difficult to interpret because, in nature, CO_2 exchange by cryptoendolithic lichens takes place very slowly through a relatively thick surface crust (Kappen & Friedmann, 1983). This may result in recycling of an internal CO_2 pool, elevating actual photosynthetic activity to levels higher than those measured under laboratory conditions (Kappen & Friedmann, 1983).

Light

Light availability to cryptoendolithic lichen phycobionts varies with rock type and exposure. Although few measurements have been made, it appears from available information that ambient light intensities in the summer are quite high (up to 1500 $\mu Em^{-2} sec^{-1}$); about one percent of this reaches the lichen zone inside the rock (Kappen, Friedmann, & Garty, 1981; Friedmann, Friedmann, & McKay, 1981).

Excessive radiation, especially UV radiation, is probably screened out by the dark-pigmented fungal layer frequently observed in cryptoendolithic lichens. The dark layer is best developed in colonies exposed to direct summer sunlight; colonies in shaded areas have poorly developed dark layers. Dark pigmentation is also characteristic of epilithic lichens from high altitudes (Kappen, 1973) and cryptoendolithic cyanobacterial colonies in hot desert rocks (Friedmann, 1971; Friedmann & Galun, 1974).

Nitrogen Economy of Endolithic Microbial Communities

The physically isolated nature of endolithic microbial communities raises questions about the source of essential elements, especially nitrogen, and the ways in which element cycling processes in endolithic systems differ from those of other systems. Friedmann and Kibler (1980) have investigated the nitrogen economics of hot and cold desert endolithic communities and have made a number of discoveries.

Their most important finding is that biological nitrogen fixation plays a very limited role in the nitrogen economy of endolithic communities. Instead, the main source of nitrogen appears to be abiotically fixed nitrogen produced by atmospheric electric discharges (lightning or aurorae), conveyed to the

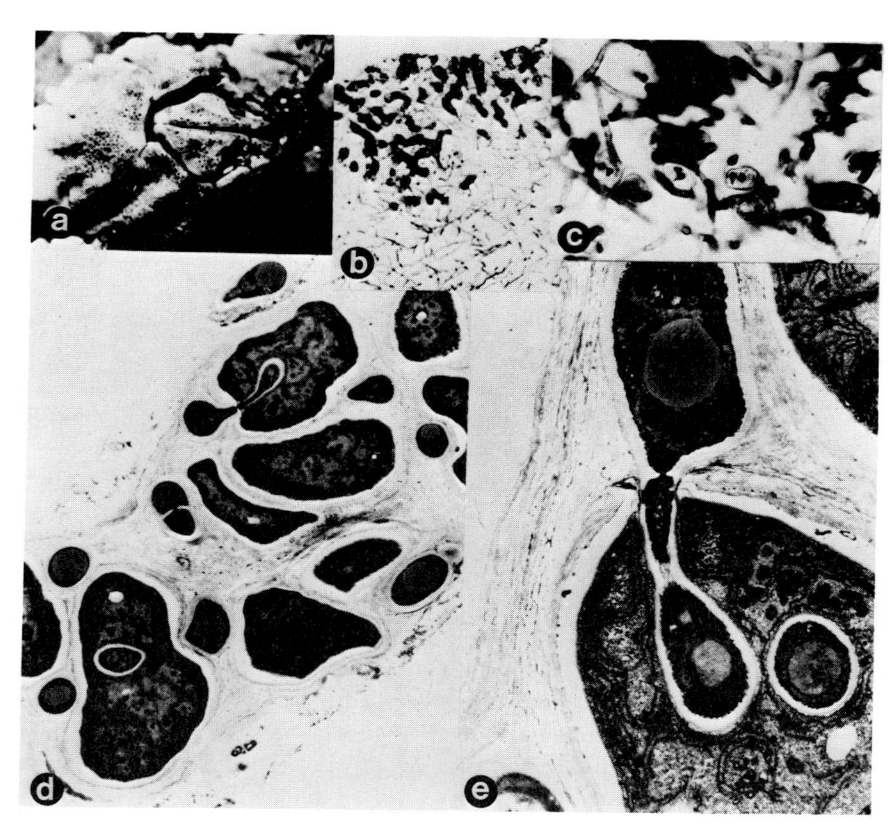

FIGURE 11.7. Structure of the endolithic lichen *Arthopyrenia halodytes*. (a) Barnacle (*Balanus glandula*) with shell heavily invaded by *A. halodytes*; dark pits are fungal reproductive structures (×5); (b) Cross section through a calcified barnacle shell invaded with *A. halodytes* (stained with GMS, ×300); (c) Detail of a cross section through a barnacle shell with *A. halodytes*; haustorial pegs are visible in all algal cells (stained with GMS, ×2000); (d) Thin section through an algal filament with closely associated fungal hyphae (×20,000); (e) Detail of haustorium shows no cytological degeneration in the prokaryotic algal cells (×60,000). Provided by David Porter.

rock by atmospheric precipitation. This contrasts sharply with the generally high levels of biological nitrogen fixation measured in other desert systems (Friedmann & Kibler, 1980). However, the low productivities and turnover rates estimated for endolithic communities suggest that nitrogen is not a limiting factor in these communities.

MARINE ENDOLITHIC LICHENS

David Porter of the University of Georgia has recently described an unusual endolithic lichen that lives on barnacles, limpets, and snails in intertidal habitats in Europe and North America. The lichen, *Arthopyrenia halodytes* Swinscow, is composed of a blue-green phycobiont, *Hyella caespitosa*, with fungal hyphae embedded in the algal filaments. An invaded barnacle (Figure 11.7a) exhibits numerous pycnidia and perithecia that pit the surface. Light microscopy of fixed and decalcified shells (Figure 11.7b & c) shows that the fungus and alga are always intimately associated in the outer shell layers. Ultrastructural analysis (Figure 11.7d & e) shows that the fungal hyphae grow primarily in the algal lamellar sheath, and that haustorial penetrations into phycobiont cells result in little cytological disturbance to the algae.

This lichen has characteristics similar to other endolithic lichens: it lives in a physically harsh environment (although not nearly as harsh as desert or Antarctic endoliths), its vegetative thallus grows inside a mineral substrate, and sexual reproductive structures are produced at the surface. Ecophysiological studies of these and other marine endolithic lichens may reveal other similarities.

12 Lichens and Pollution

Nylander (1866c) in Paris and Grindon (1859) in Manchester first documented the disappearance of lichen species and blamed it on increased air pollution levels in these cities. Since that time, sensitivity of lichens to various kinds of pollutants has stimulated interest in their use as biological monitors of environmental quality. Indeed, to many botanists unfamiliar with lichen biology, all lichen research is thought to somehow involve the study of pollution. This is obviously an exaggerated impression, but it accurately reflects the important contributions that lichen pollution studies have made to the advancement of our knowledge of basic lichen biology.

In this chapter are reviewed the various areas of lichen-pollution research, emphasizing recent advances and techniques. Interested readers will find some excellent reviews of the subject by Hawksworth (1971), Deruelle (1978), Ferry, Baddeley, and Hawksworth (1973), and Nash (1976a). Also, the British Lichen Society publishes a bibliography, "Literature on air pollution and lichens," in each issue of *The Lichenologist*.

LICHENS AS BIOINDICATORS

Nash and Sigal (1981) have discussed various ecological approaches to the use of lichenized fungi as indicators of air pollution and have pointed out that there are a number of advantages to the use of lichens. The measurement of pollution levels using analytical equipment is usually costly and frequently impractical, particularly over large areas. Data on lichen distribution and physiological activity are relatively easy and inexpensive to collect over as large a geographical area as desired.

Changes that are known to occur in lichen thalli exposed to various types of pollution are listed in Table 12.1 Because they do not possess a cuticle, stomates, or other specialized protective structures, lichens are unable to limit the flow of materials from the atmosphere to the thallus. When the thallus is dry and physiologically inactive, damage from airborne pollution is not particularly severe, but when the thallus is hydrated and active, toxic pollutants in rainwater are rapidly absorbed by lichen tissues and may cause severe structural and physiological damage.

Another useful characteristic of lichen thalli is their capacity to

TABLE 12.1. Changes Observed in Lichens as a Consequence of Air Pollution

External:

- Changes in thallus color.
- Decrease in thallus size, length of lobes, thallus reflectivity, thallus adhesion, and development of soredial and isidial structures.
- Increase in thallus thickness, surface cracks, and deposition of extracellular substances in and on the thallus.

Anatomical:

- Increase in the number of dead and plasmolyzed algal cells.
- Decrease in algal cell size, the number of dividing algal cells, and the degree of contact between symbionts.

Physiological:

- Changes in the living or dead status of algal cells.
- Decrease in net CO_2 assimilation and respiration, including a possible temporary increase in respiration.
- Decrease in N_2-fixation.
- Decrease in relative growth rate.

Changes in Chemical Concentrations:

- Increase in concentrations of pollutants and phaeophytin in acid conditions.
- Decrease in total chlorophyll and the chlorophyll a/chlorophyll b ratios.
- Changes in thallus pH and conductivity.
- Potassium efflux.

Data adapted from M. Kauppi and A. Mikkonen. 1980. "Floristic versus single species analysis in the use of epiphytic lichens as indicators of air pollution in a boreal forest region, northern Finland." *Flora* 169:255–281.

accumulate materials from the ambient environment. Materials may be accumulated by active uptake of ions from rainwater, passive adsorption of ions by ion exchange, or direct incorporation of particulate materials into lichen tissues. Materials like sulfur dioxide (SO_2) and numerous metal elements that are frequently components of industrial air pollution may be accumulated by lichens to concentrations many times that found in the immediate environment at any given time. Lichens are therefore useful as indicators of long-term, low-level pollution that is difficult to monitor with mechanical monitoring devices.

Lichen propagules, both sexual and asexual, appear to be particularly sensitive to various types of pollution. The algae of lichen soredia and isidia are easily damaged by SO_2 even at very low levels (Margot, 1973; Margot & De Sloover, 1974). Lawrey and Hale (1979) found that growth rates of minute lichen thalli were severely reduced near a heavily travelled highway, whereas large thalli appeared to be more resistant to pollution stress. This is probably related to algal sensitivity to pollution since chlorosis of young thalli was frequently observed in polluted sites. Saunders (1966; 1970) found that the ascospore germination of several nonlichenzied fungi was depressed in the presence of SO_2, suggesting that fungal propagules are also sensitive to pollution stress. Although there is little information on the effects of pollution on lichen ascospore germination, Pyatt (1969) found that SO_2 pollution depressed the ascospore production and dissemination of numerous lichen species. All of these observations suggest that pollution may affect the colonizing ability of lichens and thereby influence lichen community composition in affected areas. Numerous indices of atmospheric quality take advantage of this by correlating pollution levels with changes in lichen species diversity.

There are a number of problems with using lichens, however, of which persons interested in beginning lichen pollution studies should be aware (Johnsen, 1976). First, since most preliminary lichen pollution studies involve an evaluation of lichen species distribution patterns in relation to a presumed pollution source, it is well to consider the difficulties involved in sampling and identifying specimens. A detailed knowledge of lichen taxonomy is obviously required, and this may prove difficult to acquire. Second, lichen surveys do not, by themselves, demonstrate the effects of pollution. They provide data on distribution patterns that may or may not reflect pollution stress. Field measurements of pollution levels using analytical equipment and controlled laboratory study of the effects of these levels on lichen activity are both required to document the actual causes of the patterns observed. Third, lichen species do not all respond to pollution in the same way. Some species are especially sensitive to certain types of pollution and others are especially tolerant. The distribution patterns of these indicator species are of greater value than those whose responses to pollution are more plastic and less clearly defined. A great deal of preliminary work is required before indicator species can be identified and experiments properly designed.

These difficulties notwithstanding, lichen pollution studies have been done with increasing frequency during recent years, and information from these studies has helped to document both the effects of man-caused environmental disturbances and the success of efforts to ameliorate these disturbances. It is therefore expected that the use of lichens as biological indicators will continue in the future, especially as more laboratory information becomes available on the physiological responses of lichens to various pollution types.

Mapping

Once lichen species distributions have been established, it is possible to evaluate simultaneously the individual responses of numerous species to pollution sources by producing maps. Numerous maps have been produced around cities, industrialized areas, and specific pollution sources. Many new maps find their way into the literature each year and are referenced by the British Lichen Society, along with other papers on lichen pollution studies, in each issue of *The Lichenologist*. Hawksworth (1971; 1973b), Mellanby (1978) and others have reviewed the literature on mapping techniques and the application of various methods to lichen air pollution studies.

A problem encountered by lichenologists interested in using lichen distribution data to reveal patterns in air quality is the fact that so much information is collected it becomes difficult to interpret. One of the more popular methods used to reduce distribution data to a level suitable for mapping is to use an "Index of Atmospheric Purity" (IAP) developed by De Sloover and LeBlanc (1968) and later modified by LeBlanc and De Sloover (1970) and Hoffman (1974). The index is as follows:

$$IAP = \sum_{i=1}^{n} \frac{(Q_i \times f_i)}{10}$$

where Q_i is the "ecological index" for species i of n species at a particular site and f_i is the coverage scale value, a number from 1 to 5 that indicates the degree of coverage exhibited by species i at a particular site. The entire term is divided by 10 simply to produce a smaller, more appealing number. Hoffman (1974) substituted a term for f_i that combined a "vigor-vitality" rating and a "sociability" rating; this eliminated the subjectivity in the coverage scale value of LeBlanc and De Sloover's index. The ecological index (Q_i) is calculated at each site by determining the mean number of neighboring species found for each species at the site. The assumption is that a large number of neighbors will be found cooccurring with species from unpolluted areas and that toxitolerant species from polluted areas will be found with fewer neighbors. It is also assumed that all sites are equivalent in terms of substrate and microclimate characteristics, and that the level of pollution is the only factor influencing species richness at each site. Therefore, IAP measurements that include this term are assumed to accurately reflect the relative atmospheric purity.

De Sloover and LeBlanc (1968) used this technique to map an industrial valley in northwest Belgium and the area in and around the city of Montreal (LeBlanc & De Sloover, 1970). The Montreal study resulted in production of a baseline map, an overlay of the various zones of IAP values, and several distribution maps showing the frequency (abundant, rare, or absent) and reproductive condition (sterile or fertile) of several important lichen and bryophyte species. The patterns that emerged from these maps correlated well with the known and suspected levels of SO_2 in the region.

Hoffman (1974) has also mapped IAP values calculated using epiphytic cryptogam distribution data collected from *Robinia pseudoacacia* trees around a paper pulp mill near Lewiston, Idaho. He found no evidence of a lichen desert caused by air pollution, a phenomenon frequently observed in lichen-mapping studies done in urban areas. However, there were marked alterations in the cryptogam distribution patterns in relation to the pollution source. *Xanthoria fallax* was the most abundant lichen species encountered, along with *Physconia grisea, Phaeophyscia orbicularis, Physcia adscendens*, and *X. candelaria*. These species were considered to be the most pollution tolerant and were very close to the pulp mill. Other species, particularly bryophytes, were less tolerant and found at greater distances from the mill. Therefore, cryptogam species diversity increased steadily with distance from the mill, and this increase was reflected in the calculated IAP values.

Showman (1975) surveyed the distribution of lichen species at 128 sites around a coal-fired power plant in a rural area of southeastern Ohio (Figure 12.1). The distribution patterns of several corticolous lichen species, particularly *Pseudoparmelia caperata* (Figure 12.2), were found to be useful indicators of changes in atmospheric SO_2 levels in the area. Of the 128 stations used, those nearest the power plant and downwind exhibited the highest SO_2 levels; these stations also tended to be void of sensitive species. *Pseudoparmelia caperata* was found to be most sensitive to SO_2 pollution, followed by *Parmelia rudecta; Physcia millegrana* was most resistant. Terricolous and saxicolous lichen species were apparently not as sensitive to pollution as corticolous species because they occupied more protected microhabitats than corticolous species.

One of the most ambitious lichen-mapping projects attempted to date was organized in 1971 by the Advisory Center for Education in Britain and done by 15,000 school children. Each child was provided a clean air kit with

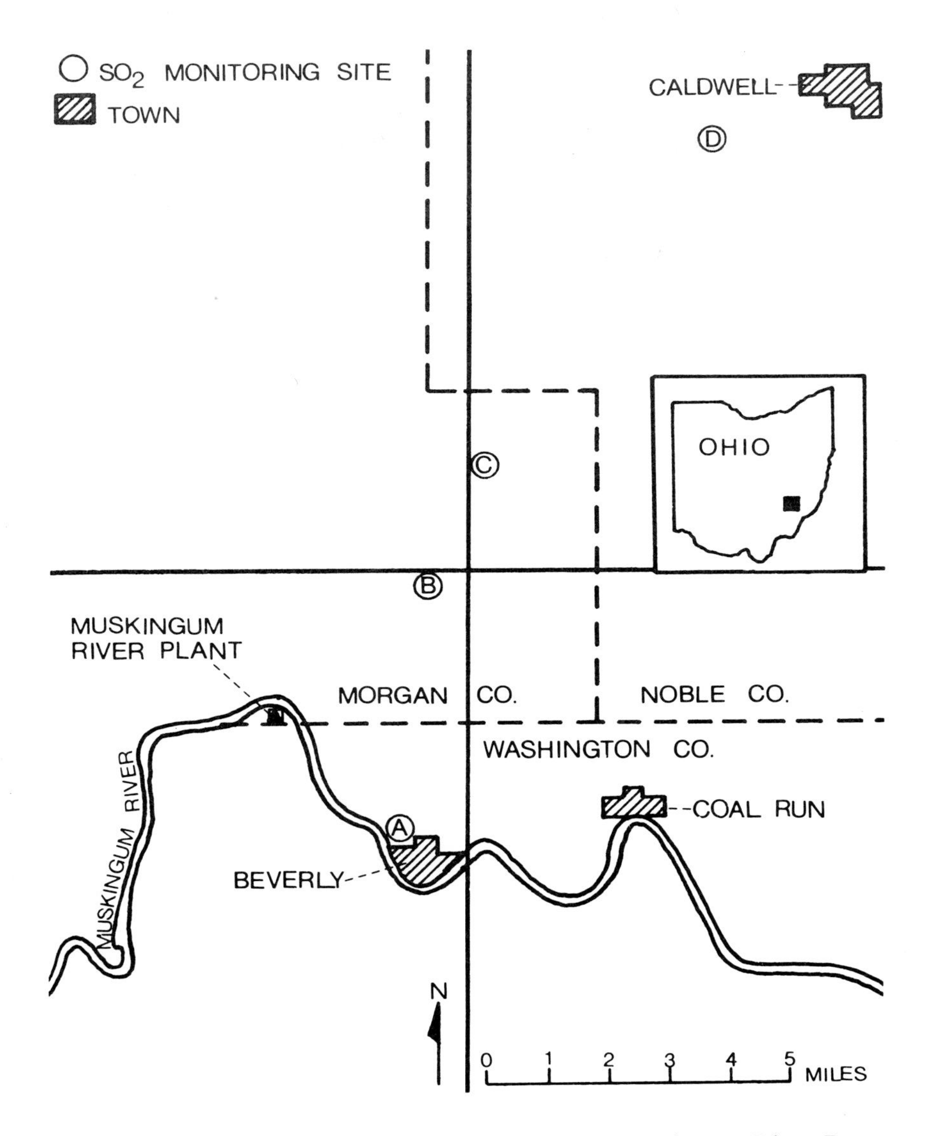

FIGURE 12.1. Base map of study area around Muskingum River Power Plant, Morgan County, Ohio. SO_2 monitoring stations A-D indicated. From R.E. Showman. 1975. "Lichens as indicators of air quality around a coal-fired power generating plant." *Bryologist* 78: 1–6. Reprinted with permission of the American Bryological and Lichenological Society and the author.

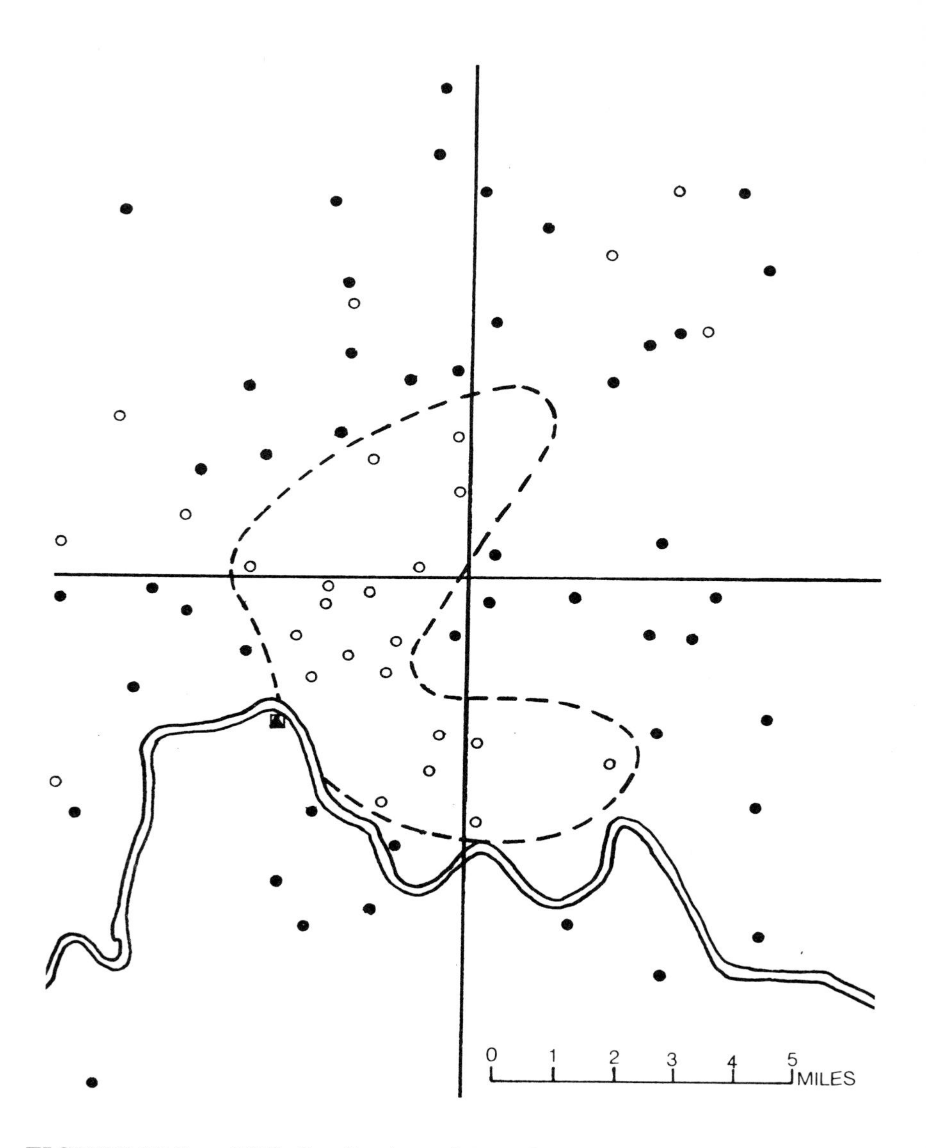

FIGURE 12.2. 1973 distribution of *Pseudoparmelia caperata* in Morgan County, Ohio study area: ● = present; ○ = absent. From R.E. Showman. 1975. "Lichens as indicators of air quality around a coal-fired power generating plant." *Bryologist* 78: 1–6. Reprinted with permission of the American Bryological and Lichenological Society and the author.

instructions for surveying his or her area, lichen species identification charts, and materials to be used in determining the pH of rainwater and lichen substrates. The identification of only a few easily identifiable and especially good indicator species (*Lecanora conizaeoides, Xanthoria parietina, Evernia prunastri, Usnea* species, etc.) was required. The results (Figure 12.3) pinpointed clearly the most heavily polluted areas in Great Britain.

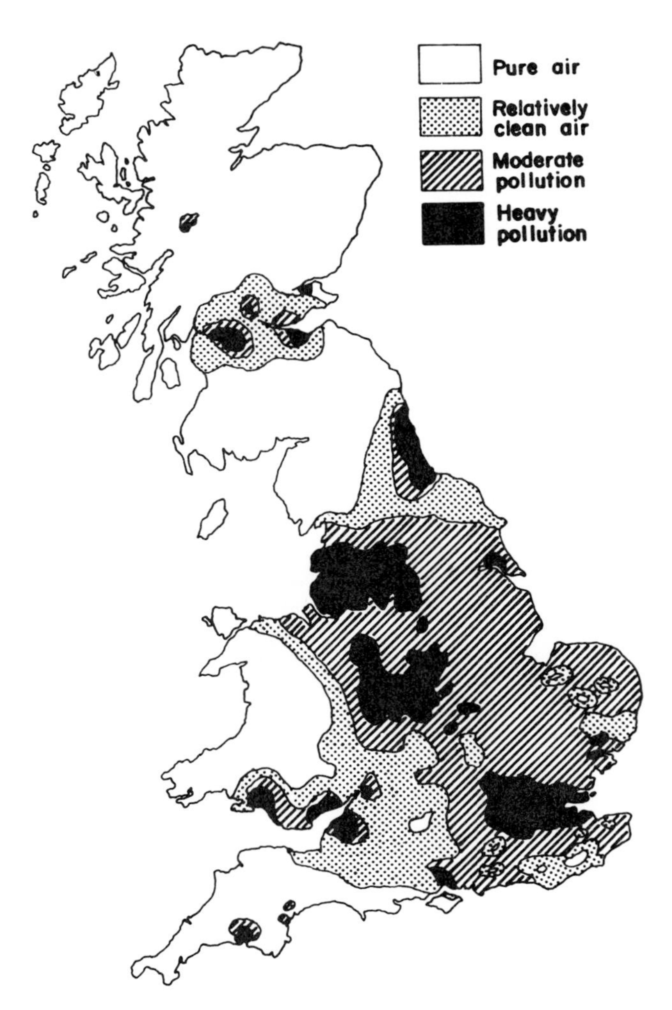

FIGURE 12.3. Air quality of Britain assessed by 15,000 school children who examined the distribution of several lichen species. Redrawn with permission of O.L. Gilbert.

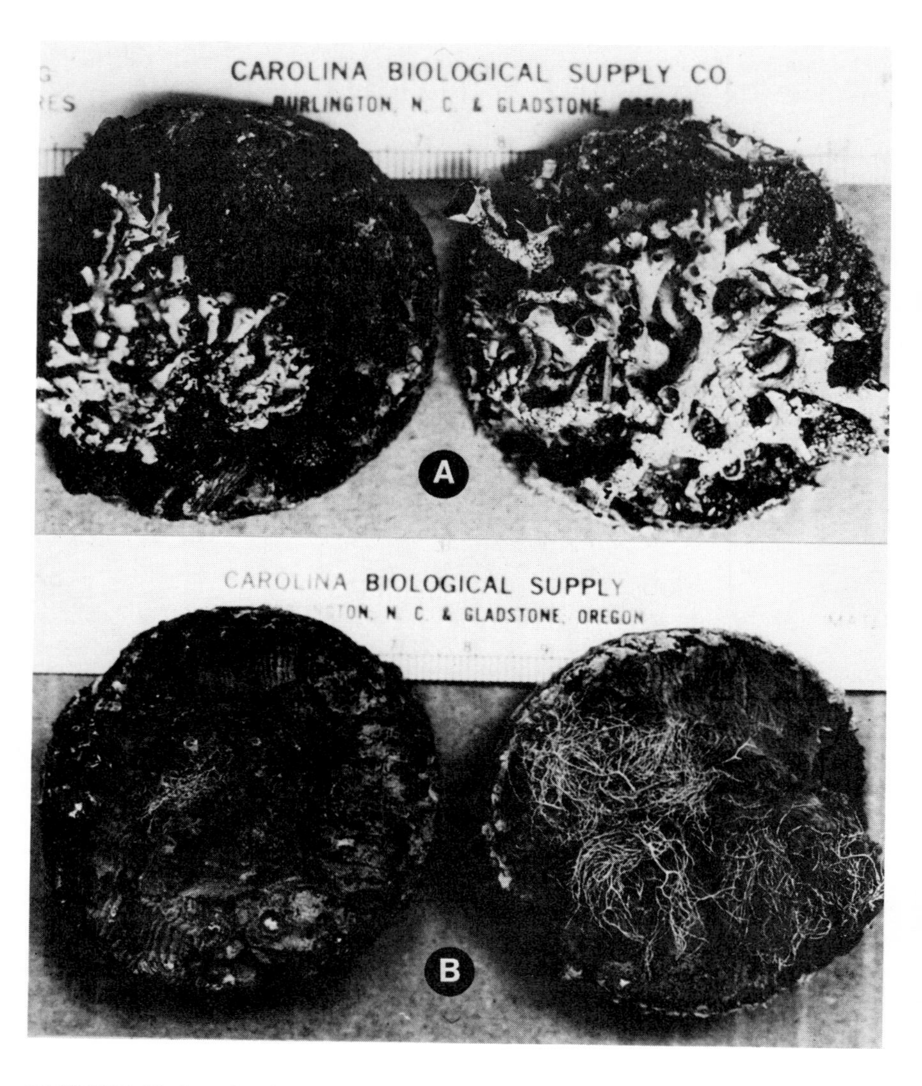

FIGURE 12.4. Bark transplants with lichens. (a) *Hypogymnia physodes*; (b) *Alectoria sarmentosa*. Left: Transplants to within one mile of the Bunker Hill smelter at Kellogg, Idaho; retrieved one year later. Right: Controls transplanted to area north of smelter where epiphytic growth on Douglas fir appeared normal. Unpublished photograph provided by G.R. Hoffman.

Transplant Studies

The estimation of pollution levels from lichen distribution data requires some knowledge of the individual responses of lichens to pollutants. Until recently, information of this sort has not been available; however, more lichenologists are beginning to document the structural and physiological changes that lichen species exhibit in the presence of varying levels of particular pollutants (e.g., Brodo, 1966; Schönbeck, 1969; Klee, 1971; LeBlanc & Rao, 1973). Field experiments generally make use of lichen transplants from relatively unpolluted to polluted areas. The responses of these transplanted lichens are then compared to controls that are transplanted from unpolluted to other unpolluted areas. Examples of transplanted lichens are shown in Figure 12.4. Techniques have been developed to obtain transplants of corticolous (Brodo, 1961a; Schönbeck, 1969; Hoffman, 1971), saxicolous (Ikonen & Kärenlampi, 1976; Armstrong, 1977), and terricolous species (Hale, 1954; Schubert & Fritsche, 1965; Kallio & Varheenmaa, 1974; Kauppi, 1976). Care must obviously be taken to insure that environmental factors other than pollution (microclimate, substrate, etc.) are the same in the transplant sites as in the sites of origin. This is often an experimental condition difficult to meet (Farrar, 1973) and should be considered carefully when interpreting results.

Søchting and Johnsen (1978) found that lichen transplants could be used to indicate various levels of SO_2 air pollution in Copenhagen. The mounted bark transplant disks bearing thalli of *Hypogymnia physodes* on wooden boards and placed these on poles at various sites throughout the city. Since this technique does not depend on the availability of trees in the transplant area, investigators were able to locate transplants throughout the city. Sulfur dioxide levels were also obtained for each site from the Greater Copenhagen Air Pollution Committee. Using a qualitative estimate of pollution damage that combined the degree of thallus discoloration with thallus growth measurements taken after approximately six months, the authors found a highly significant correlation between pollution damage and SO_2 levels (Figure 12.5). The transplantation procedure itself was found not to have had any effect on the responses observed for *H. physodes*. The authors suggested that these relatively inexpensive transplanted lichen monitors could be used to supplement mechanical monitoring devices in city environments requiring extensive pollution monitoring efforts.

LeBlanc and Rao (1973) observed a number of structural and physiological changes in lichens and mosses exposed to varying degrees of SO_2 pollution in Sudbury, Ontario. Five zones of decreasing average SO_2 concentrations were established at varying distances from three SO_2 emitting smelters. Lichen and moss disks were transplanted from unexposed areas to sites in each of these pollution zones and, after one year of exposure, were removed to the laboratory for analysis. Plants exposed to the highest levels of SO_2 generally

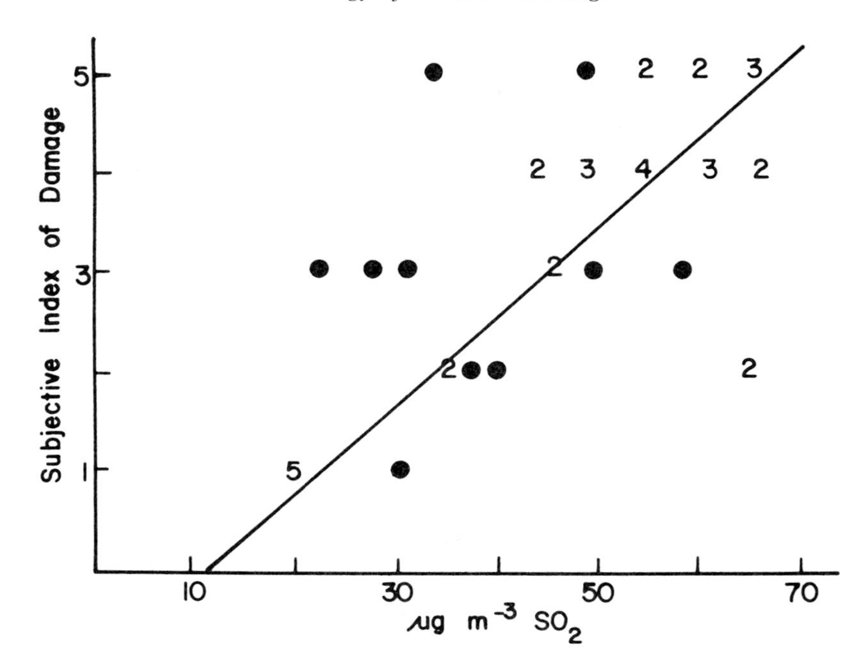

FIGURE 12.5. The correlation between SO_2 levels at transplant sites in Copenhagen and the extent of transplant thallus damage. Numbers refer to numbers of coinciding points. Source: Adapted from U. Søchting & J. Johnsen. 1978. "Lichen transplants as biological indicators of SO_2 air pollution in Copenhagen," *Bull. Environ. Contam. Toxicol.* 19:1–7, with permission of the publisher.

exhibited the lowest thallus pH and the highest sulfur concentrations. Algal cells were found to be most frequently plasmolyzed and least frequently dividing in these polluted zones (Table 12.2). Chromatographic separation of photopigments also showed a reduction in absorbance peaks for chlorophyll a in lichens from zones closest to the SO_2 sources and an associated increase in the peaks for phaeophytin a, indicating an intense phaeophytinization in lichen algae from these zones.

The physiological response of lichens transplanted from unpolluted to polluted habitats has been studied by several investigators (Schönbeck, 1968; Schumm & Kreeb, 1979). Thalli of *Hypogymnia physodes* transplanted from a rural area to various sites in the city of Stuttgart were found to exhibit decreasing photosynthetic activity (Figure 12.6) with time (Schumm & Kreeb, 1979). This response in physiological activity has been observed in other transplant studies (Gilbert, 1970; Ferry & Coppins, 1979) as well as laboratory studies (Baddeley, Ferry, & Finegan, 1973; Hill, 1974; Türk, Wirth, & Lange, 1974). Ferry and Coppins (1979) observed a number of differences in the

TABLE 12.2. Summary of Changes in Algal Cells of Lichen Thalli Transplanted to Various Pollution Zones in the Sudbury, Ontario Region. Zone I Was the Most Polluted; Zone V Was Least Polluted. Approximately 1500 Cells Were Observed for Each Estimate (See LeBlanc & Rao (1973) for Experimental Details and Discussion of Study Areas).

Zone	*Dead Cells, %*	*Plasmolyzed Cells, %*	*Divided Cells, %*
I	100	—	—
II	90–95	100	—
III	70–80	70–80	5–10
IV	5–10	10–20	20–25
V	2–5	0–5	30–40

Partial data adapted from F. LeBlanc & D.N. Rao. 1973. "Effects of sulfur dioxide on lichen and moss transplants." *Ecology* 54:612–617.

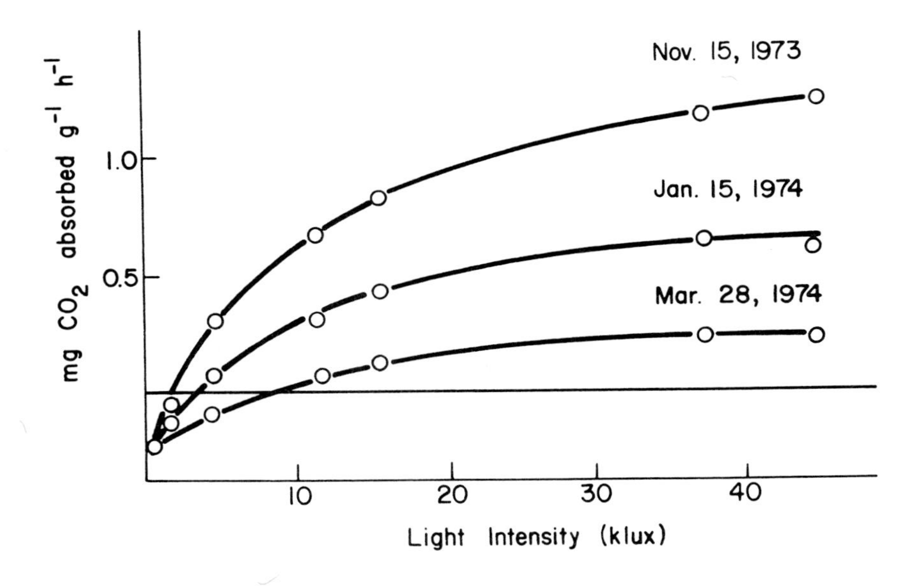

FIGURE 12.6. Light curves of *Hypogymnia physodes* transplants from unpolluted rural areas to the city of Stuttgart. Gas-exchange measurements were taken at various times after the initial transplantation. From F. Schuum & K.H. Kreeb. 1979. "Die Nettphotosynthese von Flechtentransplanten als mass für die Immissionsbelastung der Luft." *Angew. Bot.* 53: 31–39. Redrawn with permission of the Society of Angewandte Botanik.

response of transplanted thalli to pollution when acid bark species were compared with basic bark species. These differences were not explained but are of obvious importance in the experimental design and data evaluation of mapping and transplant studies. Although most of these studies have also demonstrated that removal of lichen thalli from polluted habitats results in a relatively rapid recovery of physiological activity, Ferry and Coppins (1979) found that back-transplanted thalli of *Pseudoparmelia caperata* from polluted to unpolluted control sites exhibited very little sign of recovery. They suggested that this may either be due to differences in individual species responses to pollution or the result of differences in the conditions used for recovery of the lichens. Clearly, more information is needed on the variability in species-specific responses to pollution and the degree to which species are able to recover following different levels of pollution exposure.

Recovery and Reestablishment

The successful use of lichens as indicators of environmental degradation suggests that they may also be used successfully to document improvements in environmental quality. Unfortunately, few regions have been adequately studied for sufficiently long periods of time to permit such comparative studies. One region where the lichen flora has been studied since the early seventeenth century also happens to be a region in which atmospheric quality changes have been documented for many years. This is the area in and around the city of London. Rose and Hawksworth (1981) recently surveyed the lichen flora at 29 sites north and west of Greater London and found that a number of species reported to have been rare or extinct in earlier surveys (Laundon, 1967; 1970) have extended their ranges considerably, many within the past three to seven years. The authors suggested that decreasing SO_2 levels in the city during the past two decades may be the principal reason for the improvements they observed in the lichen flora. They expressed the hope that further improvements will occur but cautioned that drastic reductions in SO_2 levels will be required before substantial improvements in lichen diversity should be expected.

Showman's (1975) study of lichen distributions near coal-fired power plants in southeastern Ohio provided the baseline floristic information necessary to document recovery and reestablishment of sensitive species as the air quality gradually improved. Since the original sampling was done in 1973, Showman has twice revisited each of his 128 sites, once in 1976 and again in 1980. Evidence for recolonization by pollution-sensitive *Pseudoparmelia caperata* in previously void areas (Figure 12.7) was interpreted to be the result of air quality improvements around the Muskingum power plant. However, mean annual SO_2 measurements taken before and after tall stacks were added to the plant were not significantly reduced. Henderson-Sellers and Seaward

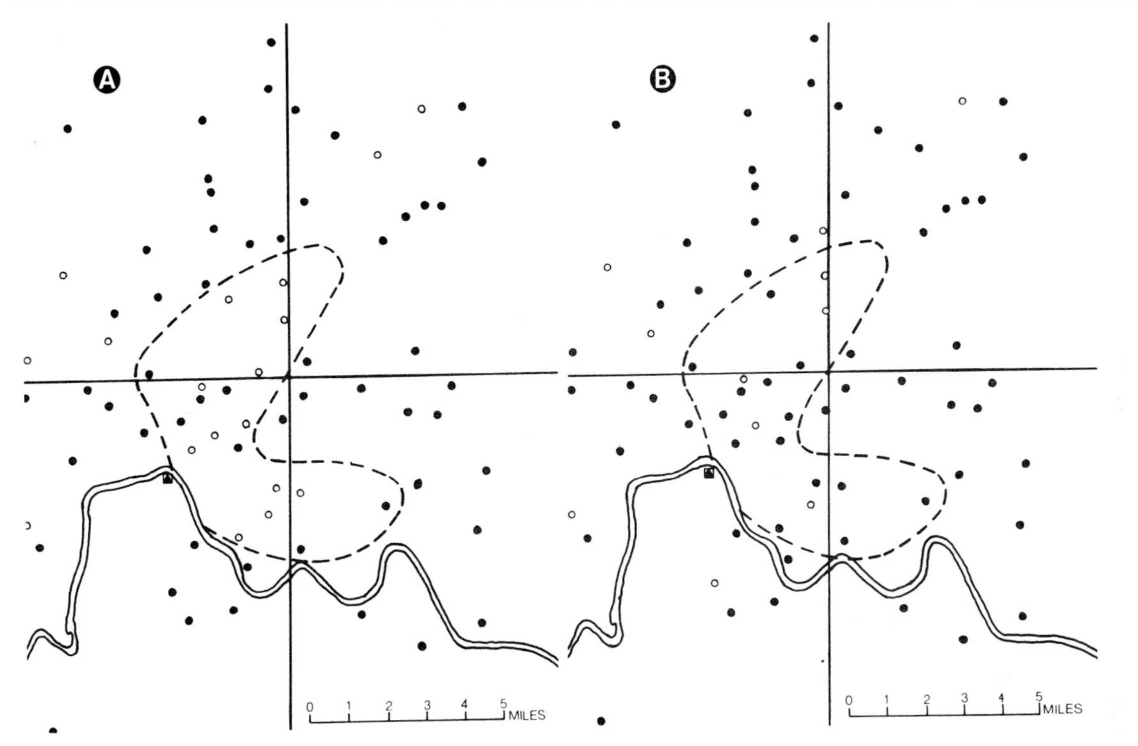

FIGURE 12.7. Reestablishment of *Pseudoparmelia caperata* around Muskingum River Power Plant, Morgan County, Ohio. (a) 1976 distribution of *P. caperata*; (b) 1980 distribution of *P. caperata*; ● = present; ○ = absent. From R.E. Showman. 1981. "Lichen recolonization following air quality improvement." *Bryologist* 84: 492–497. Reprinted with permission of the American Bryological and Lichenological Society and the author.

(1979) were similarly unable to show a clear pattern between atmospheric sulfur dioxide levels in West Yorkshire and reinvasion of *Lecanora muralis*. Showman (1981) suggested that the concentration and duration of peak SO_2 exposures are more important determinants of lichen recolonization than long-term averages. These studies demonstrated the importance of using biological monitors to document long-term changes in air quality.

SULFUR DIOXIDE

Of all the pollutants known to affect the distribution of lichens in disturbed environments, sulfur dioxide perhaps has been studied most frequently. Numerous mapping and transplant studies of lichen distribution patterns in relation to gradients of atmospheric SO_2 pollution are available in the literature. However, until recently, little experimental work had been done to demonstrate the actual phytotoxic effects of SO_2 on lichens that are the cause of these patterns.

Recent laboratory studies have demonstrated that SO_2 exposure of lichens results in a number of physiological responses, some reversible and others irreversible. Specific adverse effects of SO_2 are now known and are reviewed by Nieboer, Puckett, and Grace (1976b).

SO_2 Uptake and Thallus Moisture Content

Thallus moisture content is one of the most important factors to consider when evaluating the effect of SO_2 on lichens. Dry thalli are not injured by gaseous SO_2, but when thalli are hydrated, SO_2 damage can be quite severe (Türk, Wirth, & Lange, 1974). Therefore, the microclimatic conditions required by lichens may influence their susceptibility to SO_2 pollution. For example, the sensitivity to SO_2 that *Lobaria* and *Sticta* species exhibit may be a result of their generally moist habitat requirements (Nieboer et al., 1976c); it should also explain the resistance of other more xeric species.

Differences in SO_2 toxicity at different thallus moisture contents are also related to the physicochemical equilibrium between SO_2 in gaseous and aqueous states. Until recently, investigators who exposed lichens to gaseous SO_2 were not able to relate their results to those of others who used aqueous concentrations of SO_2. The problem with gaseous fumigations is that the effective SO_2 concentration depends to a great extent on thallus moisture content. Nieboer et al. (1977c) and Tomassini et al. (1977) have discussed this problem and experimentally evaluated the following theoretical relationship:

$$SO_2 \text{ ppm (wt/wt, aqueous)} = 10.3\sqrt{SO_2 \text{ ppm (v/v, in air)}}$$

Experiments confirmed this relationship for the extremely low gaseous SO_2 concentrations normally thought to limit lichen distributions in nature. This relationship can therefore be used to compare lichen fumigation studies involving aqueous and gaseous SO_2 when the thallus moisture contents are known.

Ferry and Baddeley (1976) considered a number of factors, in addition to thallus hydration, that influence SO_2 uptake by lichen thalli. They measured the uptake of labelled ^{35}S by lichens placed in various solutions containing sodium sulfite. They observed a significant uptake by dead lichen tissues suggesting that passive uptake into intracellular sites and/or adsorption onto cell walls was responsible, at least in part, for the accumulation of sulfite in the thallus. Uptake increased with rising temperature (Table 12.3) and was always greater in live than in dead tissues. They also observed a relationship between solution pH and sulfite uptake by both living and dead lichen tissues in buffered solutions; uptake was greatest at pH 4.2 and 5.2 and markedly reduced at pH 3.2 and 6.2. This relationship was presumably the result of pH-induced changes to the passive ion adsorption process and was probably only partially responsible for the observed phytotoxic effects of SO2. Türk and Wirth (1975) have observed a negative relationship between SO_2, pH and toxicity; the lower the pH, the greater the damage caused by SO_2.

There is some evidence that SO_2 is actively absorbed by living tissues. This may be associated with the generally efficient nutrient accumulation capacity of lichens (Smith, 1960a; 1960b). Ferry and Baddeley (1976) treated lichen tissue with various metabolic inhibitors and observed decreases in sulfite absorption. They inferred from these preliminary experiments that active sulfite uptake is at least partially responsible for observed uptake levels. However, they cautioned against drawing too many inferences from these results since effects of inhibitors on presumably passive uptake by dead tissues were also observed.

TABLE 12.3. The Influence of Temperature on Sulfite Uptake by Living and Dead Tissues of *Cladonia impexa*

	Sulfite Uptake at Various Temperatures (°C)				
	0	*5*	*15*	*25*	*35*
Live	29.0*	35.0	46.5	63.5	71.0
Dead	12.5	15.0	24.0	32.0	33.5

*μg SO_2 absorbed per g wet lichen over four hours in dark at pH 4.2, from nonbuffered solution.

B.W. Ferry & M.S. Baddeley. 1976. "Sulfur dioxide uptake in lichens." In *Lichenology: Progress and Problems* (D.H. Brown, D.L. Hawksworth, & R.H. Bailey, eds.), pp. 407–418. London: Academic Press. Reprinted with permission of Academic Press.

Effects of SO_2 on Rates of Lichen Photosynthesis, Respiration, and N_2-Fixation

The photosynthetic capacity of lichens is thought to be the most important physiological process influenced by sulfur dioxide. Respiration and N_2-fixation rates have also been found to be affected. Ever since the laboratory studies of Pearson and Skye (1965) and Rao and LeBlanc (1965) demonstrated the detrimental effect of SO_2 fumigations on lichen whole thallus photosynthesis, numerous studies have been done to determine the actual mechanisms involved. Richardson and Puckett (1973) reviewed much of this work on the effect of SO_2 on photosynthesis in Ferry, Baddeley, and Hawksworth's (1973) *Air Pollution and Lichens*; Baddeley, Ferry, and Finegan (1973) reviewed work on SO_2 and respiration in the same volume.

Rates of carbon-fixation have been shown to be significantly reduced in the presence of SO_2 (Puckett et al., 1973a; 1974; Hill, 1974; Nieboer et al., 1976b; Tomassini et al., 1977). In these experiments, lichen material was incubated in buffered aqueous solutions of $NaH^{14}CO_3$ to which various concentrations of sulfite were added. Increasing the concentration of sulfite generally resulted in reduced rates of ^{14}C-fixation (Table 12.4), although some species were found to be more sensitive to SO_2 exposure than others. These species were also found to be less likely to fully recover from SO_2 exposure, although Türk et al. (1974) found different recovery patterns when gaseous rather than aqueous SO_2 exposures were used. Table 12.5 shows that recovery of ^{14}C-fixation rates is correlated with light and moisture conditions (Puckett et al., 1974); optimal light and moisture conditions resulted in almost complete recovery, especially for SO_2-tolerant species, whereas the sub-optimal conditions resulted in reduced rates of recovery.

Türk, Wirth, and Lange (1974) investigated the SO_2 sensitivity of 12 lichen species with different growth forms using CO_2 exchange of exposed thalli as a measure of viability. Rates of net photosynthesis were found to be significantly reduced in sensitive species at atmospheric SO_2 concentrations as

TABLE 12.4. The Effect of Increasing the Amount of Sulfite at Constant Concentration for 24 Hours on $H^{14}CO_3^-$ Fixation by *Usnea* Sp.

µmol Sulfite	*Ratio of ^{14}C Fixation (Sulfite/No Sulfite)*
0.0	1.0
0.75	0.74
1.5	0.45
3.0	0.61
6.0	0.44

D.J. Hill. 1974. "Some effects of sulphite on photosynthesis in lichens." *New Phytol.* 73:1183–1205. Reprinted with permission of Academic Press.

low as 0.5 milligrams / cubic meter. The photosynthetic capacity of one of these species, *Lobaria pulmonaria,* was found to be irreversibly damaged at this concentration. Resistant species exhibited reduced rates of net photosynthesis, but no permanent damage at exposure concentrations of 4 milligrams of SO_2 per cubic meter. Sensitive species appeared to be mainly fruticose forms, but a later study (Wirth & Türk, 1975) indicated that there was no particularly strong correlation between lichen growth form and SO_2 sensitivity.

The effects of SO_2 (and HF as well) on lichen cell membrane integrity were recently investigated by L.C. Pearson and colleagues (Pearson & Henriksson, 1981; Pearson & Rodgers, 1982). Lichen species differing in SO_2 tolerance were tested for SO_2-induced membrane damage in field and laboratory experiments by placing lichen tissue in distilled deionized water for a few minutes and then measuring the water's conductivity. If the membranes are damaged by SO_2 treatment, there should be increased conductivities indicative of significant electrolyte leakage from tissues. Responses of *Evernia prunastri*, an SO_2-sensitive species, were compared with those of *Hypogymnia physodes* and *Peltigera canina*, two species relatively tolerant of SO_2 pollution (Table 12.6). In *E. prunastri*, increased electrolyte leakage was observed at all SO_2 treatment levels above 0.32 parts per million. In *H. physodes*, very high SO_2 concentrations (1000 ppm) elicited a significant response; however, no response was observed at lower concentrations. In *P. canina*, no electrolyte release responses to SO_2 were observed at all. Similar responses to HF fumigation were observed. These results demonstrated that SO_2 sensitivity exhibited by lichens may involve membrane damage as much or more than damage to respiration or photosynthetic mechanisms.

Showman (1972) has been one of the few investigators to compare physiological responses of isolated lichen symbionts to SO_2 with the responses

TABLE 12.5. The Effect of Light and Moisture on the Recovery of Photosynthetic ^{14}C Fixation in *Umbilicaria muhlenbergii*

	^{14}C Fixation*			
	Light and Moist	*Light and Dry*	*Dark and Moist*	*Dark and Dry*
Control	22.5 ± 2.0	35.6 ± 2.4	71.8 ± 4.0	60.7 ± 3.2
Sulfur dioxide	22.3 ± 3.6	11.3 ± 1.3	44.7 ± 4.3	6.5 ± 0.5
Percent recovery	99.0	31.7	62.2	10.7

*Amount of ^{14}C incorporated (cpm × 10^4) by ten 5-mm disks subsequent to periods of exposure (15 min in 75 ppm SO_2 at pH 3) and recovery (24 hours, under different conditions of light and moisture). For experimental details, see Puckett et al. 1974.

K.J. Puckett et al. 1974. "Photosynthetic ^{14}C fixation by the lichen *Umbilicaria muhlenbergii* (Ach.) Tuck. following short exposures to aqueous sulfur dioxide." *New Phytol.* 73:1183–1192.

TABLE 12.6. Effects of Sulfur Dioxide on Membrane Leakage in Three Lichen Species (*Evernia prunastri, Hypogymnia physodes,* and *Peltigera canina*). Leakage Was Measured as Conductivity in μmho/gram Lichen Tissue/ml Water Following Immersion of Pieces of Thallus Averaging 90 mg in 50 ml Deionized Distilled Water for Two Minutes. Leakage Was Measured Following 65 Days Exposure to SO_2

	Conductivity (μmho/g/ml)		
SO_2 (ppm)	E. prunastri	H. physodes	P. canina
Control (est. 0.02)	6.33 ab*	3.95 a	25.2 a
0.03	5.15 a	3.12 a	27.8 a
0.32	10.04 c	4.05 a	33.3 a
10.0	7.91 bc	4.67 a	30.6 a
30.0	8.26 bc	4.04 a	17.7 a
100.0	12.25 d	6.09 a	14.9 a
300.0	14.78 e	13.07 b	43.4 a

*Conductivity values followed by the same letter are not significantly different from each other at $p = 0.05$.

L.C. Pearson & E. Henriksson. 1981. "Air pollution damage to cell membranes in lichens. II. Laboratory experiments." *Bryologist* 84:515–520. Reprinted with permission of the American Bryological and Lichenological Society.

of whole lichen thalli. The results of his studies are particularly important because, unlike most investigations of lichen physiological responses to pollution, they demonstrated significant individualistic responses of lichen symbionts to the same pollution levels. In the three lichen species used, Showman observed net photosynthetic reductions both in whole thalli and in cultured phycobionts. Net photosynthesis was not observed at all at high SO_2 levels (138 ppm SO_2). It was interesting that no chlorophyll damage was observed in phycobionts exposed to low levels of SO_2 even though there were significant decreases in net photosynthesis and dark respiration. At high SO_2 levels, chlorophyll damage was evident. Isolated lichen mycobionts exhibited reduced rates of dark respiration with increasing SO_2 concentration (Table 12.7). However, whole thallus dark respiration increased when exposed to low levels of SO_2 and only exhibited a marked decrease when the SO_2 levels were increased to 138 parts per million (Table 12.7). Showman attributed these unexpected responses to morphological and/or physiological difference between isolated mycobionts and lichenized thalli.

Since Rao and LeBlanc (1965) first suggested that observed reductions in lichen photosynthesis in response to SO_2 were the result of chlorophyll degradation, numerous photopigment studies have been done to test this hypothesis. The results of Rao and LeBlanc's (1965) original study, which

TABLE 12.7. Mean Dark Respiratory Rates of Whole Thalli and Isolated Mycobionts of *Cladonia cristatella* Exposed to Various Concentrations of Atmospheric SO_2. Rates Were Measured at 20° C, 1.3 percent CO_2

Treatment	Cladonia cristatella *Podetia*	Cladonia cristatella *Mycobiont*
Control	1.68 ± 0.55*	—
2 ppm SO_2	0.58 ± 0.06	—
Control	0.89 ± 0.19	0.24 ± 0.005
4 ppm SO_2	0.58 ± 0.11	0.45 ± 0.036
Control	—	0.22 ± 0.020
6 ppm SO_2	—	0.35 ± 0.019
Control	0.80 ± 0.12	0.30 ± 0.027
138 ppm SO_2	0.19 ± 0.05	0.03 ± 0.007

*Mean net μl 0_2 consumed per mg dry wt. per hr ± S.D. Sample size was three.

R.E. Showman. 1972. "Residual effects of sulfur dioxide on the net photosynthetic and respiratory rates of lichen thalli and cultured lichen symbionts." *Bryologist* 75:335–341. Reprinted with permission of the American Bryological and Lichenological Society.

appeared to demonstrate SO_2-induced chlorophyll a degradation to phaeophytin, have been criticized by Hill (1971) who suggested that the observed chlorophyll breakdown was a result of experimental techniques and not necessarily a result of SO_2 effects. Showman's (1972) data, discussed previously, appear to support this idea, as do data obtained by Punz (1979). However, there have been a number of studies that have demonstrated a reduction in lichen chlorophyll content as a result of SO_2 exposure. Türk, Wirth, and Lange (1974) observed decreases in chlorophyll concentration in 12 lichen species as a result of gaseous SO_2 exposure; also, Henriksson and Pearson (1981) measured significant decreases in chlorophyll content in thalli of *Peltigera canina* exposed to SO_2 gas at levels ranging from 0.1 to 500 parts per million. Nash (1973) fumigated thalli of eight lichen species with low levels of SO_2 (0–4 ppm) and observed decreased concentrations of chlorophyll with increasing SO_2 concentration in all species. *Physcia millegrana* and *Cladonia furcata* were found to be most tolerant; *Pseudoparmelia caperata* was most sensitive. There was also a relationship between chlorophyll degradation and thallus hydration; saturated thalli were found to be more susceptible to SO_2 damage than dry thalli.

Degradation of lichen chlorophylls to phaeophytins has also been observed by comparing chlorophyll spectra before and after SO_2 exposure (Rao and LeBlanc, 1965; Puckett et al., 1973a; Nieboer et al., 1976b). A blue shift up to ten nanometers results from phaeophytinization, a process that results from a removal of the MG^{++} ion from the chlorophyll molecule in an acidic medium. Such changes in lichen photopigment chemistry are valuable indicators of pollution damage.

Lichen chlorophyll content has recently been shown to be significantly correlated with sensitivity to pollution stress. Beltman et al. (1980) determined the total chlorophyll (chlorophyll a + b) content of 13 epiphytic lichen species and related this characteristic to pollution sensitivity based on Barkman's (1969) and De Wit's (1976) field observations. They found a significant positive correlation between total chlorophyll content and pollution sensitivity; species with relatively high chlorophyll concentrations were more sensitive to pollution than species with low concentrations. They also found a positive correlation between the degree of unsaturation of numerous fatty acids and pollution sensitivity, although this correlation was less significant. From an analysis of their results, they suggested that lichen pollution sensitivity is related to the energetic dependency of the lichen mycobiont on the phycobiont; the greater this dependency, the more sensitive a species will be to pollution stress. They suggested that pollution-induced damage to the phycobiont may drain resources that would normally go to the mycobiont. Therefore, if the mycobiont requires less photosynthate during stress periods, then the association is able to withstand longer and more severe periods of pollution stress. This hypothesis needs to be further tested in environmentally stressed lichen populations from natural as well as man-influenced habitats.

The recognized importance of lichen-mediated N_2-fixation in many ecosystems has focused attention on the effects of SO_2 pollution on lichen N_2-fixation. Henriksson and Pearson (1981) exposed thalli of *Peltigera canina* to sulfur dioxide gas at levels from 0.1 to 500 parts per million and observed generally decreased rates of N_2-fixation with increased SO_2 concentrations. Denison et al. (1977) reported reduced N_2-fixation rates in *Lobaria oregana* and *L. pulmonaria* as a result of acid precipitation.

Sheridan (1979) has been studying the impact of atmospheric emissions from coal-fired power plants in southeastern Montana on the lichen flora there. He collected thalli of *Collema tenax* and a *Lecidea* species from the region and treated these with either an aqueous solution of $NaHSO_3$ or a combination of NaF and $PbCl_2$. These are expected to be the major pollutants emitted once new power plants are added to those already operating in the area. He then measured rates of N_2-fixation using an acetylene reduction technique. Both lichens exhibited a reduction in N_2-fixing rates in the presence of pollutants. *Collema tenax* was most sensitive to SO_2, with N_2-fixing capacity reduced to one-half of the control rate at 0.1 parts per million SO_2 (Figure 12.8). Sheridan (1979) suggested that this response in N_2-fixing capacity to pollution from power plant emissions represented an important

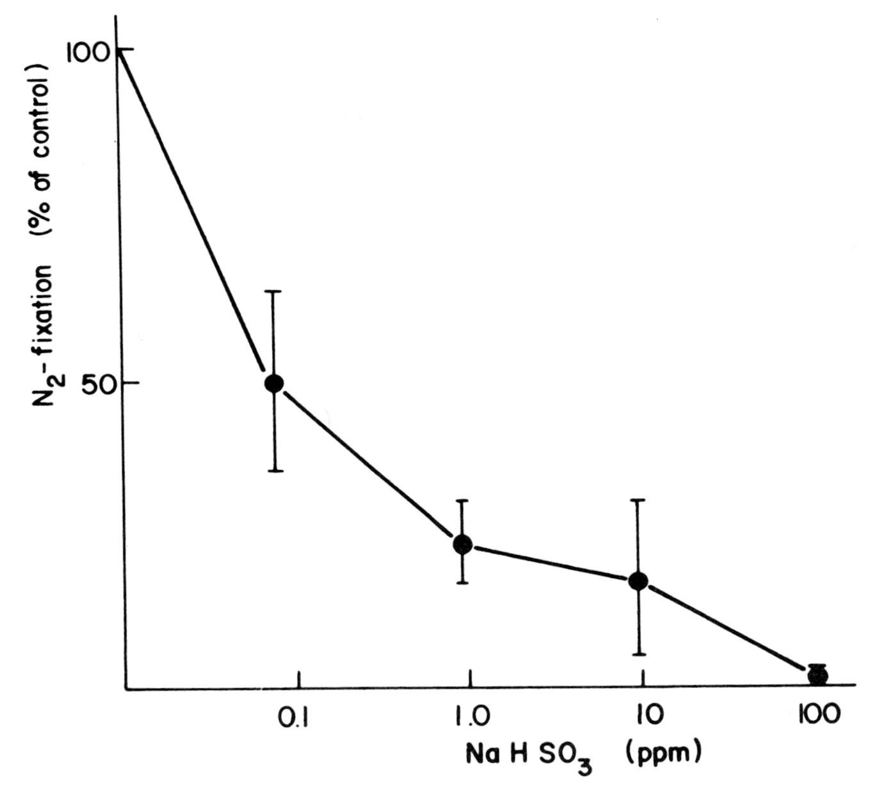

FIGURE 12.8. Sensitivity of *Collema tenax* nitrogen fixation to SO_2, as percent of control; 48 h treatment time, pH 7; bars show 95 percent confidence limits. From R.P. Sheridan. 1979. "Impact of emissions from coal-fired electricity generating facilities on N_2-fixing lichens." *Bryologist* 82: 54–58. Redrawn with permission of the American Bryological and Lichenological Society.

environmental disturbance in an ecosystem that depends so heavily on free-living, diazotrophic organisms like N_2-fixing lichens.

Potassium Efflux

Puckett et al. (1974) discovered that exposure of lichen thallus disks to aqueous SO_2 induced leakage of potassium ions (K^+) into the incubation medium. Exposure to high concentrations of SO_2 resulted in continued losses of K^+ even after lichen material was transferred to distilled water, suggesting that membrane integrity was severely disrupted (Puckett et al., 1977). The

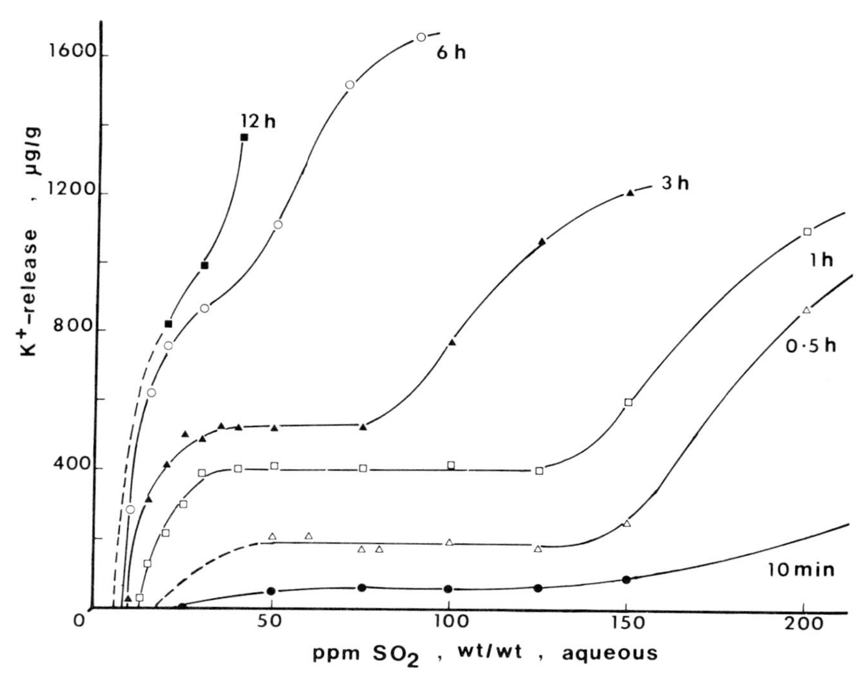

FIGURE 12.9. Potassium efflux by 150 mg samples of *Cladina rangiferina* when exposed to 12 ml of aqueous solutions of SO_2 for the exposure period indicated. From F.D. Tomassini et al. 1977. "The effect of time exposure to sulfur dioxide on potassium loss from and photosynthesis in the lichen *Cladina rangiferina* (L.) Harm." *New Phytol.* 79: 147–155. Reprinted with permission of Academic Press.

TABLE 12.8. Comparison of Induced Changes in K^+ Efflux, Photosynthetic ^{14}C-Fixation, and Photosynthetic Pigments in *Umbilicaria muhlenbergii*

Aqueous SO_2 Concentration (ppm)	*% Loss of Total K^+ Content**	*% Reduction in ^{14}C-Fixation**	*Color of Disks*
0	1	0	Olive green
35	22	10	Olive green
75	51	88	Brown
150	73	100	Brown

*Fifty 5 mm disks were incubated in 10 ml SO_2 solution for three hrs, then washed and left for 24 hrs to recover; subsequently they were washed for 60 min in 10 ml distilled water. The disks were then split into five groups of 10 disks and each group was placed in 4 ml distilled water containing 5µCi $NaH^{14}CO_3$ for 60 min. See Puckett et al. (1977) for experimental details.

K.J. Puckett et al. 1977. "Potassium efflux by lichen thalli following exposure to aqueous sulphur dioxide." *New Phytol.* 79:135–145. Reprinted with permission of Academic Press.

magnitude of K^+ loss is related to a number of factors, particularly duration of SO_2 exposure and concentration of SO_2 in solution. Figure 12.9 also shows that the K^+ efflux response to SO_2 is biphasic, suggesting that an initial K^+ release occurs at relatively low SO_2 concentrations and a massive release occurs later, presumably the result of prolonged SO_2 exposure (see Puckett et al., 1977 for an interesting speculative discussion of membrane dynamics and K^+ loss in relation to SO_2 exposure).

The relative SO_2 sensitivity of several lichen species was determined by extrapolation of K^+-efflux curves to zero K^+ release (Puckett et al., 1977; Tomassini et al., 1977). Relative sensitivities determined by this method correlated well with measured reduction in ^{14}C-fixation rates for the same lichen species (Table 12.8), suggesting that the level of K^+ loss in response to SO_2 is an accurate index of lichen metabolic damage. Further work is required to determine how damage to the fungus-alga association influences K^+ loss, perhaps by measuring rates of K^+ loss by isolated lichen mycobionts and phycobionts.

ACID RAIN

Recent annual reports of the U.S. Council on Environmental Quality (C.E.Q.) have consistently identified acid rain as "one of the two most serious global environmental problems associated with the combustion of fossil fuels (carbon dioxide build-up is the other)" (Council on Environmental Quality, 1980, p. 173). Oxides of sulfur and nitrogen are released in large quantities by the burning of fossil fuels, and reactions that form sulfuric and nitric acids in the atmosphere reduce the pH of rainfall below what it is in unpolluted air (between 5.5 and 6.5). In the 1970s, the pH of precipitation in the northeastern United States was less than 4, and one rain with a measured pH of 2.1 was recorded (National Academy of Sciences, 1975). The pH of precipitation over much of northern Europe is below 4. These problems are apparently being exacerbated by the continued production of tall smoke stacks designed to reduce the impact of pollutants in a given monitoring area. Levels of SO_2 emission in the United Kingdom are currently six million metric tons per year, 40–50 percent of which is released from tall stacks (Gilbert, 1980). Pollutants released from tall stacks often rise high into the atmosphere where they can remain for days and form aerosols that can combine with water vapor to produce acids.

The effects of acid precipitation on lichens are only now beginning to be investigated. Robitaille, LeBlanc, and Rao (1977) have shown that acid precipitation changes the chemical properties of corticolous moss and lichen epiphytes and their bark substrates by acidifying stemflow, increasing toxic bisulfite ion concentrations in stemflow, decreasing the buffering capacity of

FIGURE 12.10. Thallus of *Umbilicaria mammulata* exposed to acid rain showing concentric necrotic rings. Unpublished photograph provided by Lorene Sigal.

bark, increasing the heavy metal concentration in epiphytes, and reducing the chlorophyll concentration of mosses and lichen phycobionts. Denison et al. (1977) found that *Lobaria oregana* and *L. pulmonaria*, important N_2-fixing epiphytes in western U.S. coniferous forests, exhibited significantly reduced rates of N_2-fixation as a result of acid precipitation. Another important lichen species in northern ecosystems, the caribou lichen, *Cladina stellaris*, has been shown to be sensitive to acid precipitation. Lechowicz (1982b) wetted thalli of *C. stellaris* with simulated acid precipitation and observed reduced photosynthetic rates and an impaired ability to respond to rewetting, suggesting that growth of this lichen may be much reduced in areas affected by acid precipitation. Lorene Sigal (unpublished data) has also obtained some preliminary results of simulated acid rain studies involving lichens. So far, she has observed thallus bleaching and reduced photosynthesis in *Usnea* species and *Pseudoparmelia caperata*, and necrosis of the margins of *Umbilicaria*

TABLE 12.9. Elemental Values from Necrotic Regions of *Umbilicaria mammulata* Thalli Compared to Apparently Uninjured Portions of the Same Thallus Exposed to Acid Precipitation

	Element Concentration, μg/g dry wt.			
	Al	*Fe*	*K*	*Mg*
Uninjured	540	210	5800	520
	530	230	6000	500
Necrotic	1500	390	4300	430
	2100	440	4500	350

Provided by L. Sigal.

mammulata. An interesting pattern of concentric necrotic rings on *Umbilicaria* thalli (Figure 12.10) has been observed in the field and can be produced in the laboratory by exposing lichens to simulated rain acidified to a pH of 2.3. Even more interesting, necrotic portions of the acidified thalli appear from preliminary ICP (Inductively Coupled Plasmaspectrophotometric) analyses to contain higher levels of some heavy metals (notably Al and Fe) and lower levels of essential elements (notably K and Mg) than uninjured portions (Table 12.9). Further field and laboratory studies of lichens influenced by acid precipitation are obviously needed; it is expected, however, that lichenologists will be able to contribute as much to our understanding of acid rain impacts as they have to the biological effects of other types of pollution.

OXIDANTS

Oxidants, particularly ozone and carbon monoxide, continue to be the pollutants most often measured in the unhealthful range in the United States (Council on Environmental Quality, 1980). Nitrous oxides and peroxyacetylnitrates (PAN) are also problem pollutants in many urban areas. Nash (1976a) commented on the relative lack of field or laboratory investigations of the effect of these pollutants on lichens. Since then, a number of important studies have demonstrated the value of lichens as indicators of oxidant pollution effects.

Nash (1976b) fumigated samples of four lichen species with various concentrations of nitrogen dioxide and found significant reductions in total chlorophyll concentration after six hours at eight parts per million of NO_2 for all species. This was the first published study to suggest that there was a relationship between lichen physiological activity and NO_2 pollution. Further studies of this particular group of pollutants on lichen physiological activity are required.

The information presently available concerning lichen photosynthetic response to ozone is somewhat contradictory. Brown and Smirnoff (1978) fumigated thalli of *Cladina rangiformis* with ozone in concentrations ranging from zero to six parts per million for 2.5 hours and measured ^{14}C-fixation rate. They found no significant reduction in ^{14}C-fixation rates in fumigated samples, nor was ribitol-induced ^{14}C-release affected by ozone. The only effect of ozone they observed in their experiments with *C. rangiformis* was an ability of ozone to somehow overcome the ribitol-induced depression of ^{14}C-fixation observed in controls. The reason for this was not readily apparent.

These results contrast sharply with those of Nash and Sigal (1979), who observed significant reductions in gross photosynthesis in response to short-term (12 hour) fumigations of ozone at concentrations ranging from 0 to 0.8 parts per million. Fumigated thalli of *Parmelia sulcata* exhibited a greater depression in photosynthetic activity than did *Hypogymnia physodes*, a result that is consistent with observed distributions of these two species in areas of southern California influenced by ozone pollution. Nash and Sigal suggested that the earlier results of Brown and Smirnoff (1978) differed from theirs because measurements were taken 2.5 hours after exposure, and this was not enough time for ozone to penetrate in sufficient quantities to affect photosynthesis. Nash and Sigal were unable to detect responses in their samples until at least three hours after exposure. They attributed this lag time in response to ozone to lichen anatomy (presence of epicortical pores, thickness and density of cortical layers, etc.). Very little experimental work has considered lichen anatomical differences in relation to pollution sensitivity; however, it appears to be an important direction for future research.

Rosentreter and Ahmadjian (1977) reported an increased chlorophyll content for thalli of *Cladonia arbuscula* and a *Trebouxia* phycobiont isolated from *Cladina stellaris* exposed to 0.1 parts per million ozone for one week. Exposure to ozone concentrations above 0.1 parts per million caused no significant changes in chlorophyll concentration. The results of this study prompted Nash and Sigal (1979) to suggest that measurements of chlorophyll content are not as useful as indicators of initial ozone damage as measurements of photosynthesis. However, a better explanation for Rosentreter and Ahmadjian's results may be that of Brown (1980), who pointed out that the chlorophyll extraction procedure used by Rosentreter and Ahmadjian probably resulted in a significant chlorophyll degradation due to phaeophytinization and oxidation caused by lichen acids. Therefore, measurements of chlorophyll concentrations in these lichen samples probably did not accurately reflect true chlorophyll levels regardless of the presence of ozone.

An important group of photochemical oxidants in urban areas includes PAN and related compounds. These pollutants are known to cause considerable damage to vegetation in southern California (Taylor, 1968; 1969). Sigal and Taylor (1979) have conducted a number of laboratory studies of the effects of PAN fumigations on photosynthetic response of lichens. These studies suggested that long-term, low-level PAN fumigations significantly reduced gross photosynthetic rates in *Parmelia sulcata; Hypogymnia enteromorpha* exhibited a more variable response and *Collema nigrescens* showed little response to PAN. Acute high doses of PAN (200 ppb for 1 hr) resulted in increased photosynthetic rates in *Peltigera rufescens, P. sulcata*, and *Collema nigrescens*; no response was observed for *H. enteromorpha*. This result could not be explained.

HYDROGEN FLUORIDE AND OTHER FLUORINE COMPOUNDS

Hydrogen fluoride is generally a localized pollution problem associated with particular industries, such as aluminum reduction, rare earth metals and phosphorus production, ceramics manufacture, and fertilizer production (Nash, 1976a). Because of their ability to accumulate elements from the ambient environment, lichens are frequently used to indicate levels of fluoride release from these industrial centers. Lichen sensitivity to fluoride pollution has been documented by a number of investigators (Martin and Jacquard, 1968; Barkman, 1969; Gilbert, 1971; 1973; Nash, 1971; LeBlanc, Comeau, & Rao, 1971; Horntvedt, 1976, Olech et al., 1981). Most of these studies have demonstrated that lichen thalli near pollution sources accumulate fluorine to significantly higher concentrations than those distant from pollution sources. Evidence of thallus damage and reduced abundance is also frequently collected.

Perkins, Millar, and Neep (1980) conducted a study of corticolous and saxicolous *Ramalina* species in an area near Holyhead, North Wales, before and after an aluminum reduction plant was built. Fluorine concentrations in species nearest the plant increased to almost 1000 percent after the plant began operations; many of these lichens ultimately died. Increased F concentrations were also observed in lichens further removed from the aluminum plant (Figure 12.11). Corticolous species accumulated F more rapidly than saxicolous species and tended to exhibit obvious damage more frequently. Similar results were obtained by Asta and Garrec (1980) for lichen species in a polluted alpine valley in southeastern France. They observed significant correlations between seasonal patterns of F concentrations in corticolous species and rainfall. Concentrations of fluorine in lichens from unpolluted regions were attributed to natural sources of fluorine, a point also made by Takala, Kauranen, and Fagerstén (1979).

Recent studies have also been done of lichen fluorine accumulation around phosphate plants. The F concentration in the terricolous lichens *Cladina rangiferina* and *C. stellaris* were inversely correlated with distance from a phosphorus plant in Newfoundland, Canada (Roberts & Thompson,

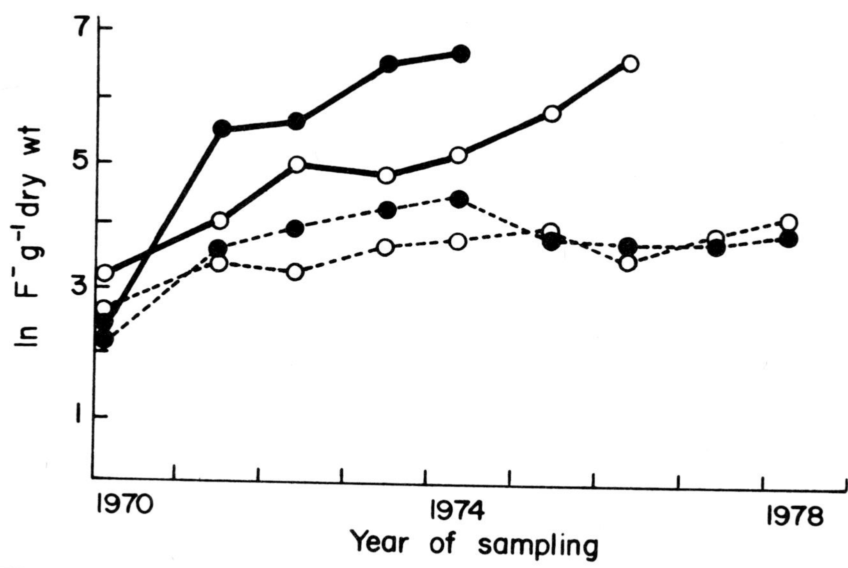

FIGURE 12.11. Concentration of fluoride ($\mu g\ F^{-}\ g^{-1}$) found in corticolous (●) and saxicolous (○) lichens sampled in successive years either 0.55-1.0 km (—) or 3.2-4.9 km (---) from an aluminum reduction plant in Anglesey, North Wales. Redrawn from D.F. Perkins, R.O. Millar & P.E. Neep. 1980. "Accumulation of airborne fluoride by lichens in the vicinity of an aluminum reduction plant." *Environ. Pollut.* (Series A) 21:155–168. Redrawn with permission from Applied Science Publishers, Inc.

1980). Concentrations ranged from 2830 parts per million dry weight to 15.5 parts per million. Thallus discoloration and structural damage were observed in lichens exhibiting a concentration of only 25 parts per million, suggesting that lichens are useful indicators of low-level fluoride pollution effects.

HEAVY METALS

The well-known ability of lichen thalli to accumulate and concentrate elements from their environment has stimulated many investigations of lichen element concentration in relation to pollution sources. Heavy metals released to the environment from automobile exhaust emissions or industrial processes of various sorts have been profitably studied by subjecting lichens collected near pollution sources to elemental analysis. The biological response of lichens to heavy metal pollution has also proven valuable in studies of the long-term environmental effects of heavy metal pollution.

Numerous reviews of lichen heavy metal accumulation are available. Lounamaa (1956; 1965) first presented overwhelming evidence that lichens

TABLE 12.10. Elevated and Background Concentrations of Various Trace Metals in Lichens

Element	*Background Value**	*Reference*	*Elevated Values**	*Reference*
Cadmium	0.05	Sergy (1978)	334	Nash (1975)
Chromium	0.5	Jaakkola et al. (1966)	70	Schutte (1977)
Copper	0.7	Scotter & Miltimore (1973)	5000	Poelt & Huneck (1968)
Lead	0.5	Puckett (1978)	1131	Lawrey & Hale (1979)
Mercury	1.0	Sheridan (1979)	2.5	Connor (1979)
Nickel	1.0	Tomassini et al. (1976)	310	Tomassini (1976)
Zinc	6.0	Puckett (1978)	3500	Maquinay et al. (1961)

*All values in ppm (dry weight).

K.J. Puckett & M.A.S. Burton. 1980. "The effect of trace elements on lower plants." In N.W. Lepp, ed. *Effect of Heavy Metal Pollution on Plants*, pp. 213–238. Vol. 2. *Metals in the Environment*. London: Applied Science Publishers, Inc. Reprinted with permission from Applied Science Publishers, Inc.

have the capacity to accumulate high concentrations of ions, particularly metals, from their environment. James (1973) and Tuominen and Jaakkola (1973) reviewed the vast, primarily descriptive literature on lichen metal accumulation that appeared following the early work of Lounamaa and others. The biological effect of heavy metals on lichens and bryophytes has been reviewed by Rao, Robitaille, and LeBlanc (1977) and Puckett and Burton (1981); and the mechanisms of uptake have been considered by Brown (1976), Johnsen (1976), Nieboer, Richardson, and Tomassini (1978), Richardson and Nieboer (1980), and Nieboer and Richardson (1981). A bibliographic synthesis of the field has been compiled by Margot and Romain (1976). These sources should be consulted for background information by the serious student since only selected topics of recent interest will be considered here.

Elevated and Background Levels of Trace Metals in Lichens

The concentrations of trace metals found in lichen thalli vary widely. Some representative background and elevated trace metal levels from lichens are given in Table 12.10. It is clear from a survey of the literature that numerous factors influence trace metal accumulation in lichens. These factors include species, location, thallus age, and substrate type. However, it is difficult to distinguish effects of these individual factors in field-collected material. It is therefore sometimes difficult to say with certainty that some anthropogenic pollution source is the cause of elevated lichen element concentrations in the field without controlled laboratory analyses.

There is a critical need for studies of normal trace metal concentrations in lichens, levels required for normal thallus growth. Additionally, knowledge about the accumulation capacity of different lichen species for different elements is necessary to distinguish elevated levels in lichens from background levels. Obviously, the more thoroughly the element accumulation processes of lichens are understood, the more reliable environmental assessments based on lichen elemental analyses will be.

Mechanisms of Uptake

There have been recent reviews (Nieboer, Richardson, & Tomassini, 1978; Richardson & Nieboer, 1980; Nieboer & Richardson, 1981) of the principal mechanisms of lichen element accumulation, namely, particulate trapping, ion exchange, passive uptake, and active uptake. Brown (1976) also discussed these mechanisms.

Particulate Trapping

Particulate trapping has been profitably studied by comparing element-to-titanium concentration ratios in lichens to ratios measured in underlying rock substrates (Nieboer, Richardson, & Tomassini, 1978). If these ratios are observed to be significantly different, an important environmental contamination is the likely cause. In Table 12.11, Nieboer, Richardson & Tomassini (1978) discovered that iron, chromium, vanadium, and nickel contents of

TABLE 12.11. Background Element Levels in *Cladonia* Spp. (See Nieboer et al., 1978 for Species and Details Concerning Data Collection).

			Observed Values	
Element	*Element-to-Titanium Ratio*	*Expected Element Level (ppm)**	*Arctic (ppm)*	*Sudbury (ppm)*
K	2	30		2300 ± 400
Fe	6.5	100	150 ± 60	120
Ti	1.00	16	16 ± 9	13
S	0.15	2	210 ± 70	500
Cr	0.02	0.3	2 ± 1	3 ± 1
V	0.02	0.3	1 ± 1	2 ± 1
Ni	0.01	0.2	1 ± 1	0
Cu	0.01	0.2	10 ± 10	10
Pb	0.002	0.03	4 ± 3	23
Zn	0.01	0.2	30	~25

*Based on the particulate content of lichen material having Ti at 16 ppm and using ratios in column 2.

E. Nieboer, D.H. Richardson, & F.D. Tomassini. 1978. "Mineral uptake and release by lichens: An overview." *Bryologist* 81:226–246. Reprinted with permission of the American Bryological and Lichenological Society.

lichens from rural areas in Canada were probably due to rock particulate trapping, whereas lead, copper, zinc, sulfur, and potassium levels in lichens were much higher, indicating a source other than rock particulates.

Particulates from pollution sources are also known to be incorporated by lichens. Garty et al. (1977) found that lichens accumulated Ni-rich particulates that were derived mainly from automobile and industrial sources. They also did a study involving scanning and transmission electron microscopy and energy-dispersive, X-ray analysis that demonstrated the particulate nature of metallic depositions accumulated by lichens (Garty, Galun, & Kessel, 1979). In Sendai, Japan, Saeki et al. (1977) compared the element concentrations of filter-trapped particulates with those of lichen material and observed a significant positive correlation, suggesting that the high concentrations found in lichens were due mainly to particulate trapping.

Ion Exchange

Passive uptake of ions is extracellular and is assumed to occur at anionic sites on cell wall surfaces (Puckett et al., 1973b; Nieboer et al., 1976a; Burton, Le Sueur, & Puckett, 1981). Figure 12.12 illustrates the ion-exchange surface of a lichen fungal cell wall (A), the functional binding groups in zone (B), a diffusion zone (C) and a bulk solution zone (D). The greatest portion of absorbed water in a hydrated intact lichen thallus is assumed to be present in

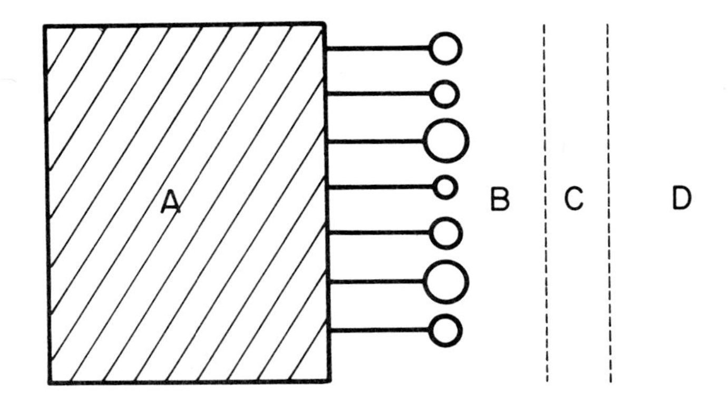

FIGURE 12.12. Diagrammatic representation of the exposed surface of a wetted or submerged lichen thallus. Zone A, ion-exchange surface (e.g., hyphal cell wall); zone B, reaction zone with functional groups or binding sites; zone C, undisturbed diffusion layer; zone D, bulk solution zone. From E. Nieboer, K.J. Puckett, & B. Grace. 1976. "The uptake of nickel by *Umbilicaria muhlenbergii:* a physicochemical process." *Can. J. Bot.* 54: 724–733. Redrawn with permission of the National Research Council of Canada.

zone (D), and the rate of ion-exchange is expected to be limited by diffusion of ions across film (C) and the binding properties associated with the cell wall (B). The functional groups involved in ion exchange are not known, but are thought to be anions of carboxylic or hydrocarboxylic acids (Nieboer & Richardson, 1981). Potentially toxic metals bound to these sites are assumed not to be of metabolic importance since they do not enter fungal or algal cells (Brown, 1976).

Metal ion uptake by lichens from solution can be very rapid. In laboratory studies of lichen metal accumulation, the process appears to be biphasic, with approximately one-half of the total absorption occurring in the first few minutes and the remainder taking from 15 to 30 minutes (Nieboer & Richardson, 1981).

The capacity for metal accumulation is determined by a number of factors, including concentration of the metal ion available in solution, thallus pH, and number and position of binding sites (Puckett et al., 1973b; Tuominen, 1967; Nieboer et al., 1977a). The variability of these factors in nature helps to explain the wide range of lichen metal concentrations reported in the literature.

Intracellular Uptake

The accumulation of elements in the cytoplasm may occur by a passive or an active process. Essential cation uptake is likely to be passive; however, anions may require active transport, at least to some extent. Feige (1977) provides evidence suggesting that sulfate and phosphate uptake by isolated phycobionts of *Peltigera aphthosa* is active and requires energy.

Brown and Slingsby (1972) showed that low concentrations of nickel induced slight potassium losses from *Cladonia rangiferina*, but sharply increased losses occurred at higher nickel concentrations. Puckett (1976) observed similar potassium effluxes by *Umbilicaria muhlenbergii* in the presence of high concentrations of nickel (Figure 12.13) and other metals. Since a portion of the total thallus potassium is intracellular, these results suggest that metals are capable of penetrating cells and accumulating to toxic levels in the cytoplasm. This is likely to be a passive process or the result of membrane damage, not an active process.

Toxicity of Metal Ions

The measurement of potassium efflux by lichen thalli in response to increasing concentrations of various elements has been used to assess the relative toxicities of various metal ions (Nieboer et al., 1979; Richardson et al., 1979; Puckett, 1976; Burton, Le Sueur, & Puckett, 1981). An example is the K^+ efflux response of *U. muhlenbergii* thallus disks to high concentrations of Ni (Figure 12.13). These studies have demonstrated that two classes of metal ions, designated class A and B by Nieboer and Richardson (1981), differ in

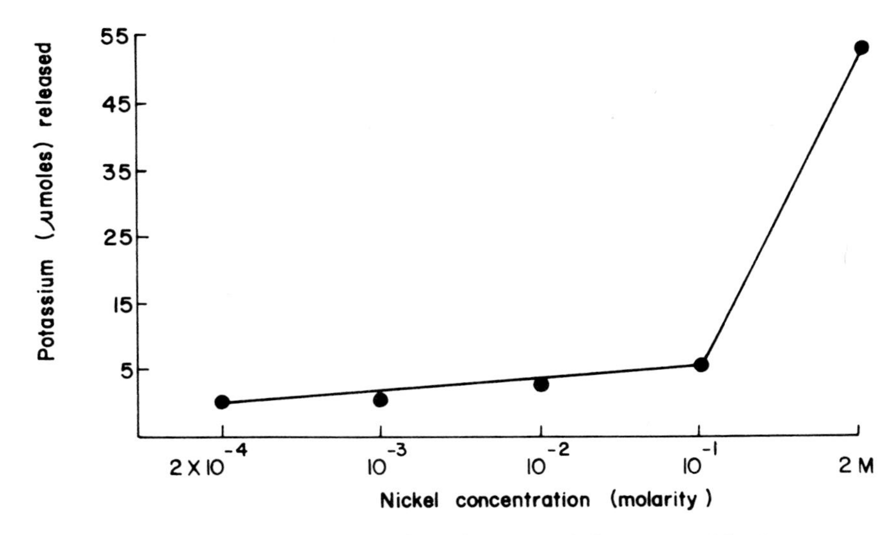

FIGURE 12.13. Potassium efflux from *Umbilicaria muhlenbergii* as a function of nickel concentration. Thallus disks were incubated 3 h in solutions of nickel chloride. From K.J. Puckett. 1976. "The effect of heavy metals on some aspects of lichen physiology." *Can J. Bot.* 54: 2695–2703. Redrawn with permission of the National Research Council of Canada.

their relative toxicities. Class A ions (e.g., K^+, Ca^{++}, Sr^{++}) are characterized by a strong preference for oxygen-containing binding sites, whereas class B ions (e.g., Ag^+, Hg^{++}, Cu^+) prefer nitrogen and sulfur-containing binding sites. Intermediate metal ions (e.g., Zn^{++}, Ni^{++}, Cu^{++}, Pb^{++}) show no strong preferences for either kind of binding site. Nieboer et al. (1979) found that borderline ions with strong class B characteristics exhibited a marked ability to induce K^+ effluxes from *U. muhlenbergii* thallus disks. Class A and borderline ions with class A characteristics induced no K^+ effluxes. Puckett (1976) also showed that class B metal ions (Ag^{++} and Hg^{++}) were extremely toxic to lichens compared to other ions tested. Recently, Burton, Le Sueur, and Puckett (1981) found that K^+ efflux from thallus disks of *Cladina rangiferina* was initiated at low concentrations of Cu^{++} and Ti^+, whereas relatively high concentrations of Mn^{++} and Ni^{++} were required to elicit the same response. These results are also consistent with Nieboer and Richardson's (1981) hypothesis.

Another interesting method used to assess the relative toxicities of metal ions to lichens is through enzyme activity assays. Two recent studies (Lane & Puckett, 1979; Le Sueur & Puckett, 1980) showed that phosphatase activity in thalli of *Cladina rangiferina* is inhibited by uranyl and vanadyl cations and a number of anions. Some cations (copper, nickel, and silver) enhanced enzyme activity.

Retrospective Trace Metal Studies

Lichen trace metal concentrations reflect the availability of trace metals in a particular place for a particular interval of time. Unless studies are designed so that lichen material can be collected before and after a potential pollution source is established in a particular location, it is difficult to say with certainty that the pollution source is responsible for the concentrations of elements found in lichen material.

One solution to this problem is to analyze herbarium lichen material and compare element concentrations of these lichens to levels found under present-day conditions. These retrospective studies not only demonstrate the effects of specific pollution sources on lichen trace element concentrations but also help to document gradual, long-term accumulation trends for relatively large geographic regions.

Rühling and Tyler (1968) were among the first to demonstrate elevated lead content of a number of moss species collected from various locations in Sweden during an interval from 1860 to 1968. They found a distinct increase in lead levels at the turn of the century and another after the second World War that they attributed to increased use of leaded automobile fuel. Since this study was published, other retrospective studies have demonstrated that the elemental analysis of herbarium moss material can be extremely useful in the estimation of long-term trace metal availability (Rasmussen, 1977; Johnsen & Rasmussen, 1977; Rao, Robitaille, & LeBlanc, 1977; Rühling and Tyler, 1969; 1971; 1973; Tyler, 1971).

Although relatively few retrospective studies of lichen trace metal concentrations have been done, the few that have appeared in the literature demonstrate that lichens are as effective as bryophytes as long-term integrators of trace metal accumulation patterns (Table 12.12). In central Sweden, Persson, Holm, and Lidén (1974) analyzed field-collected *Cladina stellaris* and unidentified *Cladonia* species from herbarium collections dating to 1882 for radiolead (Pb-210) and stable lead. They found a significantly decreased ratio of Pb-210 to stable Pb after 1940, indicating an increased deposition of stable Pb at that time. Schutte (1977) compared chromium (Cr) content of field-collected materials to that of herbarium specimens of *Pseudoparmelia caperata* and *Parmelia rudecta* from several locations in Ohio and West Virginia. She found an increased Cr content in the most recently collected lichens that she attributed to air pollution effects. Lawrey and Hale (1981) analyzed lead (Pb) contents of field-collected and herbarium specimens of *Pseudoparmelia baltimorensis, Xanthoparmelia conspersa, and Cladina subtenuis* collected from numerous locations in the northeastern United States. They found increased Pb concentrations in recent material collected near heavily travelled roads. Perhaps even more interesting, they observed steadily increasing Pb levels in herbarium material with time that could not be attributed to specific pollution sources. They suggested that these steady,

TABLE 12.12. Selected Lichen Trace Metal Concentration Values from Retrospective Studies

Species	*Location*	*Element*	*Collection Date*	*Concentration*	*Source*
Pseudoparmelia caperata	Franklin County, OH	Cr	1893	4.8*	Schutte (1977)
			1961	8.8	"
			1976	22.0	"
P. baltimorensis	Plummers Island, MD	Pb	1907	82.3	Lawrey & Hale (1981)
			1938	127.8	"
			1958	342.9	"
			1978	1893.5	"
Xanthoparmelia conspersa	Plummers Island, MD	Pb	1907	82.9	"
			1978	1647.5	"
Cladina subtenuis	Plummers Island, MD	Pb	1907	35.0	"
			1933	57.0	"
			1978	272.0	"

*Data are in μg/g dry wt. See original reports for sample sizes.

long-term increases in lichen Pb concentration reflect an increasing background level of atmospheric lead in the region, a phenomenon that cannot be accurately assessed without comparing lichen material collected in the same location over lengthy intervals of time.

Retrospective studies of long-term trace metal accumulation trends in lichens are limited in scope and design only by the length of time and amount of material collected in a given area. Imaginative investigators will undoubtedly find numerous applications for such retrospective analyses in the future.

RADIONUCLIDE ACCUMULATION

Lichens-Reindeer/Caribou-Man

Shortly after large-scale atmospheric nuclear weapons testing began in 1956, cryptogams were found to contain higher concentrations of radionuclides, notably ^{238}Pu, ^{239}Pu, ^{137}Cs, ^{90}Sr, ^{210}Pb, ^{55}Fe, and ^{22}Na, than vascular plants as shown by Table 12.13 (Gorham, 1958; Aarkrog & Lippert, 1959; Gorham, 1959; Hvinden & Lillegraven, 1961; Lidén, 1961; Rickard et al., 1965; Watson, Hanson, & Davis, 1964). Since reindeer and caribou (each considered a subspecies of *Rangifer tarandus*) are known to feed almost exclusively on

TABLE 12.13. Radionuclides in Certain Plants in the Vicinity of Enare, Finnish Lapland (from Rahola & Miettinen 1969–70)

	Radionuclide Concentration (nCi/kg dry wt)			
Species	^{137}Cs	^{59}Mn	^{106}Ru	^{125}Sb
Cryptogams				
Cladonia spp.	29.3	0.6	—	—
Pleurotium schreberi	25.5	0.6	—	—
Polytrichum commune	3.1	0.4	5.3	1.9
Equisetum sp.	6.5	0.1	—	—
Phanerogams				
Betula nana	3.3	1.1	3.1	0.2
B. pubescens	2.9	0.2	0.7	—
Betula sp.	1.2	0.4	0.8	—
Salix sp.	1.5	0.7	—	—
Vaccinium vitis-idaea	3.1	0.4	—	—
Carex sp.	1.4	0.1	—	—
Deschampsia flexuosa	2.0	—	2.8	—

Partial data adapted from T. Rahola & J.K. Miettinen. 1969–1970. *Radionuclides in plants and reindeer meat in Lapland—Preliminary report in radioactive foodchains in the subarctic environment*, Ann. Rep. Dept. Radiochemistry, Helsingfors Univ., Helsinki, Finland, 1969–70.

lichens during the winter, and these animals are important food sources for certain human populations (e.g., Eskimo populations in Alaska and Lapp populations in Fennoscandia), numerous studies were undertaken to determine the potential health effects of radionuclide fallout.

In 1957, reindeer bones from Norway were found to contain high levels of ^{137}Cs and ^{90}Sr (Richardson, 1975); in 1961, Lidén demonstrated a higher-than-average ^{137}Cs body burden in Swedish Lapps. Since then, numerous studies in arctic and subarctic regions have been done, all of which strongly suggest a bioaccumulation of radionuclides through the relatively simple arctic and subarctic lichen-reindeer/caribou-man ecosystem.

The importance of lichens in this process was emphasized in a study by Book, Connolly, and Longhurst (1972) at the University of California's Hopland Field Station located in the Mayacamus Mountains 90 miles north of San Francisco. Levels of ^{137}Cs were determined in rumen contents and muscle from 68 wild Columbian black-tailed deer over a two-year period. Samples from deer collected from oak woodland habitats contained higher ^{137}Cs concentrations than those from deer living in chaparral. It was subsequently discovered that deer from oak woodland habitats ate appreciable amounts of lichens (mostly *Usnea* and *Ramalina* species, unidentified) all year round, whereas chaparral deer had a more varied diet. Furthermore, analysis of all forage species revealed that lichens contained up to 140 times the ^{137}Cs concentration of other forage types.

Apparently, not all radionuclides are absorbed or retained by lichens the same way. For example, ^{90}Sr is distributed homogeneously throughout the lichen thallus, and ^{137}Cs is concentrated in thallus growing tips (Salo & Miettinen, 1964; Hanson, Watson, & Perkins, 1967; Nifontova & Kulikov, 1977; Mattsson, 1975a). Mattsson (1975b) found that numerous other radionuclides (^{155}Eu, ^{144}Ce, ^{137}Cs, ^{106}Ru, ^{95}Zr, ^{54}Mn, and ^{7}Be) exhibited highest concentrations in the uppermost layers of *Cladonia alpestris* carpets. In northern Alaska, Hanson, Watson, and Perkins (1967) found seasonal patterns in lichen radionuclide concentrations that apparently correlated well with the level of snow cover; evidently, snow cover prevented fallout deposition on most species. An exception was *Cornicularia divergens*, a species that inhabits windswept ridges. This lichen usually contained the highest fallout concentration. Concentrations of ^{137}Cs in all species generally tended to increase with time regardless of snow cover, however, suggesting that this particular radionuclide may be retained from snow melt. The importance of these different accumulation patterns is obvious when one considers that radionuclides in the top layers of the lichen carpet are most likely to be ingested by caribou and reindeer and thereafter passed onto humans.

During the 1961–1965 interval when atmospheric weapons testing was at its peak, the body burden of radioactivity in Lapps due to ^{137}Cs levels was

TABLE 12.14. Concentration of ^{55}Fe in Blood Samples Obtained from Reindeer and Humans in Relation to the ^{137}Cs and Stable Fe Content (from Persson 1969)

	$^{55}Fe/^{137}Cs$	$nCi\ ^{55}Fe/g$ *Stable Fe*
Lapps		
Male	4 ± 1	23 ± 2
Female	20 ± 10	55 ± 23
Non-Lapps		
Male	1 ± 0.5	4 ± 2
Female	6 ± 1	16 ± 4
Reindeer	43 ± 9	360 ± 40

Partial data adapted from R.B.R. Persson. 1969. "Iron-55 in northern Sweden: Relationships and annual variation from 1956 until 1967 in lichen and reindeer as well as uptake and metabolism in man." *Health Physics* 16:69–78.

observed to be fed up to five times higher than in non-Lapps from the same area who did not consume reindeer meat (Lidén & Gustafsson, 1966). There was also an obvious seasonal pattern in the ^{137}Cs levels observed for Lapps; lower levels were observed in summer months when foods other than reindeer meat were most frequently consumed. Persson (1969) observed similar differences between Lapps and non-Lapps in the concentration of ^{55}Fe in blood samples (Table 12.14); this difference was more pronounced in females. Despite these observations, however, the health effects of elevated levels of radionuclide exposure have been considered slight. Since atmospheric nuclear weapons testing has ceased for the most part, the concentrations of radionuclides in lichens, reindeer/caribou and man have all declined steadily, thus reducing further any health hazard that might once have been considered possible.

POLLUTION-INDUCED ULTRASTRUCTURAL MODIFICATIONS

Although the effects of pollutants on lichen fine structure have received little attention, the few investigations done to date indicate that pollution damage can be observed in lichens at the cellular level long before morphological damage is apparent. This is of obvious importance in assessing the long-term, low-level effects of pollution on lichen biological activity.

Ultrastructural pollution damage is most apparent in the phycobiont. The following are among the changes that have been observed:

1. Thylakoid swelling and increased number of starch grains (Ikonen & Kärenlampi, 1976; Slocum, 1977; Silva-Pando & Ascaso, 1982; Holopainen, 1983).

2. Rounded chloroplasts, indicative of chloroplast inactivation (Holopainen, 1983).

3. Unidentified dark vacuolar bodies observed most frequently in algal cells from polluted sites, although they are also found occasionally in control lichens (Holopainen, 1983).

4. General swelling of algal mitochondria accompanied by poor resolution of mitochondrial cristae (Holopainen, 1983).

Pollution-induced ultrastructural modifications to mycobionts are less frequently observed. Holopainen (1983) found that mycobionts of both the pollution-sensitive *Bryoria capillaris* and the pollution-tolerant *Hypogymnia physodes* showed cellular modifications in thalli collected near two pollution sources in central Finland. These changes were particularly apparent in cortical tissues. The fungi were heavily vacuolated, perhaps an indication of enhanced lysosomatic activity. Also, the vacuoles contained unidentified dark accumulations, never seen in healthy cells from control thalli, but frequently seen in degenerating hyphae from both sites. Other organelles, including mitochondria and concentric bodies, were apparently unaffected by pollution; the frequency of haustorial penetration was also unchanged.

In an ultrastructural investigation of the *Nostoc* phycobiont of *Peltigera canina*, Sharma et al. (1982) discovered a number of changes induced by laboratory fumigations with SO_2. The most important was a decreased number of carboxysomes, indicating a disturbance in the algal photosynthetic capacity. Other changes were apparently stress-related; these included an increased number of cyanophycin granules and an unusually common initation of akinetes.

Many of these ultrastructural responses to pollution, particularly changes in the type and density of storage organelles, are observed in lichens from environmentally extreme habitats. This suggests that, from the lichen's point of view, pollution is a physiological stress to which a generalized response can be made. Depending on the severity and duration of the pollution load, this type of response may be sufficient to maintain normal metabolic activity. However, when the pollution level exceeds that which can be tolerated by the thallus, normal activity ceases and a certain amount of cellular breakdown probably occurs, resulting in structural aberrations.

Further ultrastructural study of lichens exposed to known amounts of pollution will undoubtedly make us more aware of the normal ability of lichens to cope with stress as well as the cellular basis for recognized patterns of pollution tolerance and sensitivity in various lichen species.

Bibliography

Aarkrog, A. & J. Lippert. 1959. *Environmental radioactivity at Risö, 1 April 1958–31 March 1959.* Report 9, Risö, Denmark: Danish Atomic Energy Commission.

des Abbayes, H. 1939. "Revision monographique des *Cladonia* du sous-genre *Cladina* (Lichens)." *Bull. Soc. Scient. (Bretagne)* 16 (Ser. 2): pp. 1–154.

———. 1951. "Traité de Lichenologie." *Encycl. Biol. (Paris)* 41: pp. 1–217.

Adams, D.B. & P.G. Risser. 1971a. "The effect of host specificity on the interspecific associations of bark lichens." *Bryologist* 74: pp. 451–457.

———. 1971b. "Some factors influencing the frequency of bark lichens in north central Oklahoma." *Am. J. Bot.* 58: pp. 752–757.

Adams, M.S. 1971. "Temperature response of CO_2 exchange of *Cladonia rangiferina* from the Wisconsin Pine Barrens, and comparisons with an arctic population." *Am. Midl. Nat.* 86: pp. 224–227.

Ahmadjian, V. 1960a. "The taxonomy and physiology of lichen algae and problems of lichen synthesis." Ph.D. dissertation, Harvard University, Cambridge, MA.

———. 1960b. "Some new and interesting species of *Trebouxia,* a genus of lichenized algae." *Am. J. Bot.* 47: pp. 677–683.

———. 1961. "Studies on lichenized fungi." *Bryologist* 64: pp. 168–179.

———. 1962a. "Investigations on lichen synthesis." *Am. J. Bot.* 49: pp. 277–283.

———. 1962b. "Lichens." In *Physiology and Biochemistry of Algae,* edited by R.A. Lewin, pp. 817–822. New York: Academic Press.

———. 1963. "Studies of lichen synthesis and the lichen symbiosis." *Am. J. Bot.* 50: p. 624.

———. 1964. "Further studies on lichenized fungi." *Bryologist* 67: pp. 87–98.

———. 1966a. "Artificial reestablishment of the lichen *Cladonia cristatella.*" *Science* 151: pp. 199–201.

———. 1966b. "Lichens." In *Symbiosis,* vol. I, edited by S.M. Henry, pp. 35–97. New York: Academic Press.

———. 1967a. *The Lichen Symbiosis.* Waltham, MA: Blaisdell.

———. 1967b. "A guide to the algae occurring as lichen symbionts: isolation, culture, cultural physiology, and identification." *Phycologia* 6: pp. 128–160.

———. 1970. "The lichen symbiosis: its origin and evolution." *Evol. Biol.* 4: pp. 163–184.

———. 1973a. "Resynthesis of lichens." In *The Lichens,* edited by V. Ahmadjian and M.E. Hale, Jr., pp. 565–579. New York: Academic Press.

———. 1973b. "Methods of isolating and culturing lichen symbionts and thalli." In *The Lichens,* edited by V. Ahmadjian and M.E. Hale, Jr., pp. 653–659. New York: Academic Press.

———. 1977. "Qualitative requirements and utilization of nutrients: lichens." In (M. Rechcigl, Jr., ed.) *CRC Handbook Series in Nutrition and Food.* Section D. *Nutritional Requirements.* Vol. I. *Comparative and Qualitative Requirements,* pp. 203–215. Cleveland: CRC Press.

———. 1978. "Culture media (natural and synthetic): lichens." In (M. Rechcigl, Jr., ed.) *CRC Handbook Series in Nutrition and Food,* Section G. *Diets, Culture Media, Food Supplements.* Vol. III. *Culture Media for Microorganisms and Plants,* pp. 505–506. Cleveland: CRC Press.

———. 1980a. "Separation and artificial synthesis of lichens." In *Cellular Interactions in Symbiosis and Parasitism,* edited by C.B. Cook, P.W. Pappas, and E.D. Rudolph, pp. 3–25. Columbus: Ohio State University Press.

———. 1980b. "Guide to culture collections of lichen symbionts." *Intern. Lichenol. Assoc. Newsletter* 13: pp. 13–16.

———. 1982a. "Algal/fungal symbioses." In *Progress in Phycological Research,* edited by F.C. Round and D.J. Chapman, pp. 179–233. Amsterdam: Elsevier Biomedical Press.

———. 1982b. "The nature of lichens." *Natural History* 91: pp. 30–37.

Ahmadjian, V. & H. Heikkilä. 1970. "The culture and synthesis of *Endocarpon pusillum* and *Staurothele clopima.*" *Lichenologist* 4: pp. 259–267.

Ahmadjian, V. & J.B. Jacobs. 1970. "The ultrastructure of lichens. III. *Endocarpon pusillum.*" *Lichenologist* 4: pp. 268–270.

———. 1981. "Relationship between fungus and alga in the lichen *Cladonia cristatella* Tuck." *Nature* 289: pp. 169–172.

———. 1982. "Artificial reestablishment of lichens. III. Synthetic development of *Usnea strigosa.*" *J. Hattori Bot. Lab.* 52: pp. 393–399.

———. 1983. "Algal-fungal relationships in lichens: recognition, synthesis, and development." In *Algal Symbioses: A Continuum of Interaction Strategies,* edited by L.J. Goff, pp. 147–172. New York: Cambridge University Press.

Ahmadjian, V., J.B. Jacobs, & L.A. Russell. 1978. "Scanning electron microscope study of early lichen synthesis." *Science* 200: pp. 1062–1064.

Ahmadjian, V. & J.T. Reynolds. 1961. "Production of biologically active compounds by isolated lichenized fungi." *Science* 133: pp. 700–701.

Ahmadjian, V., L.A. Russell, & K.C. Hildreth. 1980. "Artificial reestablishment of lichens. I. Morphological interactions between the phycobionts of different lichens and the mycobionts *Cladonia cristatella* and *Lecanora chrysoleuca.*" *Mycologia* 72: pp. 73–89.

Ahti, T. 1959. "Studies on the caribou lichen stands of Newfoundland." *Ann. Bot. Soc. "Vanamo"* 30: pp. 1–44.

———. 1961. "Taxonomic studies on reindeer lichens (*Cladonia,* subgenus *Cladina*)." *Ann. Bot. Soc. Zool. Bot. Fenn. "Vanamo"* 32: pp. 1–160.

Albersheim, P. & A.J. Anderson-Prouty. 1975. "Carbohydrates, proteins, cell surfaces, and the biochemistry of pathogenesis." *Ann. Rev. Plant Physiol.* 26: pp. 31–52.

Albersheim, P. & B.S. Valent. 1978. "Host-pathogen interactions in plants." *J. Cell Biol.* 78: pp. 627–643.

Alexander, M. 1971. *Microbial Ecology.* New York: Wiley.

Alexopoulos, C.J. 1962. *Introductory Mycology.* 2nd ed. New York: Wiley.

Almborn, O. 1948. "Distribution and ecology of some south Scandanavian lichens." *Bot. Notis. Suppl.* 1: pp. 1–354.

Akvin, K.L. 1960. "Observations on the lichen ecology of South Haven Peninsula, Studland Heath, Dorset," *J. Ecol.* 48: pp. 331–339.

Am Ende, I. 1950. "Zur Ernährungsphysiologie des Pilzes der *Xanthoria parietina.*" *Arch. Mikrobiol.* 15: 185–202.

Andersen, J.L. & J.L. Sollid. 1971. "Glacial chronology and glacial geomorphology of the glaciers Mitdalsbreen and Nigardsbreen, South Norway." *Norsk. Geogr. Tidskr.* 25: pp. 1–38.

Anderson, M.C. 1966. "Ecological groupings of plants." *Nature* 212: pp. 54–56.

Andrews, J.T. & P.J. Webber. 1969. "Lichenometry to evaluate changes in glacial mass budgets: As illustrated from north-central Baffin Island, N.W.T." *Arctic & Alpine Res.* 1: pp. 181–194.

Aplin, P.S. & D.J. Hill. 1979. "Growth analysis of circular lichen thalli." *J. Theoret. Biol.* 78: pp. 347–363.

Archibald, P.A. 1975. "*Trebouxia* de Puymaly (Chlorophyceae, Chlorococcales) and *Pseudotrebouxia* gen. nov. (Chlorophyceae, Chlorosarcinales)." *Phycologia* 14: pp. 125–137.

———. 1977. "Physiological characteristics of *Trebouxia* (Chlorophyceae, Chlorococcales) and *Pseudotrebouxia* (Chlorophyceae, Chlorosarcinales)." *Phycologia* 16: pp. 295–300.

Armesto, J.J. & L.C. Contreras. 1981. "Saxicolous lichen communities: Nonequilibrium systems?" *Am. Nat.* 118: pp. 597–604.

Armstrong, R.A. 1973. "Seasonal growth and growth rate-colony size relationships in six species of saxicolous lichens." *New Phytol.* 72: pp. 1023–1030.

———. 1974. "Growth phases in the life of a lichen thallus." *New Phytol.* 73: pp. 913–918.

———. 1975. "The influence of aspect on the pattern of seasonal growth in the lichen *Parmelia glabratula* ssp. *fuliginosa* (Fr. ex Duby) Laund." *New Phytol.* 75: pp. 245–251.

———. 1976. "Studies on the growth rates of lichens." In *Lichenology: Progress and Problems,* edited by D.H. Brown, D.L. Hawksworth, and R.H. Bailey, pp. 309–322. London: Academic Press.

———. 1977. "The response of lichen growth to transplantation to rock surfaces of different aspect." *New Phytol.* 78: pp. 473–478.

———. 1978. "The colonization of a slate rock surface by a lichen." *New Phytol.* 81: pp. 85–88.

———. 1981. "Field experiments on the dispersal, establishment and colonization of lichens on a slate rock surface." *Environ. Exp. Bot.* 21: pp. 115–120.

———. 1982 "Competition between three saxicolous species of *Parmelia* (lichens)." *New Phytol.* 90: pp. 67–72.

Asahina, Y. 1934. "Uber die Reaktion von Flechten-Thallus." *Acta Phytochim.* 8: pp. 47–64.

Asahina, Y. & S. Shibata. 1954. *Chemistry of Lichen Substances.* Tokyo: Japan Society for the Promotion of Science.

Ascaso, C. 1978. "Ultrastructural modifications in lichens induced by environmental humidity." *Lichenologist* 10: pp. 209–219.

Ascaso, C. & J. Galvan. 1975. "Concentric bodies in three lichen species." *Arch. Microbiol.* 105: pp. 129–130.

———. 1976a. "The ultrastructure of the symbionts of *Rhizocarpon geographicum, Parmelia conspersa* and *Umbilicaria pustulata* growing under dryness conditions." *Protoplasma* 87: pp. 409–418.

———. 1976b. "Studies on the pedogenetic action of lichen acids." *Pedobiologia* 16: pp. 321–331.

Asta, J. & J.P. Garrec. 1980. "Etude de l'accumulation du fluor dans les lichens d'une vallée alpine polluée." *Environ. Pollut.* (Series A) 21: pp. 267–286.

Asta, J. & B. Lachet. 1978. "Analyses de relations entre la teneur en carbonate de calcium des substrats et divers groupements phytosociologiques de lichens saxicoles." *Oecol. Plant.* 13: pp. 193–206.

Bachelor, F.W. & G.G. King. 1970. "Chemical constituents of lichens: Aphthosin, a homologue of peltigerin." *Phytochem.* 9: pp. 2587–2589.

Bachmann, E. 1915. "Kalklösende Algen." *Ber. Deut. Bot. Ges.* 33: pp. 44–57.

Bachmann, F.M. 1912. "New type of spermogonium and fertilization in *Collema.*" *Ann. Bot.* 26: pp. 747–760.

Baddeley, M.S., B.W. Ferry, & E.J. Finegan. 1973. "Sulphur dioxide and respiration in lichens." In *Air Pollution and Lichens,* edited by B.W. Ferry, M.S. Baddeley, and D.L. Hawksworth, pp. 299–313. Toronto: University of Toronto Press.

Bailey, R.H. 1966. "Studies on the dispersal of lichen soredia." *J. Linn. Soc. Bot.* 59: pp. 479–490.

———. 1976. "Ecological aspects of dispersal and establishment in lichens." In *Lichenology: Progress and Problems,* edited by D.H. Brown, D.L. Hawksworth, and R.H. Bailey, pp. 215–247. London: Academic Press.

Bailey, R.H. & R.M. Garrett. 1968. "Studies of the discharge of ascospores from lichen apothecia." *Lichenologist* 4: pp. 57–65.

Bannister, P. 1976. *Introduction to Physiological Plant Ecology.* New York: John Wiley & Sons.

Barber, J. 1968. "Measurement of the membrane potential and evidence for active transport of ions in *Chlorella pyrenoidosa.*" *Biochim. Biophys. Acta* 150: pp. 618–625.

Barkman, J. 1958. *On the Ecology of Cryptogamic Epiphytes.* The Hague: Van Gorcum & Co.

———. 1969. "The influence of air pollution on bryophytes and lichens." In *Air Pollution, Proc. 1st European Congr. Influence of Air Pollut. on Plants and Animals,* Wageningen 1969, pp. 197–209. The Netherlands: Centre for Agricultural Publishing and Documentation.

Bates, J.W. 1975. "A quantitative investigation of the saxicolous bryophyte and lichen vegetation of Cape Clear Island, County Cork." *J. Ecol.* 63: pp. 143–162.

Bauer, H. & E. Sigarlakie. 1973. "Cytochemistry on ultrathin frozen sections of yeast cells." *J. Microsc.* 99: p. 205.

Baur, E. 1898. "Zur Frage nach der Sexualität der Collenaceen." *Ber. Deut. Bot. Ges.* 16: pp. 363–367.

Bazzaz, F.A. 1979. "The physiological ecology of plant succession." *Ann. Rev. Ecol. Syst.* 10: pp. 351–371.

Bednar, T.W. 1963. "Physiological studies on the isolated components of the lichen *Peltigera aphthosa.*" Ph.D. dissertation, University of Wisconsin, Madison.

Bednar, T.W. & O. Holm-Hansen. 1964. "Biotin liberation by the lichen alga *Coccomyxa* sp. and by *Chlorella pyrenoidosa.*" *Plant Cell Physiol.* (*Tokyo*) 5: pp. 297–303.

Bednar, T.W. & B.E. Juniper. 1964. "Microfibrillar structure in the fungal portions of the lichen *Xanthoria parietina* (L.) Th. Fr." *Proc. Exp. Cell Res.* 36: pp. 680–683.

Bednar, T.W. & D.C. Smith. 1966. "Studies in the physiology of lichens. VI. Preliminary studies of photosynthesis and carbohydrate metabolism of the lichen *Xanthoria aureola.*" *New Phytol.* 65: pp. 211–220.

Bellemère, M.-A. 1973. "Observation de 'corps concentriques' semblables à ceux des lichéns, dans certaines cellules de plusieurs ascomycètes non lichénisants." *C.R. Hebd. Séances Acad. Sci. Ser. D. Sci. Nat.* 276D: pp. 949–952.

Bellemère, A. & M.-A. Letrouit-Galinou. 1981. "The lecanoralean ascus: An ultrastructural preliminary study." In *Ascomycete Systematics: The Luttrellian Concept,* edited by D.R. Reynolds, pp. 54–70. New York: Springer-Verlag.

Beltman, I.H., L.J. deKok, P.J.C. Kuiper, & P.R. van Hasselt. 1980. "Fatty acid composition and chlorophyll content of epiphytic lichens and a possible relation to their sensitivity to air pollution." *Oikos* 35: pp. 321–326.

Benedict, J.B. 1967. "Recent glacial history of an alpine area in the Colorado Front Range, U.S.A. I. Establishing a lichen-growth curve." *J. Glaciol.* 6: pp. 817–832.

Ben-Shaul, Y., N. Paran, & M. Galun. 1969. "The ultrastructure of the association between phycobiont and mycobiont in three ecotypes of the lichen *Caloplaca aurantia* var. *aurantia.*" *J. Microsc.* 8: pp. 415–422.

Berenbaum, M. 1980. "Adaptive significance of midgut pH in larval Lepidoptera." *Am. Nat.* 115: pp. 138–146.

Bergerud, A.T. 1972. "Food habits of Newfoundland caribou." *J. Wildlife Management* 36: pp. 913–923.

Bergerud, A.T. & M.J. Nolan. 1970. "Food habits of hand-reared caribou *Rangifer tarandus* L. in Newfoundland." *Oikos* 21: pp. 348–350.

Berner, L. 1970. "Y a-t-il une compétition entre lichens et mousses?" *Rev. Bryol. Lichénol.* 37: pp. 397–383.

Bertsch, A. & H. Butin. 1967. "Die Kultur der Erdflechte *Endocarpon pusillum* im Labor." *Planta* 72: pp. 29–42.

Beschel, R.E. 1950. "Flechten als Altersmasstat rezenter Moränen." *Z. Gletscherkd. Glazialgeol.* 1: pp. 152–161.

———. 1958. "Flechtenvereine der Städte, Stadflechten und ihre Wachstum." *Ber. Naturwiss. Med. Ver. Innsbruck* 52: pp. 1–158.

———. 1961. "Dating rock surfaces by lichen growth and its application to glaciology and physiography (lichenometry)." In *Geology of the Arctic, Vol. 2,* edited by G.O. Raasch, pp. 1044–1062. Toronto: University of Toronto Press.

Bewley, J.C. 1979. "Physiological aspects of desiccation tolerance." *Ann. Rev. Plant Physiol.* 30: pp. 195–238.

Billings, W.D. & W.B. Drew. 1938. "Bark factors affecting the distribution of corticolous bryophyte communities." *Am. Midl. Nat.* 20: pp. 302–333.

Bloomer, J.L., W.R. Eder, & W.F. Hoffman. 1970a. "Biosynthesis of (+)-protolichesterinic acid in *Cetraria islandica.*" *J. Chem. Soc. "C"* 1970: pp. 1848–1850.

———. 1970b. "Some problems in lichen metabolism: Studies with the mycobionts *Cetraria islandica* and *Cladonia papillaria.*" *Bryologist* 73: pp. 586–591.

Blum, O.B. 1973. "Water relations." In *The Lichens,* edited by V. Ahmadjian and M.E. Hale, Jr., pp. 381–400. New York: Academic Press.

Boardman, N.K. 1977. "Comparative photosynthesis of sun and shade plants." *Ann. Rev. Plant Physiol.* 28: pp. 355–377.

Bogusch, E.R. 1944. "Isolation in unialgal culture of lichen gonidia by a simple plasmolysis technique." *Plant Physiol.* 19: pp. 559–561.

Boissière, M.-C. 1972. "Cytologie du *Peltigera canina* (L.) Willd. en microscopie électronique: 1. Premières observations." *Rev. Gén. Bot.* 79: pp. 167–185.

———. 1979. "Cytologie du *Peltigera canina* (L.) Willd. en microscopie électronique:

le mycobionte à l'état végétatif." *Rev. Mycol.* 43: pp. 1–49.

———. 1982. "Cytochemical ultrastructure of *Peltigera canina*: Some features related to its symbiosis." *Lichenologist* 14: pp. 1–27.

Bonnier, G. 1888. "Germination des spores des Lichens sur les protonemas des Mousses et sur des Algues différents des gonidies du Lichen." *Compt. Rend. Soc. Biol. (Paris)* 40: pp. 541–543.

———. 1899. "Germination des Lichens sur les protonemas des Mousses." *Rev. Gén.* Bot. 1: pp. 165–169.

Book, S.A., G.E Connolly, & W.M. Longhurst. 1972. "Fallout ^{137}Cs accumulation in two adjacent populations of northern California deer." *Health Physics* 22: pp. 379–385.

Botkin, D.B., J.F. Janak, & J.R. Wallis. 1972. "Some ecological consequences of a computer model of forest growth." *J. Ecol.* 60: pp. 849–872.

Boudet, A. & P. Gadal. 1965. "Sur l'inhibition des enzymes par les tannins des feuilles de *Quercus sessilis* Ehrl. Isolement des tannins." *C. R. Acad. Sci. (Paris)* 260: pp. 4057–4060.

Bowler, P.A., & P.W. Rundel. 1975. "Reproductive strategies in lichens." *J. Linn. Soc. Bot.* 70: pp. 325–340.

———. 1978. "The *Ramalina farinacea* complex in North America: Chemical, ecological and morphological variation." *Bryologist* 81: pp. 386–403.

Brightman, F.H. 1959. "Some factors influencing lichen growth in towns." *Lichenologist* 1: pp. 104–108.

Brightman, F.H. & M.R.D. Seaward. 1977. "Lichens of man-made substrates." In *Lichen Ecology*, edited by M.R.D. Seaward, pp. 253–293. London: Academic Press.

Broadhead, E. 1958. "The psocid fauna of larch trees in northern England. An ecological study of mixed species populations exploiting a common resource." *J. Animal Ecol.* 27: pp. 217–263.

Broadhead, E. & I.W.B. Thornton. 1955. "*Elipsocus mclachlani* feeding on lichens." *Oikos* 6: pp. 1–50.

Brock, T.D. 1973. "Primary colonization of Surtsey, with special reference to the blue-green algae." *Oikos* 24: pp. 239–243.

———. 1975. "The effect of water potential on photosynthesis in whole lichens and in their liberated algal components." *Planta* 124: pp. 13–23.

Brodie, H.J. 1951. "The splash cup dispersal mechanism in plants." *Can. J. Bot.* 29: 224–234.

Brodie, H.J. & P.H. Gregory. 1953. "The action of wind in the dispersal of spores from cup-shaped plant structures." *Can. J. Bot.* 31: pp. 402–410.

Brodo, I.M. 1961a. "Transport experiments with corticolous lichens using a new technique." *Ecology* 42: pp. 838–841.

———. 1961b. "A study of lichen ecology in central Long Island, New York." *Am. Midl. Nat.* 65: pp. 290–310.

———. 1966. "Lichen growth and cities: A study on Long Island, New York." *Bryologist* 69: pp. 427–449.

———. 1973. "Substrate ecology." In *The Lichens*, edited by V. Ahmadjian and M.E. Hale, Jr., pp. 401–441. New York: Academic Press.

Brodo, I.M. & D.H.S. Richardson. 1978. "Chimeroid associations in the genus *Peltigera.*" *Lichenologist* 10: pp. 157–170.

Broughton, W.J. 1978. "Control of specificity in legume-*Rhizobium* associations." *J. Appl. Bacteriol.* 45: pp. 165–194.

Brown, D.H. 1976. "Mineral uptake by lichens." In *Lichenology: Progress and Problems,* edited by D.H. Brown, D.L. Hawksworth, and R.H. Bailey, pp. 419–439. London: Academic Press.

———. 1980. "Notes on the instability of extracted chlorophyll and a reported effect of ozone on lichen algae." *Lichenologist* 12: pp. 151–154.

Brown, D.H. & G.W. Buck. 1979. "Desiccation effects and cation distribution in bryophytes." *New Phytol.* 82: pp. 115–125.

Brown, D.H. & T.N. Hooker. 1977. "The significance of acidic substances in the estimation of chlorophyll and phaeophytin in lichens." *New Phytol.* 78: pp. 617–624.

Brown, D.H. & D.R. Slingsby. 1972. "The cellular location of lead and potassium in the lichen *Cladonia rangiferina* (L.) Hoffm." *New Phytol.* 71: pp. 297–305.

Brown, D.H. & N. Smirnoff. 1978. "Observations on the effect of ozone on *Cladonia rangiformis.*" *Lichenologist* 10: pp. 91–94.

Brown, R.M. & R. Wilson. 1968. "Electron microscopy of the lichen *Physcia aipolia* (Ehrh.) Nyl." *J. Phycol.* 4: pp. 230–240.

Brown, R.T. & P. Mikola. 1974. "The influence of fruticose soil lichens upon the mycorrhizae and seedling growth of forest trees." *Acta Forest. Fenn.* 141: pp. 1–23.

Bubrick, P. & M. Galun. 1980a. "Symbiosis in lichens: Differences in cell wall properties of freshly isolated and cultured phycobionts." *FEMS Microbiol. Lett.* 7: pp. 311–313.

———. 1980b. "Proteins from the lichen *Xanthoria parietina* which bind to phycobiont cell walls. Correlation between binding patterns and cell wall cytochemistry." *Protoplasma* 104: pp. 167–173.

Bubrick, P., M. Galun, & A. Frensdorff. 1981. "Proteins from the lichen *Xanthoria parietina* which bind to phycobiont cell walls: Localization in the intact lichen and cultured mycobiont." *Protoplasma* 105: pp. 207–211.

Buck, G.W. & D.H.Brown. 1979. "The effect of desiccation on cation location in lichens." *Ann. Bot.* 44: pp. 265–277.

Buffard, Y. & Y. Rondon. 1977. "Effets de l'extrait du lichen *Parmelia conspersa* (Ehrh.) Ach. sur la rhizogenese de boutures de *Peperomia magnoliaefolia* Dietr." *Bull. Soc. Bot. France* 124: pp. 161–166.

Burkholder, P.R. & A.W. Evans. 1945. "Further studies on the antibiotic activity of lichens." *Bull. Torrey Bot. Club* 72: pp. 157–164.

Burkholder, P.R., A.W. Evans, I. McVeigh, & H.K. Thornton. 1944. "Antibiotic activity of lichens." *Proc. Nat. Acad. Sci. U.S.A.* 30: 250–255.

Burton, M.A.S., P. Le Sueur, & K.J. Puckett. 1981. "Copper, nickel, and thallium uptake by the lichen *Cladonia rangiferina.*" *Can. J. Bot.* 59: pp. 91–100.

Burzlaff, D.F. 1950. "The effect of extracts from the lichen *Parmelia molliuscula* upon seed germination and upon the growth rate of fungi." *J. Colorado-Wyoming Acad. Sci.* 4: p. 56 (Abstr.).

Butin, H. 1954. "Physiologisch-ökologische Untersuchungen über den Wasserhaushalt und die Photosynthese bei Flechten." *Biol. Zentralbl.* 73: pp. 459–502.

Capriotti, A. 1959. "The effects of "Usno" on yeasts." *G. Microbiol.* 7: pp. 187–206.

———. 1961. "The effects of "Usno" on yeasts isolated from the excretions of tuberculosis patients." *Antibiot. Chemother.* 11: pp. 409–410.

Černhorsky, Z. 1950. "Fluorescence of lichens in ultra-violet light. Genus *Parmelia* Ach." *Stud. Bot. Cech.* 11: pp. 98–100.

Chadefaud, M. 1960. "Les végétaux non vasculaires." In *Traité de Botanique Systematique,* Vol. 1., edited by M. Chadefaud and L. Emberger, pp. 429–686. Paris: Masson.

Chambers, S., M. Morris, & D.C. Smith. 1976. "Lichen physiology XV. The effect of digitonin and other treatments on biotrophic transport of glucose from alga to fungus in *Peltigera polydactyla.*" *New Phytol.* 76: pp. 485–500.

Chervin, R.E., G.E. Baker, & H. R. Hohl. 1968. "The ultrastructure of phycobiont and mycobiont in two species of *Usnea.*" *Can. J. Bot.* 46: pp. 241–245.

Childress, S. & J.B. Keller. 1980. "Lichen growth." *J. Theor. Biol.* 82: pp. 157–165.

Chilvers, G.A., M. Ling-Lee, & A.E. Ashford. 1978. "Polyphosphate granules in the fungi of two lichens." *New Phytol.* 81: pp. 571–574.

Cirillo, V.P. 1961. "Sugar transport in microorganisms." *Ann. Rev. Microbiol.* 15: pp. 197–218.

Clements, F.E. 1916. "Plant succession: An analysis of the development of vegetation." Carnegie Inst. Wash. Publ. 242.

———. 1936. "Nature and structure of the climax." *J. Ecol.* 24: pp. 252–284.

Clements, F.E. & V.E. Shelford. 1939. *Bioecology.* New York: Wiley.

Cobb, F., M. Krstic, E. Zavarin, & H. Barber. 1968. "Inhibitory effects of volatile oleoresin components on *Fomes annosus* and four *Ceratocystis* species." *Phytopathology* 58: pp. 1327–1335.

Coker, P.D. 1967. "Damage to lichens by gastropods." *Lichenologist* 3: pp. 428–429.

Coley, P.D. 1980. "Effects of leaf age and plant life history patterns on herbivory." *Nature* 284: pp. 545–546.

Collins, C.R. & J.F. Farrar. 1978. "Structural resistances to mass transfer in the lichen *Xanthoria parietina.*" *New Phytol.* 81: pp. 71–83.

Cooper, R. & E.D. Rudolph. 1953. "Role of lichens in soil formation and plant succession." *Ecology* 34: pp. 805–807.

Council on Environmental Quality. 1980. *Environmental Quality.* 11th Annual Report of the Council on Environmental Quality. Washington, D.C.: U.S. Government Printing Office.

Cowan, D.A., T.G.A. Green, & A.T. Wilson. 1979. "Lichen metabolism. 1. The use of tritium labelled water in studies of anhydrobiotic metabolism in *Ramalina celastri* and *Peltigera polydactyla.*" *New Phytol.* 82: pp. 489–503.

Culberson, C.F. 1969. *Chemical and Botanical Guide to Lichen Products.* Chapel Hill: University of North Carolina Press.

———. 1970. "Supplement to *Chemical and Botanical Guide to Lichen Products.*" *Bryologist* 73: pp. 177–377.

———. 1972a. "High speed liquid chromatography of lichen extracts." *Bryologist* 75: pp. 54–62.

———. 1972b. "Improved conditions and new data for the identification of lichen products by a standardized thin-layer chromatographic method." *J. Chromatogr.* 72: pp. 113–125.

Culberson, C.F. & V. Ahmadjian. 1980. "Artificial reestablishment of lichens. II. Secondary products of resynthesized *Cladonia cristatella* and *Lecanora chrysoleuca.*" *Mycologia* 72: pp. 90–109.

Culberson, C.F. & W.L. Culberson. 1977. "Chemosyndromic variation in lichens." *Systematic Bot.* 1: pp. 325–339.

Culberson, C.F., W.L. Culberson, & A. Johnson. 1977a. *Second supplement to Chemical and Botanical Guide to Lichen Products.* St. Louis, MO: American Bryological and Lichenological Society, Missouri Bot. Garden.

———. 1977b. "Nonrandom distribution of an epiphytic *Lepraria* on two species of *Parmelia.*" *Bryologist* 80: pp. 201–203.

———. 1981. " A standardized TLC analysis of β-orcinol depsidones." *Bryologist* 84: pp. 16–29.

Culberson, C.F. & A. Johnson. 1976. "A standardized two-dimensional thin-layer chromatographic method for lichen products." *J. Chromatogr.* 128: pp. 253–259.

Culberson, C.F. & H. Kristinsson. 1970. "A standardized method for the identification of lichen products." *J. Chromatogr.* 46: pp. 85–93.

Culberson, W.L. 1955a. "The corticolous communities of lichens and bryophytes in upland forests." *Ecol. Monogr.* 25: pp. 215–231.

———. 1955b. "Qualitative and quantitative studies on the distribution of corticolous lichens and bryophytes in Wisconsin." *Lloydia* 18: pp. 25–36.

———. 1969. "Norstictic acid as a hymenial constituent of *Letharia.*" *Mycologia* 61: pp. 731–736.

———. 1970. "Chemosystematics and ecology of lichen-forming fungi." *Ann. Rev. Ecol. Syst.* 1: pp. 153–170.

———. 1972. "Disjunctive distributions in the lichen-forming fungi." *Ann. Mo. Bot. Garden* 59: pp. 165–173.

———. 1973. "The *Parmelia perforata* group: Niche characteristics of chemical races, speciation by parallel evolution, and a new taxonomy." *Bryologist* 76: pp. 20–29.

Culberson, W.L. & C.F. Culberson. 1967. "Habitat selection by chemically differentiated races of lichens." *Science* 158: pp. 1195–1197.

———. 1970. "A phylogenetic view of chemical evolution in lichens." *Bryologist* 73: pp. 1–31.

———. 1973. "Parallel evolution in lichen-forming fungi." *Science* 180: pp. 196–198.

———. 1982. "Evolutionary modification of ecology in a common lichen species." *Syst. Bot.* 7: pp. 158–169.

Cuthbert, J.B. 1934. "Further notes on the physiology of *Teloschistes flavicans.*" *Trans. Roy. Soc. South Africa* 22: pp. 35–54.

Czeczuga, B. 1979a. "Investigations on carotenoids in lichens. I. The presence of carotenoids in representatives of certain families." *Nova Hedwigia* 31: pp. 337–347.

———. 1979b. "Investigations on carotenoids in lichens. II. Members of the Usneaceae family." *Nova Hedwigia* 31: pp. 349–356.

———. 1980a. "Investigations on carotenoids in lichens. III. Species of *Peltigera*

Willd." *Cryptog., Bryol. Lichénol.* 1980: pp. 189–196.

———. 1980b. "Investigations on carotenoids in lichens. IV. Representatives of the Parmeliaceae family." *Nova Hedwigia* 32: pp. 105–111.

Czygan, F.C. 1976. "Carotenoid-Garnitur und -Stoffwechsel der Flechte *Haematomma ventosum* (L.) Massal. s. str. und ihres Phycobionten." *Z. Pflanzenphysiol.* 79: pp. 438–445.

Dalvi, R.R., B. Singh, & D.K. Salunkhe. 1972. "Physiological and biochemical investigation on the phytotoxicity of usnic acid." *Phyton (Buenos Aires)* 29: pp. 63–72.

Degelius, G. 1954. "The lichen genus *Collema.*" *Symb. Bot. Upsal.* 13: pp. 1–499.

———. 1964. "Biological studies of the epiphytic vegetation on twigs of *Fraxinus excelsior.*" *Acta Horti Gothob.* 27: pp. 11–55.

———. 1978. "Further studies on the epiphytic vegetation on twigs." *Bot. Gothob.* 7: pp. 1–58.

Denison, R., B. Caldwell, B. Bormann, L. Eldred, C. Swanberg, & S. Anderson. 1977. "The effects of acid rain on nitrogen fixation in western Washington coniferous forests." *Water, Air & Soil Pollution* 8: pp. 21–34.

Denison, W.C. 1973. "Life in tall trees." *Sci. Am.* 228: pp. 74–80.

Denton, G.H. & W. Karlén. 1973. "Lichenometry: Its application to holocene moraine studies in southern Alaska and Swedish Lapland." *Arctic & Alpine Res.* 5: pp. 347–372.

Deruelle, S. 1978. "Les lichens et la pollution atmosphérique." *Bull. Ecol.* 9: pp. 87–128.

De Sloover, J. & F. LeBlanc. 1968. "Mapping of atmospheric pollution on the basis of lichen sensitivity." In *Proc. Symp. on Recent Adv. Trop. Ecol.,* edited by R. Misra and B. Golpal, pp. 42–56, Varanasi, India.

Des Meules, P. & J. Heyland. 1969. "Contribution to the study of the food habits of caribou. Part I—Lichen preferences." *Natural. Canad.* 96: pp. 317–331.

De Wit, A. 1976. *Epiphytic Lichens and Air Pollution in the Netherlands. Bibl. Lichenologica,* Bd. 5, Vaduz: Cramer.

Dibben, M.J. 1971. "Whole-lichen culture in a phytotron." *Lichenologist* 5: pp. 1–10.

Di Benedetto, G. & F. Furnari. 1962. "Sull crescita di *Trebouxia albulescens* e *T. humicola* trattate con acido β-indol-acetico e con acido gibberellico." *Boll. dell'Ist. di Bot. dell'Univ. di Catania* 3: pp. 34–38.

Diner, B., V. Ahmadjian, & H. Rosenkrantz. 1964. "Preliminary fractionation of pigments from the lichen fungus *Acarospora fuscata.*" *Bryologist* 67: pp. 363–368.

Drew, E.A. & D.C. Smith. 1967a. "Studies in the physiology of lichens. VII. The physiology of the *Nostoc* symbiont of *Peltigera polydactyla* compared with cultured and free-living forms." *New Phytol.* 66: pp. 379–388.

———. 1967b. "Studies in the physiology of lichens. VIII. Movement of glucose from alga to fungus during photosynthesis in the thallus of *Peltigera polydactyla.*" *New Phytol.* 66: pp. 389–400.

Drury, W.H., Jr. & I.C.T. Nisbet. 1973. "Succession." *J. Arnold Arboretum* 54: pp. 331–368.

Du Rietz, G.E. 1931. "Studier över vinddriften på snöfält i de skandinaviska fjallen." *Bot. Notiser* 1931: pp. 31–46.

———. 1932. "Zur Vegetationsökologie der östschwedische Kustenfelsen." *Beih. Bot. Zbl.* 49: pp. 61–112.

Durrell, L.W. 1967. "An electron miscroscope study of algal hyphal contact in lichens." *Mycopath. Mycol. Appl.* 31: pp. 273–286.

Ejiri, H. & S. Shibata. 1974. "Zeorin from the mycobiont of *Anaptychia hypoleuca.*" *Phytochem.* 13: pp. 2871–2873.

Ellis, E.A. 1975. "Observations on the ultrastructure of the lichen *Chiodecton sanguineum.*" *Bryologist* 78: pp. 471–476.

Ellis, E.A. & R.M. Brown. 1972. "Freeze-etch ultrastructure of *Parmelia caperata* (L.) Ach." *Trans. Am. Microsc. Soc.* 91: pp. 411–421.

Emerson, F.W. 1947. *Basic Botany.* Philadelphia: The Blakiston Co.

Ercegović, A. 1925. "Litofitska vegetacija vapnenaca i dolomita u Hrvatskoj. (La végétation des lithophytes sur les calcaires et les dolomites en Croatie)." *Acta Bot. Inst. Bot. R. Univ. Zagrebensis* 1: pp. 64–114.

Ertl, L. 1951. "Über die Lichtverhaltnisse in Laubflechten." *Planta* 39: pp. 245–270.

Esslinger, T.L. 1977. "A chemosystematic revision of the brown *Parmeliae.*" *J. Hattori Bot. Lab.* 42: pp. 1–211.

———. 1978. "Studies in the lichen family Physciaceae. II. The genus *Phaeophyscia* in North America." *Mycotaxon* 7: pp. 283–320.

Evodokimova, T.I. 1962. "On the toxicity of several mosses and lichens with respect to *Azobacter* in Karelian soils." *Pochvovedenie* 1962: pp. 88–92.

Fabiszewski, J. 1975. "East Canadian peat bog ecosystems and the biological role of their lichens." *Phytocoenosis* (Warszawa-Białowieza) 4: pp. 3–94.

Fahselt, D. 1975. "Gas-liquid chromatography of the lichen substance usnic acid." *Bryologist* 78: pp. 452–455.

———. 1976. "Growth rates in vegetative and reproductive tissue of *Parmelia cumberlandia* after relocation to a new environment." *Can. Field-Natur.* 91: pp. 134–140.

———. 1979. "Lichen substances of transplanted thallus segments of *Parmelia cumberlandia.*" *Can. J. Bot.* 57: pp. 23–25.

———. 1981. "Lichen products of *Cladonia stellaris* and *C. rangiferina* maintained under artificial conditions." *Lichenologist* 13: pp. 87–91.

Farrar, J.F. 1973. "Physiological lichen ecology." Ph.D. dissertation, The University of Oxford, England.

———. 1974. "A method for investigating lichen growth rates and succession." *Lichenologist* 6: pp. 151–155.

———. 1976a. "Ecological physiology of the lichen *Hypogymnia physodes.* I. Some effects of constant water saturation." *New Phytol.* 77: pp. 93–103.

———. 1976b. "Ecological physiology of the lichen *Hypogymnia physodes.* II. Effects of wetting and drying cycles and the concept of 'physiological buffering'." *New Phytol.* 77: 105–113.

———. 1976c. "The uptake and metabolism of phosphate by the lichen *Hypogymnia physodes.*" *New Phytol.* 77: pp. 127–134.

———. 1976d. "The lichen as an ecosystem: Observation and experiment." In *Lichenology: Progress and Problems,* edited by D.H. Brown, D.L. Hawksworth, and R.H. Bailey, pp. 385–406. London: Academic Press.

Farrar, J.F. & D.C. Smith. 1976. "Ecological physiology of the lichen *Hypogymnia physodes.* III. The importance of the rewetting phase." *New Phytol.* 77: pp. 115–125.

Fawcett, C.H. & D.M. Spencer. 1969. "Natural antifungal compounds." In *Fungicides,* Vol. 2, edited by D.C. Torgeson, pp. 637–639.

Fedoseev, K.G. & P.A. Yakimov. 1960. "Preparation of usnic acid from lichens. 1. A study of conditions of chemical leaching of usnic acid from *Cladonia.*" *Farm. Zh.* (Leningrad) 9: pp. 139–149.

Feeny, P.P. 1968. "Effect of oak leaf tannins on larval growth of the winter moth *Operophtera brumata.*" *J. Insect Physiol.* 14: pp. 805–817.

———. 1969. "Inhibitory effect of oak leaf tannins on the hydrolysis of proteins by trypsin." *Phytochem.* 8: pp. 2119–2126.

———. 1970. "Seasonal changes in the oak leaf tannins and nutrients as a cause of spring feeding by winter moth caterpillars." *Ecology* 51: pp. 565–581.

———. 1976. "Plant apparency and chemical defense." *Rec. Adv. Phytochem.* 10: pp. 1–40.

Feige, G.B. 1973. "Untersuchungen zur Ökologie und Physiologie der marinen Blaualgenflechte *Lichina pygmaea* Ag. II. Die Reversibilität der Osmoregulation." *Z. Pflanzenphysiol.* 68: pp. 415–421.

———. 1977. "Untersuchungen über die Aufnahme von Sulfat und Phosphat durch isolierte Phycobionten der Flechte *Peltigera aphthosa* (L.) Willd." *Z. Pflanzenphysiol.* 82: pp. 347–354.

Ferry, B.W. & M.S. Baddeley. 1976. "Sulphur dioxide uptake in lichens." In *Lichenology: Progress & Problems,* edited by D.H.Brown, D.L. Hawksworth, and R.H. Bailey, pp. 407–418. London: Academic Press.

Ferry, B.W., M.S. Baddeley, & D.L. Hawksworth. 1973. *Air Pollution and Lichens.* London: University of London, Athlone Press.

Ferry, B.W. & B.J. Coppins. 1979. "Lichen transplant experiments and air pollution studies." *Lichenologist* 11: pp. 63–73.

Fisher, K.A. & N.J. Lang. 1971. "Comparative ultrastructure of cultured species of *Trebouxia.*" *J. Phycol.* 7: pp. 155–165.

Fisher, P.J. & M.C.F. Proctor. 1978. "Observations on a season's growth in *Parmelia caperata* and *P. sulcata* in South Devon." *Lichenologist* 10: pp. 81–90.

Fisher, R.F. 1979. "Possible allelopathic effects of reindeer-moss (*Cladonia*) on jack pine and white spruce." *Forest Sci.* 25: pp. 256–260.

Fletcher, A. 1973a. "The ecology of marine (littoral) lichens on some rocky shores of Anglesey." *Lichenologist* 5: pp, 368–400.

———. 1973b. "The ecology of maritime (supralittoral) lichens on some rocky shores of Anglesey." *Lichenologist* 5: pp. 401–422.

———. 1976. "Nutritional aspects of marine and maritime lichen ecology." In *Lichenology: Progress and Problems,* edited by D.H. Brown, D.L. Hawksworth, and R.H. Bailey, pp. 359–384. London: Academic Press.

Follmann, G. 1960. "Die Durchlässig Keitseigenshaften der Protoplasten von Phycobionten aus *Cladonia furcata* (Huds.) Schrad." *Naturwiss.* 47: pp. 405–406.

———. 1961. "Lichenometrische Alterbestimmungen an Vorchristlichen Steinsetzungen der Polunesischen Osterinsel." *Naturwiss.* 48: pp. 627–628.

———. 1965. "Das Alter der Steinriesen auf der Osterinsel, Flechtenstudien als Hilfsmittel der Datierung." *Umschau in Wiss. & Tech.* 65: pp. 374–377.

Follmann, G. & M. Nakagava. 1963. "Keimhemmung von Angiospermensamen durch Flechtenstoffe." *Naturwiss.* 50: pp. 696–697.
Follmann, G. & R. Peters. 1966. "Flechtenstoffe und Bodenbildung." *Z. Naturf.* 21: pp. 386–387.
Follmann, G. & V. Villagrán. 1965. "Flechtenstoffe und Zellpermeabilität." *Z. Naturf.* 20B: p. 723.
Foote, K.G. 1966. "The vegetation of lichens and bryophytes on limestone outcrops in the driftless area of Wisconsin." *Bryologist* 69: pp. 265–292.
Forman, R.T.T. 1975. "Canopy lichens with blue-green algae: A nitrogen source in a Colombian rain forest." *Ecology* 56: pp. 1176–1184.
Forssell, K.B. 1883. "Studier öfver cephalodierne." *Bihang K. Svensk. Vet.-Akad.* 8: pp. 1–112.
Fortin, J.-A. & J.-R. Thibault. 1972. "Présence d'auxine chez un lichen (*Cladonia alpestris*) et transformation du tryptophane en acide indolyl-acétique par trois mycobiontes." *Natural. Can.* 99: pp. 213–218.
Fox, C.H. 1966. "Experimental studies on the physiology of the phycobiont and mycobiont *Ramalina ecklonii.*" *Physiol. Plantarum* 19: pp. 830–839.
Frey, E. 1959. "Die Flechtenflora und Vegetation des Nationalparks im Unterengadin. II. Teil. Die Entwicklung der Flechtenvegetation auf photogrammetrisch kontrollierten Dauerflachen." *Ergebn. Wiss. Unters. Schweiz. Natn.-Parks* 6: pp. 237–319.
Friedmann, E.I. 1971. "Light and scanning electron microscopy of the endolithic desert algal habitat." *Phycologia* 10: pp. 411–428.
———. 1977. "Microorganisms in Antarctic desert rocks from dry valleys and Dufek Massif." *Antarct. J. U.S.* 12: pp. 26–30.
———. 1978. "Melting snow in the dry valleys is a source of water for endolithic microorganisms." *Antarct. J. U.S.* 13: pp. 162–163.
———. 1980. "Endolithic microbial life in hot and cold deserts." *Origins of Life* 10: pp. 223–235.
———. 1982. "Endolithic microorganisms in the Antarctic cold desert." *Science* 215: pp. 1045–1053.
Friedmann, E.I., O.R. Friedmann, & C.P. McKay. 1981. "Adaptations of cryptoendolithic lichens in the Antarctic desert." In *Les Ecosystèmes Terrestres Subantarctiques.* Paris: Comité National Français des Recherches Antarctiques.
Friedmann, E.I. & M. Galun. 1974. "Desert algae, lichens, and fungi." In *Desert Biology*, Vol. II, edited by G.W. Brown, pp. 165–212. New York: Academic Press.
Friedmann, E.I. & A.P. Kibler. 1980. "Nitrogen economy of endolithic microbial communities in hot and cold deserts." *Microb. Ecol.* 6: pp. 95–108.
Friedmann, E.I., Y. Lipkin, & R. Ocampo-Paus. 1967. "Desert algae of the Negev (Israel)." *Phycologia* 6: pp. 185–200.
Friedmann, E.I. & R. Ocampo. 1976. "Endolithic blue-green algae in the dry valleys: Primary producers in the Antarctic desert ecosystem." *Science* 193: pp. 1247–1249.
Fries, N. 1973. "Effects of volatile organic compounds on the growth and development of fungi." *Trans. Brit. Mycol. Soc.* 60: pp. 1–21.
Fry, E.J. 1927. "The mechanical action of crustaceous lichens on substrata of shale,

schist, gneiss, limestone and obsidian." *Ann. Bot.* 41: pp. 437–460.

Fujita, M. 1968. "Fine structure of lichens." *Misc. Bryol. Lichenol.* 4: pp. 157–160.

Furnari, F. & F. Luciani. 1962. "Esperenze sulla crescita dei micobionti in *Sarcogyne similis* e *Acarospora fuscata* in coltura pura su vari substrati." *Boll. dell'Ist. di Bot. dell'Univ. di Catania* 3: pp. 39–47.

Furuya, T, S. Shibata, & H. Iizuka. 1966. "Gas-liquid chromatography of anthraquinones." *J. Chromatogr.* 21: pp. 116–118.

Gagnon, J.D. 1966. "Le lichen *Lecidea granulosa* constitue un milieu favorable à la germination de l'épinette noire." *Natural. Can.* 93: pp. 89–98.

Galinou, M.A. 1956. "Sur la mise en évidence de quelques biocatalyseurs chez les lichens." *Proc. Int. Bot. Congr. 8th, 1954.* Sect. XVIII, pp. 2–4.

Galløe, O. 1939. "*Lobaria amplissima* L." In Galløe, *Natural History of the Danish Lichens.* Vol. 6, p. 79. Copenhagen: H. Ascheoug.

Galun, M., L. Behr, & Y. Ben-Shaul. 1974. "Evidence for protein content in concentric bodies of lichenized fungi." *J. de Microsc.* (Paris) 19: pp. 193–196.

Galun, M., Y. Ben-Shaul, & N. Paran. 1971b. "The fungus-alga association in the Lecideaceae: An ultrastructural study." *New Phytol.* 70: pp. 837–839.

Galun, M., A. Braun, A. Frensdorff, & E. Galun. 1976. "Hyphal walls of isolated lichen fungi." *Arch. Microbiol.* 108: pp. 9–16.

Galun, M., E. Kushnir, L. Behr, & Y. Ben-Shaul. 1973. "Ultrastructural investigation on the alga-fungus relation in pyrenocarpous lichen species." *Protoplasma* 78: pp. 187–193.

Galun, M., N. Paran, & Y. Ben-Shaul. 1970a. "Structural modifications of the phycobiont in the lichen thallus." *Protoplasma* 69: pp. 85–96.

Galun, M., N. Paran, & Y. Ben-Shaul. 1970b. "The fungus-alga associations in the Lecanoraceae: An ultrastructural study." *New Phytol.* 69: pp. 599–603.

Galun, M., N. Paran, & Y. Ben-Shaul. 1970c. "An ultrastructural study of the fungus-alga associations in *Lecanora radiosa* growing under different environmental conditions." *J. de Microsc.* (Paris) 9: pp. 801–806.

Galun, M., N. Paran, & Y. Ben-Shaul. 1971a. "Electron microscopic study of the lichen *Dermatocarpon hepaticum* (Ach.) Th. Fr." *Protoplasma* 73: pp. 457–468.

Galvan, J., C. Rodriguez, & C. Ascaso. 1981. "The pedogenic action of lichens in metamorphic rocks." *Pedobiologia* 21: pp. 60–73.

Gardner, C.R. & D.M.J. Mueller. 1981. "Factors affecting the toxicity of several lichen acids: Effect of pH and lichen acid concentration." *Am. J. Bot.* 68: pp. 87–95.

Garrett, R.M. 1971. "Studies on some aspects of ascospore liberation and dispersal in lichens." *Lichenologist* 5: pp. 33–44.

Garton, G.A. & W.R.H. Duncan. 1971."Fatty acid composition and intramolecular structure of triglycerides from adipose tissue of the red deer and the reindeer." *J. Sci. Food & Agric.* 22: pp. 29–33.

Garty, J., M. Gal, & M. Galun. 1974. "The relationship between physico-chemical soil properties and substrate choice of 'multisubstrate' lichen species." *Lichenologist* 6: pp. 146–150.

Garty, J., M. Galun, C. Fuchs, & N. Zisapel. 1977. "Heavy metals in the lichen *Caloplaca aurantia* from urban, suburban and rural regions in Israel (a comparative study)." *Water, Air and Soil Pollut.* 8: pp. 171–188.

Garty, J., M. Galun, & M. Kessel. 1979. "Localization of heavy metals and other elements accumulated in the lichen thallus." *New Phytol.* 82: pp. 159–168.

Gerson, U. 1973. "Lichen-arthropod associations." *Lichenologist* 5: pp. 434–443.

Gerson, U. & M.R.D. Seaward. 1977. "Lichen-invertebrate associations." In *Lichen Ecology*, edited by M.R.D. Seaward, pp. 69–119. London: Academic Press.

Gilbert, O.L. 1970. "A biological scale for the estimation of sulphur dioxide pollution." *New Phytol.* 69: pp. 629–634.

———. 1971. "The effect of airborne fluorides on lichens." *Lichenologist* 5: pp. 26–32.

———. 1973. "The effect of airborne fluorides." In *Air Pollution and Lichens*, edited by B.W. Ferry, M.S. Baddeley, & D.L. Hawksworth, pp. 176–191. London: Athlone.

———. 1980. "Acid rain threat to lichens?" *Bull. Brit. Lichen Soc.* 46: pp. 1–2.

Gilenstam, G. 1969. "Studies in the lichen genus *Conotrema*." *Ark. Bot.* 7: pp. 149–179.

Gimingham, C.H. 1964. "Dwarf shrub heaths." In *The Vegetation of Scotland*, edited by J.H. Burnett, pp. 232–287. Edinburgh: Oliver & Boyd.

Giudici Di Nicola, M. & R. Tomaselli. 1961a. "Ricerche preliminari sui pigmenti nel ficosimbionte lichenico *Trebouxia decolorans* Ahm. I. Carotenoidi." *Boll. dell'Ist. di Bot. dell'Univ. di Catania* 2: pp. 22–28.

———. 1961b. "Ricerche preliminari sui pigmenti nel ficosimbionte lichenico *Trebouxia decolorans* Ahm. II. Clorofille." *Boll. dell'Ist. di Bot. dell'Univ. di Catania* 2: pp. 29–34.

Giudici Di Nicola, M. & G. Di Benedetto. 1962. "Ricerche preliminari sui pigmenti nel ficobionte lichenico *Trebouxia decolorans* Ahm. III. Clorofille e carotenoidi." *Boll. dell'Ist. di Bot. dell'Univ. di Catania* 3: pp. 22–33.

Gjelstrup, P. & U. Søchting. 1979. "Cryptostigmatid mites (Acarina) associated with *Ramalina siliquosa* (Lichens) on Bornholm in the Baltic." *Pedobiologia* 19: pp. 237–245.

Glück, H. 1899. "Ent wurf zu einer vergleichenden Morphologie der Flechten-Spermagonien." *Verh. Naturhist. -Med. Ver. Heidelberg* [N.S.] 6: pp. 81–216.

Goldstein, J.L. & T. Swain. 1965. "The inhibition of enzymes by tannins." *Phytochem.* 4: pp. 185–192.

Golubic, S., E.I. Friedmann, & J. Schneider. 1981. "The lithobiontic ecological niche, with special reference to microorganisms." *J. Sedimentary Petrology* 51: pp. 475–478.

Gorham, E. 1958. "Accumulation of radioactive fall-out by plants in the English Lake District." *Nature* 181: pp. 1523–1524.

———. 1959. "A comparison of lower and higher plants as accumulators of radioactive fallout." *Can. J. Bot.* 37: pp. 327–329.

Gough, L.P. 1975. "Cryptogam distributions on *Pseudotsuga menziesii* and *Abies lasiocarpa* in the Front Range, Boulder County, Colorado." *Bryologist* 78: pp. 124–145.

Goyal, R. & M.R.D. Seaward. 1981. "Metal uptake in terricolous lichens. I. Metal localization within the thallus." *New Phytol.* 89: pp. 631–645.

Granett, A.L. 1974. "Ultrastructural studies of concentric bodies in the ascomycetous fungus *Venturi inaequalis*." *Can. J. Bot.* 52: pp. 2137–2139.

Green, T.G.A. 1970. "The biology of lichen symbionts." Ph.D. dissertation, The University of Oxford, England.

Green, T.G.A., J. Horstmann, H. Bonnett, A. Wilkins, & W.B. Silvester. 1980. "Nitrogen fixation by members of the Stictaceae (Lichenes) of New Zealand." *New Phytol.* 84: pp. 339–348.

Green, T.G.A. & D.C. Smith. 1974. "Lichen physiology. XIV. Differences between lichen algae in symbiosis and in isolation." *New Phytol.* 73: pp. 753–766.

Green, T.G.A. & W.P. Snelgar. 1981. "Carbon dioxide exchange in lichens. Relationship between net photosynthetic rate and CO_2 concentration." *Plant Physiol.* 68: pp. 199–201.

Green, T.G.A., W.P. Snelgar, & D.H. Brown. 1981. "Carbon dioxide exchange in lichens. Carbon dioxide exchange through the cyphellate lower cortex of *Sticta latifrons* Rich." *New Phytol.* 88: pp. 421–426.

Greenhalgh, G.N. & D. Anglesea. 1979. "The distribution of algal cells in lichen thalli." *Lichenologist* 11: pp. 283–292.

Gregory, P.H. 1961. *The Microbiology of the Atmosphere.* London: Leonard Hill.

Griffiths, H.B. & A.D. Greenwood. 1972. "The concentric bodies of lichenized fungi." *Arch. Mikrobiol.* 87: pp. 285–302.

Griffiths, H.B., A.D. Greenwood, & J.W. Millbank. 1972. "The frequency of heterocysts in the *Nostoc* phycobiont of the lichen *Peltigera canina* Willd." *New Phytol.* 71: pp. 11–13.

Grime, J.P. 1977. "Evidence for the existence of three primary strategies in plants and its relevance to ecological and evolutionary theory. *Am. Nat.* 111: pp. 1169–1194.

Grindon, L.H. 1859. *The Manchester Flora.* London: W. White.

Gross, M. & V. Ahmadjian. 1966. "The effects of *l*-amino acids on the growth of two species of lichen fungi." *Svensk Botanisk Tidskrift.* 60: pp. 74–80.

Gustafson, F.G. 1954. "A study of riboflavin, thiamine, niacin and ascorbic acid content of plants in Northern Alaska." *Bull. Torrey Bot. Club* 81: pp. 313–322.

Hakulinen, R. 1966. "Über die Wachstumgeschwindigheit einiger Laubflechten." *Ann. Bot. Fenn.* 3: pp. 167–179.

Hale, M.E., Jr. 1954. "First report on lichen growth rate and succession at Aton Forest, Connecticut." *Bryologist* 57: pp. 244–247.

———. 1955. "Phytosociology of corticolous cryptogams in the upland forests of southern Wisconsin." *Ecology* 36: pp. 45–63.

———. 1956a. "Fluorescence of lichen depsides and depsidones as a taxonomic criterion." *Castanea* 21: pp. 30–32.

———. 1956b. "Ultraviolet absorption spectra of lichen depsides and depsidones." *Science* 123: p. 671.

———. 1957. "Conidial stage of the lichen fungus *Buellia stillingiana* and its relation to *Sporidesmium folliculatum.*" *Mycologia* 49: pp. 417–419.

———. 1958. "Vitamin requirements of three lichen fungi." *Bull. Torrey Bot. Club* 85: pp. 182–187.

———. 1961. *Lichen Handbook.* Washington, D.C.: Smithsonian Institution Press.

———. 1966. "Chemistry and evolution in lichens." *Israel J. Bot.* 15: pp. 150–157.

———. 1967. *The Biology of Lichens.* London: Edward Arnold.

———. 1970. "Single-lobe growth-rate patterns in the lichen *Parmelia caperata.*" *Bryologist* 73: pp. 72–81.

———. 1972. "Natural history of Plummers Island, Maryland. XXI. Infestation of the lichen *Parmelia baltimorensis* Gyel. & For. by *Hypogastrura packardi* Folsom (Collembola)." *Proc. Biol. Soc. Washington* 85: pp. 287–296.

———. 1973a. "Growth." In *The Lichens,* edited by V. Ahmadjian and M.E. Hale, Jr., pp. 473–492. New York: Academic Press.

———. 1973b. "Fine structure of the cortex in the lichen family Parmeliaceae viewed with the scanning-electron microscope." *Smithsonian Contributions to Botany,* Vol. 10. Washington, D.C.: Smithsonian Institution Press.

———. 1974. *The Biology of Lichens.* 2nd ed. London: Edward Arnold.

———. 1975. "A monograph of the lichen genus *Relicina* (Parmeliaceae)." *Smithsonian Contributions to Botany,* No. 26. Washington, D.C.: Smithsonian Institution Press.

———. 1976. "Lichen structure viewed with the scanning electron microscope." In *Lichenology: Progress and Problems,* edited by D.H. Brown, D.L. Hawksworth and R.H. Bailey, pp. 1–15. London: Academic Press.

———. 1978. "A new species of *Ramalina* from North America (Lichenes: Ramalinaceae)." *Bryologist* 81: pp. 599–602.

———. 1979. *How to Know the Lichens.* 2nd ed. Dubuque, IA: W.C. Brown.

———. 1981. "Pseudocyphellae and pored epicortex in the Parmeliaceae: Their delimitation and evolutionary significance." *Lichenologist* 13: pp. 1–10.

———. 1983. *The Biology of Lichens.* 3rd ed. London: Edward Arnold.

Hale, M.E., Jr. & G. Vobis. "*Santessonia,* a new lichen genus from southwest Africa." *Bot. Notiser* 131: pp. 1–5.

Hallbauer, D.K. & H.M. Jahns. 1977. "Attack of lichens on quartzitic rock surfaces." *Lichenologist* 9: pp. 119–122.

Hamada, N. 1982a. "The distribution pattern of the medullary depsidone salazinic acid in the thallus of *Ramalina siliquosa* (lichens)." *Can. J. Bot.* 60: pp. 379–382.

———. 1982b. "The effect of temperature on the content of the medullary depsidone salazinic acid in *Ramalina siliquosa* (lichens)." *Can. J. Bot.* 60: pp. 383–385.

Hampton, R.E. 1973. "Photosynthetic pigments in *Peltigera canina* (L.) Willd. from sun and shade habitats." *Bryologist* 76: pp. 534–545.

Handley, R., R. Overstreet, & K. Grossenbacher. 1969. "The fine structure of *Protococcus* as the algal host of *Ramalina reticulata.*" *Nova Hedwigia* 18: pp. 581–596.

Hanson, W.C., D.G. Watson, & R.W. Perkins. 1967. "Concentration and retention of fallout radionuclides in Alaskan Arctic ecosystems." In *Radioecological Concentration Processes,* edited by B. Aberg and F.P. Hungate, pp. 233–245. New York: Pergamon Press.

Harder, R. & E. Uebelmesser. 1958. "Über die Beeinflussung niederer Erdphycomyceten durch Flechten." *Arch. Mikrobiol.* 31: pp. 82–86.

Harley, J.L. 1969. *The Biology of Mycorrhiza.* London: Leonard Hill.

Harley, J.L. & D.C. Smith. 1956. "Sugar absorption and surface carbohydrate activity of *Peltigera polydactyla* (Neck.) Hoffm." *Ann. Bot.* 20: pp. 513–543.

Harper, J.L., P.H. Lovell, & K.G. Moore. 1970. "The shapes and sizes of seeds." *Ann. Rev. Ecol. Syst.* 1: pp. 327–356.

Harper, J.L. & J. White. 1974. "The demography of plants." *Ann. Rev. Ecol. Syst.* 5: pp. 419–463.

Harris, G.P. 1971. "The ecology of corticolous lichens. II. The relation between physiology and the environment." *J. Ecol.* 59: pp. 441–452.

Harris, G.P. & K.A. Kershaw. 1971. "Thallus growth and the distribution of stored metabolites in the phycobionts of the lichens *Parmelia sulcata* and *P. physodes.*" *Can. J. Bot.* 49: p. 1367.

Hartwell, J.L. 1971. "Plants used against cancer. A survey." *Lloydia* 34: pp. 386–438.

Hasenhüttl, G. & J. Poelt. 1978. "Über die Brutkörner bei der Flechtengattung *Umbilicaria.*" *Ber. Deut. Bot. Ges.* 91: pp. 275–296.

Hawksworth, D.L. 1971. "Lichens as litmus for air pollution: A historical review." *Int. J. Environ. Studies* 1: pp. 281–296.

———. 1973a. "Ecological factors and species delimitation in the lichens." In *Taxonomy and Ecology,* edited by V.H. Heywood, pp. 31–69. London: Academic Press.

———. 1973b. "Mapping studies." In *Air Pollution and Lichens,* edited by B.W. Ferry, M.S. Baddeley, and D.L. Hawksworth, pp. 38–76. London: Athlone Press.

———. 1976. "Lichen chemotaxonomy." In *Lichenology: Progress and Problems,* edited by D.H. Brown, D.L. Hawksworth, and R.H. Bailey, pp. 139–184. London: Academic Press.

———. 1978. "The taxonomy of lichen-forming fungi: Reflections on some fundamental problems." In *Essays in Plant Taxonomy,* edited by H.E. Street, pp. 211–243. New York: Academic Press.

———. 1979. "The lichenicolous Hyphomycetes." *Bull. Br. Mus. (Natural History) Bot.* 6: pp. 183–300.

Hawksworth, D.L. & A.O. Chater. 1979. "Dynamism and equilibrium in a saxicolous lichen mosaic." *Lichenologist* 11: pp. 75–80.

Haynes, F.N. 1964. "Lichens." In *Viewpoints in Biology,* 3, edited by J.D. Carthy and C.L. Duddington, pp. 64–115. London: Butterworth & Co.

Heatwole, H. 1966. "Moisture exchange between the atmosphere and some lichens of the genus *Cladonia.*" *Mycologia* 58: pp. 148–156.

Hedlund, J.T. 1895. "Über Thallusbildung durch Pyknokonidien bei *Catillaria denigrata* (Fr.) und *C. prasina* (Fr.)." *Bot. Zbl.* 63: pp. 9–16.

Heilman, A.S. & A.J. Sharp. 1963. "A probable antibiotic effect of some lichens on bryophytes." *Rev. Bryol. Lichénol.* 32: p. 215.

Henderson-Sellers, A. & M.R.D. Seaward. 1979. "Monitoring lichen reinvasion of ameliorating environments." *Environ. Pollut.* 19: pp. 207–213.

Henningsson, B. & H. Lundström. 1970. "The influence of lichens, lichen extracts and usnic acid on wood destroying fungi." *Material & Organismen* 5: pp. 19–31.

Henriksson, E. 1951. "Nitrogen fixation by a bacteria-free, symbiotic *Nostoc* strain isolated from *Collema.*" *Physiol. Plantarum* 4: pp. 542–545.

———. 1961. "Studies in the physiology of the lichen *Collema.* IV. The occurrence of polysaccharides and some vitamins outside the cells of the phycobiont, *Nostoc* sp." *Physiol. Plantarum* 14: pp. 813–817.

———. 1963. "The occurrence of carotenoids in some lichen species belonging to the Collemataceae." *Physiol. Plantarum* 16: pp. 867–869.

Henriksson, E. & L.C. Pearson. 1968. "Carotenoids extracted from mycobionts of *Collema tenax, Baeomyces roseus,* and some other lichens." *Sven. Bot. Tidskr.* 62: pp. 441–447.

———. 1981. "Nitrogen fixation rate and chlorophyll content of the lichen *Peltigera canina* exposed to sulfur dioxide." *Am. J. Bot.* 68: pp. 680–684.

Henssen, A. 1981. "The lecanoralean centrum." In *Ascomycete Systematics: The Luttrellian Concept*, edited by D.R. Reynolds, pp. 138–234. New York: Springer-Verlag.

Henssen, A. & H.M. Jahns. 1974. *Lichenes. Eine Einführung in die Flechtenkunde.* Stuttgart: Georg Thieme Verlag.

Hess, D. 1959a. "Untersuchungen über die hemmende Wirkung von Extrakten aus Flechtenpilzen auf das Wachstum von *Neurospora crassa*." *Z. für Bot.* 48: pp. 136–142.

———. 1959b. "Untersuchungen über die Bildung von Phenolkörpen durch isolierte Flechtenpilze." *Z. Naturforsch.* 14B: pp. 345–347.

Hessler, R. & E. Peveling. 1978. "Die Lokalization von ^{14}C-Assimilaten in Flechtenthalli von *Cladonia incrassata* Floerke und *Hypogymnia physodes* (L.) Ach." *Z. Pflanzenphysiol.* 86: pp. 287–302.

Higinbotham, N. 1973. "Electropotentials of plant cells." *Ann. Rev. Plant Physiol.* 24: pp. 25–46.

Hildreth, K.C. & V. Ahmadjian. 1981. "A study of *Trebouxia* and *Pseudotrebouxia* isolates from different lichens." *Lichenologist* 13: pp. 65–86.

Hilitzer, A. 1925. "La vègètation epiphyte de la Bohème." *Publ. Fac. Sc. Univ. Charles, Prague,* Čislo 41: pp. 1–200.

Hill, D.J. 1971. "Experimental study of the effects of sulphite on lichens with reference to atmospheric pollution." *New Phytol.* 70: pp. 831–836.

———. 1972. "The movement of carbohydrate from the alga to the fungus in the lichen *Peltigera polydactyla*." *New Phytol.* 71: pp. 31–39.

———. 1974. "Some effects of sulphite on photosynthesis in lichens." *New Phytol.* 73: pp. 1193–1205.

———. 1976. "The physiology of lichen symbiosis." In *Lichenology: Progress and Problems,* edited by D.H. Brown, D.L. Hawksworth, and R.H. Bailey, pp. 457–496. London: Academic Press.

———. 1981. "The growth of lichens with special reference to the modelling of circular thalli." *Lichenologist* 13: pp. 265–287.

Hill, D.J. & V. Ahmadjian. 1972. "Relationship between carbohydrate movement and the symbiosis in lichens with green algae." *Planta* 103: pp. 267–277.

Hill, D.J. & D.C. Smith. 1972. "Lichen physiology. XII. The 'inhibition technique'." *New Phytol.* 71: pp. 15–30.

Hill, D.J. & H.W. Woolhouse. 1966. "Aspects of the autecology of *Xanthoria parietina* agg." *Lichenologist* 3: pp. 207–214.

Hintikka, V. 1970. "Selective effect of terpenes on wood-decomposing Hymenomycetes." *Karstenia* 11: pp. 28–32.

Hitch, C.J.B. & J.W. Millbank. 1975. "Nitrogen metabolism in lichens. VI. The blue-green phycobiont content, heterocyst frequency and nitrogenase activity in *Peltigera* spp." *New Phytol.* 74: pp. 473–476.

Hoder, D. & A. Mathey. 1980. "Color cathodoluminescence (CCL) applied to the study of the distribution of lichen products." In *Scanning Electron Microscopy* 1980/III, pp. 555–562. Chicago, IL: SEM Inc., AMF O'Hare.

Hoffman, G.R. 1971. "An ecologic study of epiphytic bryophytes and lichens on

Pseudotsuga menziesii on the Olympic Peninsula, Washington. II. Diversity of the vegetation." *Bryologist* 74: pp. 413–427.

———. 1974. "The influence of a paper pulp mill on the ecological distribution of epiphytic cryptogams in the vicinity of Lewiston, Idaho and Clarkston, Washington." *Environ. Pollut.* 7: pp. 283–301.

Hoffman, G.R. & D.M. Gates. 1970. "An energy budget approach to the study of water loss in cryptogams." *Bull. Torrey Bot. Club* 97: pp. 361–366.

———. 1971. "Transpirational water loss and energy budgets of selected plant species." *Oecol. Plant.* 6: pp. 115–131.

Hoffman, G.R. & R.G. Kazmierski. 1969. "An ecologic study of epiphytic bryophytes and lichens on *Pseudotsuga menziesii* on the Olympic Peninsula, Washington. I. A description of the vegetation." *Bryologist* 72: pp. 1–19.

Holleman, D.F. & J. R. Luick. 1977. "Lichen species preference by reindeer." *Can. J. Zool.* 55: pp. 1368–1369.

Holopainen, T.H. 1982. "Summer versus winter condition of the ultrastructure of the epiphytic lichens *Bryoria capillaris* and *Hypogymnia physodes* in Central Finland." *Ann. Bot. Fenn.* 19: pp. 39–52.

———. 1983. "Ultrastructural changes in epiphytic lichens, *Bryoria capillaris* and *Hypogymnia physodes,* growing near a fertilizer plant and a pulp mill in central Finland." *Ann. Bot. Fenn.* 20: pp. 169–185.

Honneger, R. & U. Brunner. 1981. "Sporopollenin in the cell walls of *Coccomyxa* and *Myrmecia* phycobionts of various lichens: An ultrastructural and chemical investigation." *Can. J. Bot.* 59: pp. 2713–2734.

Hooker, T.N. 1980. "Lobe growth and marginal zonation in crustose lichens." *Lichenologist* 12: pp. 313–323.

Hooker, T.N. & D.H. Brown. 1977. "A photographic method for accurately measuring the growth of crustose and foliose saxicolous lichens." *Lichenologist* 9: pp. 65–75.

Horn, H.S. 1974. "The ecology of secondary succession." *Ann. Rev. Ecol. Syst.* 5: pp. 25–37.

———. 1975. "Markovian properties of forest succession." In *Ecology and Evolution of Communities,* edited by M.L. Cody and J.M. Diamond, pp. 196–211. Cambridge, MA: Belknap Press of Harvard University Press.

Horntvedt, R. 1976. "The response of epiphytic lichens to fluoride air pollution." In *Proc. Kuopio Meeting on Plant Damages Caused by Air Pollution,* edited by L. Kärenlampi, pp. 93–94. The University of Kuopio. Kuopio, Finland.

Huneck, S. 1968. "Lichen substances." *Progr. Phytochem.* 1: pp. 223–346.

———. 1971. "Chemie und Biosynthese der Flechtenstoffe." *Forschr. Chem. Org. Naturst.* 29: pp. 209–306.

———. 1973. "Nature of lichen substances." In *The Lichens,* edited by V. Ahmadjian and M.E. Hale, Jr., pp. 495–522. New York: Academic Press.

Huneck, S. & K. Schreiber. 1972. "Wachstumsregulatorische Eigenschaften von Flechten- und Moos-Inhaltstoffen." *Phytochem.* 11: pp. 2429–2434.

Hvinden, T. & A. Lillegraven. 1961. "Cesium-137 and strontium-90 in precipitation, soil and animals in Norway." *Nature* 192: pp. 1144–1146.

Ikekawa, N., S. Natori, H. Ageta, K. Iwata, & M. Matsui. 1965. "Gas chromatography of triterpenes. X. Hopanezeorinane and onocerane groups." *Chem. Pharm. Bull.*

(Tokyo) 13: pp. 320–325.

Ikonen, S. & L. Kärenlampi. 1976. "Physiological and structural changes in reindeer lichens transplanted around a sulphite pulp mill." In *Proc. Kuopio Meeting on Plant Damages Caused by Air Pollution,* edited by L. Kärenlampi, pp. 37–45. The University of Kuopio, Kuopio, Finland.

Iskandar, I.K. & J.K. Syers. 1971. "Solubility of lichen compounds in water: Pedogenetic implications." *Lichenologist* 5: pp. 45–50.

———. 1972. "Metal-complex formation by lichen compounds." *J. Soil Sci.* 23: pp. 255–265.

Jaag, O. 1929. "Recherches expérimentales sur les gonidies des lichins appartenant aux genres *Parmelia* et *Cladonia.*" *Bull. Soc. Bot.* Genève 21: pp. 1–119.

Jackson, T.A. & W.D. Keller. 1970. "A comparative study of the role of lichens and inorganic processes in the chemical weathering of recent Hawaiian lava flows." *Am. J. Sci.* 269: pp. 446–466.

Jacobs, J.B. & V. Ahmadjian. 1969. "The ultrastructure of lichens. I. A general survey." *J. Phycol.* 5: pp. 227–240.

———. 1971a. "The ultrastructure of lichens. II. *Cladonia cristatella:* The lichen and its isolated symbionts." *J. Phycol.* 7: pp. 71–82.

———. 1971b. "The ultrastructure of lichens. IV. Movement of carbon products from alga to fungus as demonstrated by high resolution radiography." *New Phytol* 70: pp. 47–50.

———. 1973. "The ultrastructure of lichens. V. *Hydrothyria venosa,* a freshwater lichen." *New Phytol.* 72: pp. 155–160.

Jahns, H.M. 1970. "Untersuchungen zur Entwicklungsgeschichte der Cladoniaceen." *Nova Hedwigia* 20: pp. 1–177.

———. 1972. "Die Entwicklung von Flechten-Cephalodien aus *Stigonema*-Algen." *Ber. Deut. Bot. Ges.* 85: pp. 615–622.

———. 1973. "Anatomy, morphology and development." In *The Lichens,* edited by V. Ahmadjian and M.E. Hale, Jr., pp. 3–58. New York: Academic Press.

———. 1982. "The cyclic development of mosses and the lichen *Baeomyces rufus* in an ecosystem." *Lichenologist* 14: pp. 261–265.

Jahns, H.M., D. Mollenhauer, M. Jenninger, & D. Schönborn. 1979. "Die Neubesiedlung von Baumrinde durch Flechten. I." *Natur. & Mus.* 109: pp. 40–51.

Jahns, H.M. & S. Ott. 1982. "Flechtenentwicklung an dicht benachbarten Standorten." *Herzogia* 6: pp. 201–241.

James, P.W. 1973. "The effect of air pollutants other than hydrogen fluoride and sulphur dioxide on lichens." In *Air Pollution and Lichens,* edited by B.W. Ferry, M.S. Baddeley, and D.L. Hawksworth, pp. 143–175. London: Athlone Press.

———. 1975. "Lichen chimeras." *Rep. Br. Mus. Nat. Hist.* 1972–74: pp. 37–43.

James, P.W. & A. Henssen. 1976. "The morphological and taxonomic significance of cephalodia." In *Lichenology: Progress and Problems,* edited by D.H. Brown, D.L. Hawksworth, and R.H. Bailey, pp. 27–77. London: Academic Press.

Janzen, D.H. 1974. "Tropical blackwater rivers, animals, and mast fruiting by the Dipterocarpaceae." *Biotropica* 6: pp. 69–103.

———. 1975. *Ecology of Plants in the Tropics.* London: Edward Arnold.

Jatimliansky, J.R. & E.M. Sivori. 1969. "Inhibición del crecimiento de *Scenedesmus obliquus* por alcaloides." *Anales Soc. Cient. Argent.* 187: pp. 49–57.

Jesberger, J.A. & J.W. Sheard. 1973. "A quantitative study and multivariate analysis of corticolous lichen communities in the southern boreal forest of Saskatchewan." *Can. J. Bot.* 51: pp. 185–201.

Jochimsen, M. 1973. "Does the size of lichen thalli really constitute a valid measure for dating glacial deposits?" *Arctic & Alpine Res.* 5: pp. 417–424.

Joenje, W. & H.J. During. 1977. "Colonisation of a desalinating Wadden-polder by bryophytes." *Vegetatio* 35: pp. 177–185.

Johnsen, I. 1976. "Problems in relation to the use of plants as monitors of air pollution with metals." In *Proc. Kuopio Meeting on Plant Damages Caused by Air Pollution,* edited by L. Karënlampi, pp. 110–114. The University of Kuopio, Kuopio, Finland.

Johnsen, I. & L. Rasmussen. 1977. "Retrospective study (1944–1976) of heavy metals in the epiphyte *Pterogonium gracile* collected from one phorophyte." *Bryologist* 80: pp. 625–629.

Jones, D., M.J. Wilson, & W.J. McHardy. 1981. "Lichen weathering of rock-forming minerals: Application of scanning electron microscopy and microprobe analysis." *J. Microsc.* 124: pp. 95–104.

Jones, J.M. & R.B. Platt. 1969. "Effects of ionizing radiation, climate, and nutrition on growth and structure of a lichen *Parmelia conspersa* (Ach.). Ach." *Proc. Int. Symp. on Radioecol. Concentr. Processes,* 2nd, 1969, edited by D.L. Nelson and F.C. Evans, pp. 111–119. Ann Arbor: University of Michigan.

Jordan, W.P. 1970. "The internal cephalodia of the genus *Lobaria.*" *Bryologist* 73: pp. 669–681.

Jordan, W.P. & F.R. Rickson. 1971. "Cyanophyte cephalodia in the lichen genus *Nephroma.*" *Am. J. Bot.* 58: pp. 561–568.

Kallio, S. & T. Varheenmaa. 1974. "On the effect of air pollution on nitrogen fixation in lichens." *Rep. Kevo Subarct. Res. Sta.* 11: pp. 42–46.

Kappen, L. 1973. "Response to extreme environments." In *The Lichens,* edited by V. Ahmadjian and M.E. Hale, Jr., pp. 311–380. New York: Academic Press.

Kappen, L. & E.I. Friedmann. 1983. "Ecophysiology of lichens in the dry valleys of southern Victoria Land, Antarctica. II. CO_2 gas exchange in cryptoendolithic lichens." *Polar Biol.* 1: pp. 227–232.

Kappen, L., E.I. Friedmann, & J. Garty. 1981. "Ecophysiology of lichens in the dry valleys of southern Victoria Land, Antarctica. I. Microclimate of the cryptoendolithic lichen habitat." *Flora* 171: pp. 216–235.

Kappen, L., O.L. Lange, E.-D. Schulze, M. Evenari, & U. Buschbom. 1979. "Ecophysiological investigations on lichens of the Negev Desert. VI. Annual course of photosynthetic production of *Ramalina maciformis* (Del.) Bory." *Flora* 168: pp. 85–108.

Kärenlampi, L. 1966. "The succession of the lichen vegetation in the rocky shore geolittoral and adjacent parts of the epilittoral in the southwestern archipelago of Finland." *Ann. Bot. Fenn.* 3: pp. 79–85.

———. 1970. "Distribution of chlorophyll in the lichen *Cladonia alpestris.*" *Rep. Kevo Subarct. Res. Sta.* 7: pp. 1–8.

———. 1971. "Studies on the relative growth rate of some fruticose lichens." *Rep. Kevo Subarct. Res. Sta.* 7: pp. 33–39.

Kauppi, M. 1976. "Fruticose lichen transplant technique for air pollution experiments." *Flora* 165: pp. 407–414.

Kauppi, M. & A. Mikkonen. 1980. "Floristic versus single species analysis in the use of epiphytic lichens as indicators of air pollution in a boreal forest region, northern Finland." *Flora* 169: pp. 255–281.

Keever, C. 1957. "Establishment of *Grimmia laevigata* on bare granite." *Ecology* 38: pp. 422–429.

Kershaw, K.A. 1964. "Preliminary observations on the distribution and ecology of epiphytic lichens in Wales." *Lichenologist* 2: pp. 263–276.

———. 1975a. "Studies on lichen-dominated systems. XII. The ecological significance of thallus color." *Can. J. Bot.* 53: pp. 660–667.

———. 1975b. "Studies on lichen-dominated systems. XIV. The comparative ecology of *Alectoria nitidula* and *Cladonia alpestris.*" *Can. J. Bot.* 53: pp. 2608–2613.

Kershaw, K.A. & G.P. Harris. 1971. "A technique for measuring the light profile in a lichen canopy." *Can. J. Bot.* 49: pp. 609–611.

Kershaw, K.A. & J.D. MacFarlane. 1980. "Physiological-environmental interactions in lichens. X. Light as an ecological factor." *New Phytol.* 84: pp. 687–702.

Kershaw, K.A. & J.W. Millbank. 1970. "Nitrogen metabolism in lichens. II. The partition of cephalodial-fixed nitrogen between the mycobiont and phycobionts of *Peltigera aphthosa.*" *New Phytol.* 69: pp. 75–79.

Kershaw, K.A., T. Morris, M.J. Tysiaczny, & J.D. MacFarlane. 1979. "Physiological-environmental interactions in lichens. VIII. The environmental control of dark CO_2 fixation in *Parmelia caperata* (L.) Ach. and *Peltigera canina* var. *praetextata* Hue." *New Phytol.* 83: pp. 433–444.

Kershaw, K.A. & M.M. Smith. 1978. "Studies on lichen-dominated systems. XXI. The control of seasonal rates of net photosynthesis by moisture, light, and temperature in *Stereocaulon paschale.*" *Can. J. Bot.* 56: pp. 2825–2830.

Kinraide, W.T.B. & V. Ahmadjian. 1970. "The effects of usnic acid on the physiology of two cultured species of the lichen alga *Trebouxia* Puym." *Lichenologist* 4: pp. 234–247.

Klee, R. 1971. "Die Wirkung von Gas und Staubformigen Immissionen auf Respiration und Inhaltstoffe von *Parmelia physodes.*" *Angew. Bot.* 44: pp. 253–261.

Klee, R. & L. Steubing. 1977. "Untersuchungen zur Bedeutung des Atranorins in Flechtenthalli." *Herzogia* 4: pp. 357–362.

Kohlmeyer, J. & E. Kohlmeyer. 1979. *Marine Mycology. The Higher Fungi.* New York: Academic Press.

Komiya, T. & S. Shibata. 1969. "Formation of lichen substances by mycobionts of lichens. Isolation of (+) usnic acid and salazinic acid from mycobionts of *Ramalina* spp." *Chem. Pharm. Bull.* (Tokyo) 17: pp. 1305–1306.

Kunkel, G. 1980. "Microhabitat and structural variation in the *Aspicilia desertorum* group (lichenized Ascomycetes)." *Am. J. Bot.* 67: pp. 1137–1144.

Kupchan, S.M. & H.L. Kopperman. 1975. "*l*-Usnic acid: Tumor inhibitor isolated from lichens." *Experimentia* 31: p. 625.

Kushnir, E. & M. Galun. 1977. "The fungus-alga association in endolithic lichens." *Lichenologist* 9: pp. 123–130.

Lambright, D.D. & S.C. Tucker. 1980. "Observations of the ultrastructure of *Trypethelium eluteriae* Spreng." *Bryologist* 83: pp. 170–178.
Lane, I. & K.J. Puckett. 1979. "Response of the phosphatase activity of the lichen *Cladina rangiferina* to various environmental factors including metals." *Can. J. Bot.* 57: pp. 1534–1540.
Lang, G.E., W.A. Reiners, & R.K. Heier. 1976. "Potential alteration of precipitation chemistry by epiphytic lichens." *Oecologia* 25: pp. 229–241.
Lange, O.L. 1953. "Hitze- und Trockenresistenz der Flechten in Beziehung zu ihrer Verbreitung." *Flora* 140: pp. 39–97.
———. 1954. "Einige Messungen zum Wärmehaushalt poikilohydrer Flechten und Moose." *Archiv. Meterorol. Geophys. Bioklimatol.* Ser. B 5: pp. 182–190.
Lange, O.L., E.-D. Schulze, & W. Koch. 1970. "Experimentell-ökologische Untersuchungen an Flechten der Negev-Wüste II. CO_2-Gaswechsel und Wasserhaushalt von *Ramalina maciformis* (Del.) Bory am naturlichen Standort-während der sommerlichen Trockenperiode." *Flora* 159: pp. 38–62.
Larson, D.W. 1979a. "Preliminary studies of the physiological ecology of *Umbilicaria* lichens." *Can. J. Bot.* 57: pp. 1398–1406.
———. 1979b. "Lichen water relations under drying conditions." *New Phytol.* 82: pp. 713–731.
———. 1980a. "Seasonal change in the pattern of net CO_2 exchange in *Umbilicaria* lichens." *New Phytol.* 84: pp. 349–369.
———. 1980b. "Patterns of species distribution in an *Umbilicaria* dominated community." *Can. J. Bot.* 58: pp. 1269–1279.
———. 1981. "Differential wetting in some lichens and mosses: The role of morphology." *Bryologist* 84: pp. 1–15.
Larson, D.W. & K.A. Kershaw. 1974. "Studies on lichen-dominated systems. VII. Interactions of the general lichen-heath with edaphic factors." *Can. J. Bot.* 52: pp. 1163–1176.
———. 1975a. "Measurement of CO_2 exchange in lichens: A new method." *Can. J. Bot.* 53: pp. 1535–1541.
———. 1975b. "Acclimation in arctic lichens." *Nature* 254: pp. 421–423.
———. 1976. "Studies on lichen-dominated systems. XVIII. Morphological control of evaporation in lichens." *Can. J. Bot.* 54: pp. 2061–2073.
Laundon, J.R. 1967. "A study of the lichen flora of London." *Lichenologist* 3: pp. 277–327.
———. 1970. "London's lichens." *London Natur.* 49: pp. 20–69.
———. 1971. "Lichen communities destroyed by psocids." *Lichenologist* 5: p. 177.
Lawrey, J.D. 1977a. "Inhibition of moss spore germination by acetone extracts of terricolous *Cladonia* species." *Bull. Torrey Bot. Club* 194: pp. 49–52.
———. 1977b. "Adaptive significance of *O*-methylated lichen depsides and depsidones." *Lichenologist* 9: pp. 137–142.
———. 1980a. "Sexual and asexual reproductive patterns in *Parmotrema* (Parmeliaceae) that correlate with latitude." *Bryologist* 83: pp. 344–350.
———. 1980b. "Calcium accumulation by lichens and transfer to lichen herbivores." *Mycologia* 72: pp. 586–594.
———. 1980c. "Correlations between lichen secondary chemistry and grazing activity by *Pallifera varia.*" *Bryologist* 83: pp. 328–334.

———. 1981. "Evidence for competitive release in simplified saxicolous lichen communities." *Am. J. Bot.* 68: pp. 1066–1073.

———. 1983. "Lichen herbivore preference: A test of two hypotheses." *Am. J. Bot.* 70: pp. 1188–1194.

Lawrey, J.D. & M.E. Hale, Jr. 1977. "Natural history of Plummers Island, Maryland. XXIII. Studies on lichen growth rate at Plummers Island, Maryland." *Proc. Biol. Soc. Washington* 90: pp. 698–725.

———. 1979. "Lichen growth responses to stress induced by automobile exhaust pollution." *Science* 204: pp. 423–424.

———. 1981. "Retrospective study of lichen lead accumulation in the northeastern United States." *Bryologist* 84: pp. 449–456.

Lazo, W.R. 1961. "Growth of green algae with myxomycete plasmodia." *Am. Midl. Natur.* 65: pp. 381–383.

———. 1964. "An experimental association between *Chlorella xanthella* and streptomycetes." *Am. J. Bot.* 51: pp. 678–679 (Abstr.).

———. 1966. "An experimental association between *Chlorella xanthella* and a *Streptomyces.*" *Am. J. Bot.* 53: pp. 105–107.

Leak, W.B. 1970. "Successional change in northern hardwoods predicted by birth and death simulation." *Ecology* 51: pp. 794–801.

LeBlanc, F., G. Comeau, & D. N. Rao. 1971. "Fluoride injury symptoms in epiphytic lichens and mosses." *Can. J. Bot.* 49: pp. 1691–1698.

LeBlanc, F. & J. De Sloover. 1970. "Relation between industrialization and the distribution and growth of epiphytic lichens and mosses in Montreal." *Can. J. Bot.* 48: pp. 1485–1496.

LeBlanc, F. & D.N. Rao. 1973. "Effects of sulphur dioxide on lichen and moss transplants." *Ecology* 54: pp. 612–617.

Lechowicz, M.J. 1982a. "Ecological trends in lichen photosynthesis." *Oecologia* (Berlin) 53: pp. 330–336.

———. 1982b. "The effects of simulated acid precipitation on photosynthesis in the caribou lichen *Cladina stellaris* (Opiz.) Brodo." *Water, Air & Soil Pollut.* 18: pp. 421–430.

———. 1983. "Age dependence of photosynthesis in the caribou lichen *Cladina stellaris.*" *Plant Physiol.* 71: pp. 893–895.

Lechowicz, M.J. & M.S. Adams. 1973. "Net photosynthesis of *Cladonia mitis* Sandst. from sun and shade sites of the Wisconsin Pine Barrens." *Ecology* 54: pp. 413–419.

———. 1974a. "Ecology of *Cladonia* lichens. I. Preliminary assessment of the ecology of terricolous lichen-moss communities in Ontario and Wisconsin." *Can. J. Bot.* 52: pp. 55–64.

———. 1974b. "Ecology of *Cladonia* lichens. II. Comparative physiological ecology of *C. mitis, C. rangiferina,* and *C. uncialis.*" *Can. J. Bot.* 52: pp. 411–422.

———. 1979. "Net CO_2 exchange in *Cladonia* lichen species endemic to southeastern North America." *Photosynthetica* 13: pp. 155–162.

Leibundgut, H. 1952. "Flechtenrasen als Hindernis für die Ansamung." *Schw. Zeitschr. f. Forstw.* 103: pp. 162–168.

Leighton, W.A. 1867. "Notulae lichenologicae. No. XII. On the *Cladoniei* in the Hookerian herbarium at Kew." *Ann. Mag. Nat. Hist.* 19 (Ser. 3): pp. 99–124.

Leonian, L.H. & V.G. Lilly. 1937. "Is heteroauxin a growth promoting substance?" *Am. J. Bot.* 24: pp. 135–139.

Le Sueur, P. & K.J. Puckett. 1980. "Effect of vanadium on the phosphatase activity of lichens." *Can. J. Bot.* 58: pp. 502–504.

Letrouit-Galinou, M.-A. 1968. "The apothecia of the Discolichens." *Bryologist* 71: pp. 297–327.

———. 1973. "Sexual reproduction." In *The Lichens,* edited by V. Ahmadjian and M.E. Hale, Jr., pp. 59–90. New York: Academic Press.

Levin, D.A. 1971. "Plant phenolics: An ecological perspective." *Am. Nat.* 105: pp. 157–181.

———. 1976. "The chemical defenses of plants to pathogens and herbivores." *Ann. Rev. Ecol. Syst.* 7: pp. 121–159.

Lewis, D.H. 1973. "Concepts in fungal nutrition and the origin of biotrophy." *Biol. Rev.* 48: pp. 261–278.

Lewis, D.H. & D.C. Smith. 1967. "Sugar alcohols (polyols) in fungi and green plants. I. Distribution, physiology and metabolism." *New Phytol.* 66: pp. 143–184.

Lidén, K. 1961. "Cesium-137 burdens in Swedish Laplanders and reindeer." *Acta Radiol.* 56: p. 237.

Lidén, K. & M. Gustafsson. 1966. "Relationships and seasonal variation of ^{137}Cs in lichen, reindeer and man in northern Sweden 1961–65." In *Radioecological Concentration Processes,* edited by B. Aberg and F.P. Hungate, pp. 193–208. Oxford: Pergamon Press.

Lindsay, D.C. 1973. "Estimates of lichen growth rates in the maritime Antarctic." *Arctic & Alpine Res.* 5: pp. 341–346.

———. 1977. "Lichens of cold deserts." In *Lichen Ecology,* edited by M.R.D. Seaward, pp. 183–209. London: Academic Press.

Link, S.O., T.J. Moser, & T.H. Nash III. 1983. "Relationships among initial rate, closed chamber, and 14-CO_2 techniques with respect to lichen photosynthetic CO_2 dependencies." *Photosynthetica* (in press).

Lockhart, C.M., P. Rowell, & W.D.P. Stewart. 1978. "Phytohaemagglutinins from the nitrogen-fixing lichens *Peltigera canina* and *P. polydactyla.*" *FEMS Microbiol. Lett.* 3: pp. 127–130.

Lounamaa, J. 1956. "Trace elements in plants growing wild on different rocks in Finland." *Ann. Bot. Soc. Zool. Bot. Fenn. "Vanamo"* 29: pp. 1–196.

———. 1965. "Studies on the content of Fe, Mn and Zn in macrolichens." *Ann. Bot. Fenn.* 2: pp. 127–137.

Lundström, H. & B. Henningsson. 1973. "The effect of ten lichens on the growth of wood-destroying fungi." *Material & Organismen* 8: pp. 233–246.

Luttrell, E.S. 1951. *Taxonomy of the Pyrenomycetes.* University of Missouri Studies No. 24, pp. 1–120.

Maass, W.S.G. 1975a. "Lichen substances V. Methylated derivatives of orsellinic acid, lecanoric acid and gyrophoric acid from *Pseudocyphellaria crocata.*" *Can. J. Bot.* 53: pp. 1031–1039.

———. 1975b. "Lichen substances VI. The phenolic constitution of *Peltigera aphthosa.*" *Phytochem.* 14: pp. 2487–2489.

———. 1975c. "Lichen substances VII. Identification of orsellinate derivatives from *Lobaria linita.*" *Bryologist* 78: pp. 178–182.

———. 1975d. "Lichen substances VIII. Phenolic constituents of *Pseudocyphellaria quercifolia.*" *Bryologist* 78: pp. 183–186.

MacArthur, R.H. 1958. "A note on stationary age distributions in single-species populations and stationary species populations in a community." *Ecology* 39: pp. 146–147.

MacArthur, R.H. & J.H. Connell. 1966. *The Biology of Populations.* New York: Wiley.

MacArthur, R.H. & E.O. Wilson. 1963. "An equilibrium theory of insular zoogeography." *Evolution* 17: pp. 373–387.

———. 1967. *The Theory of Island Biogeography.* Princeton, N.J: Princeton University Press.

McCarthy, P.M. & J.A. Healey. 1978. "Dispersal of lichen propagules by slugs." *Lichenologist* 10: pp. 131–134.

MacFarlane, J.D. & K.A. Kershaw. 1978. "Thermal sensitivity in lichens." *Science* 201: pp. 739–741.

———. 1980. "Physiological-environmental interactions in lichens. XI. Snowcover and nitrogenase activity." *New Phytol.* 84: pp. 703–710.

———. 1982. "Physiological-environmental interactions in lichens. XIV. The environmental control of glucose movement from alga to fungus in *Peltigera polydactyla, P. rufescens* and *Collema furfuraceum.*" *New Phytol.* 91: pp. 93–101.

McWhorter, F.P. 1921. "Destruction of mosses by lichens." *Bot. Gaz.* 72: pp. 321–325.

Maikawa, E. & K.A. Kershaw. 1976. "Studies on lichen-dominated systems. XIX. The postfire recovery sequence of black spruce-lichen woodland in the Abitau Lake Region, N.W.T." *Can. J. Bot.* 54: pp. 2679–2687.

Malachowski, J.A., K.K. Baker, & G.R. Hooper. 1980. "Anatomy and algal-fungal interactions in the lichen *Usnea cavernosa.*" *J. Phycol.* 16: pp. 346–354.

Malicki, J. 1965. "The effect of lichen acids on the soil microorganisms. Part I. The washing down of the acids into the soil." *Ann. Univ. Mariae Curie—Skłodowska* (Sec. C) 20: pp. 239–248.

———. 1967. "The influence of lichen acids on soil microbes. Part II. The influence of aqueous extracts from *Cladonia* species on soil bacteria." *Ann. Univ. Mariae Curie—Skłodowska* (Sec. C) 22: pp. 159–163.

———. 1970. "The influence of lichen acids on soil microorganisms. Part III. The influence of the species *Cladonia* on bacterial relations in the soil of Peucedano-Pinetum cladonietosum associations." *Ann. Univ. Mariae Curie—Skłodowska* (Sec. C) 25: p. 75.

Malinowski, E. 1911. "Mozaika porostów naskalnych." *Spraw. Posied. Tow. Nauk. Warsz.* 4: pp. 393–400.

———. 1912. "Przyczynek do biologii i ekologii porostów epilitycznych." *Bull. Int. Acad. Sci. Lett. Cracovie, Cl. Sci. Mat. Nat. B* 1911: pp. 349–390.

Malone, C.P. 1977. "Observations on *Endocarpon pusillum:* The role of rhizomorphs in asexual reproduction." *Mycologia* 69: pp. 1042–1045.

Margalef, R. 1968. *Perspectives in Ecological Theory.* Chicago, IL: University of Chicago Press.

Margot, J. 1973. "Experimental study of the effects of sulphur dioxide on the soredia of *Hypogymnia physodes.*" In *Air Pollution and Lichens,* edited by B.W. Ferry,

M.S. Baddeley, & D.L. Hawksworth, pp. 314–329. Toronto: University of Toronto Press.

Margot, J. & J. De Sloover. 1974. "La culture des sorédies lichéniques: Un test de la vitalitée des thalles soumis à la pollution. *Bull. Soc. Roy. Bot. Belg.* 197: pp. 33–40.

Margot, J. & M.-T. Romain. 1976. "Métaux lourds et cryptogames terrestres synthèse bibliographique." *Mém. Soc. Roy. Bot. Belg.* 7: pp. 25–47.

Martin, J.E. & F. Jacquard. 1968. "Influence des fumées d'usines sur la distribution des lichens dans la vallée de la Romanche (Isère)." *Pollution Atmospherique* 10: pp. 95–99.

Martin, N.M. 1938. "Some observations on the epiphytic moss flora of Argyll." *J. Ecol.* 26: pp. 82–95.

Marton, K. & M. Galun. 1976. "*In vitro* dissociation and reassociation of the symbionts of the lichen *Heppia echinulata.*" *Protoplasma* 87: pp. 135–143.

Mathey, A. 1979. "Contribution à l'étude de la famille des Trypéthéliacées (lichens pyrenomycetes)." *Nova Hedwigia* 31: pp. 917–935.

Mathey, A. & D. Hoder. 1978. "Distribution of lichen substances by means of fluorescence microscopy, cathodoluminescence in scanning electron microscope and X-ray microanalysis in *Lecanora-*, *Buellia-*, *Laurera-* and *Trypethelium-*species." *Nova Hedwigia* 30: pp. 127–138.

Matthews, J.A. 1973. "Lichen growth on an active medial moraine, Jotunheimen, Norway." *J. Glaciol.* 12: pp. 305–313.

Mattsson, L.J.S. 1975a. "^{137}Cs in the reindeer lichen *Cladonia alpestris:* Deposition, retention and internal distribution, 1961–1970." *Health Physics* 28: pp. 233–248.

———. 1975b. "Deposition, retention and internal distribution of ^{155}Eu, ^{144}Ce, ^{125}Sb, ^{106}Ru, ^{95}Zr, ^{54}Mn and ^{7}B in the reindeer lichen *Cladonia alpestris,* 1961–1970." *Health Physics* 29: pp. 27–41.

Meier, J.L. & R.L. Chapman. 1983. "Ultrastructure of the lichen *Coenogonium interplexum* Nyl." *Am. J. Bot.* 70: pp. 400–407.

Mellanby, K. 1978. "Biological methods of environmental monitoring." In *Measuring and Monitoring the Environment,* edited by J. Lenihan and W.W. Fletcher, pp. 1–13. Glasgow and London: Blackie.

Millbank, J.W. 1972. "Nitrogen metabolism in lichens. IV. The nitrogenase activity of the *Nostoc* phycobiont in *Peltigera canina.*" *New Phytol.* 71: pp. 1–10.

———. 1976. "Aspects of nitrogen metabolism in lichens." In *Lichenology: Progress and Problems,* edited by D.H. Brown, D.L. Hawksworth and R.H. Bailey, pp. 441–455. London: Academic Press.

———. 1977. "The oxygen tension within lichen thalli." *New Phytol.* 79: pp. 649–657.

Millbank, J.W. & K.A. Kershaw. 1973. "Nitrogen metabolism." In *The Lichens,* edited by V. Ahmadjian and M.E. Hale, Jr., pp. 289–307. New York: Academic Press.

Miller, A.G. 1966. "Lichen growth and species composition in relation to duration of thallus wetness." B.S. dissertation, Queen's University, Kingston, Ontario.

Mirando, M. & D. Fahselt. 1978. "The effect of thallus age and drying procedure on extractable lichen substances." *Can. J. Bot.* 56: pp. 1499–1504.

Mish, L.B. 1953. "Biological studies of symbiosis between the alga and fungus in the rock lichen *Umbilicaria papulosa.*" Ph.D. dissertation, Harvard University, Cambridge, MA.

Mitchell, B.D., A.C. Birnie, & J.K. Syers. 1966. "The thermal analysis of lichens growing on limestone." *Analyst* 91: pp. 783–789.

Mitsuno, M. 1953. "Paper chromatography of lichen substances. I." *Pharm. Bull.* 1: pp. 170–173.

Moissejeva, E.N. 1961. "Biochemical properties of lichens and their practical importance" [in Russian with English summary]. *Izd. Akad. Nauk C.C.C.R.*, (Moscow), p. 82.

Molisch, H. 1937. *Der Einfluss einer Pflanze auf die andere—Allelopathie.* Jena: Fischer.

Möller, A. 1887. Ueber die Cultur flechtenbildener Ascomyceten ohne Algen. Unters. aus Bot. Inst. K. Akad. Münster i W.

Mooney, H.A. & S.L. Gulmon. 1982. "Constraints on leaf structure and function in reference to herbivory." *BioScience* 32: pp. 198–206.

Moor, H. & K. Mühlethaler. 1963. "Fine structure in frozen-etched yeast cells." *J. Cell Biol.* 17: pp. 609–628.

Moore, R.T. & J.H. McAlear. 1960. "Fine structure of Mycota. 2. Demonstration of the haustoria of lichens." *Mycologia* 52: pp. 805–806.

———. 1961. "Fine structure of Mycota. 5. Lomasomes—previously uncharacterized hyphal structures." *Mycologia* 53: pp. 194–200.

Moreau, F. 1956. "Sur la théorie biomorphogénique des lichens." *Rev. Bryol. Lichénol.* 25: pp. 183–186.

Mosbach, K. 1973. "Biosynthesis of lichen substances. In *The Lichens,* edited by V. Ahmadjian and M.E. Hale, Jr., pp. 523–546. New York: Academic Press.

Mosbach, K. & U. Ehrensvärd. 1966. "Studies on lichen enzymes. Part I. Preparation and properties of a depside hydrolyzing esterase and of orsellinic acid decarboxylase." *Biochem. Biophys. Res. Commun.* 22: pp. 145–150.

Mosbach, K. & J. Schultz. 1971. "Studies on lichen enzymes. Purification and properties of orsellinate decarboxylase obtained from *Lasallia pustulata.*" *Eur. J. Biochem.* 22: pp. 485–488.

Moser, T.J. & T.H. Nash III. 1978. "Photosynthetic patterns of *Cetraria cucullata* (Bell.) Ach. at Anaktuvuk Pass, Alaska." *Oecologia* (Berlin) 34: pp. 37–43.

Muller, C.H. 1969. "Allelopathy as a factor in ecological process." *Vegetatio* 18: pp. 348–357.

Murty, T.K. & S.S. Subramanian. 1958. "Carotene content of *Roccella montagnei.*" *J. Sci. Ind. Res.* 17C: pp. 105–106.

Muscatine, L., R.R. Pool, & R.K. Trench. 1975. "Symbiosis of algae and invertebrates: Aspects of the symbiont surface and the host-symbiont interface." *Trans. Am. Microsc. Soc.* 94: pp. 450–469.

Nannfeldt, J.A. 1932. "Studien über die Morphologie und Systematik der nicht-lichenisierten Inoperculaten Discomycetes." *Nova Acta Regiae Soc. Sci. Upsal.* 8: pp. 1–368.

Nash, T.H, III. 1971. "Lichen sensitivity to hydrogen fluoride." *Bull. Torrey Bot. Club* 98: pp. 103–106.

———. 1973. "Sensitivity of lichens to sulfur dioxide." *Bryologist* 76: pp. 333–339.

———. 1976a. "Lichens as indicators of air pollution." *Naturwiss.* 63: pp. 364–367.

———. 1976b. "Sensitivity of lichens to nitrogen dioxide fumigations." *Bryologist* 79: pp. 103–106.

Nash, T.H., III, T.J. Moser, & S.O. Link. 1980. "Nonrandom variation of gas exchange within arctic lichens." *Can. J. Bot.* 58: pp. 1181–1186.

Nash, T.H., III, T.J. Moser, S.O. Link, L.J. Ross, A. Olafsen, & U. Matthes. 1983. "Lichen photosynthesis in relation to CO_2 concentration." *Oecologia* 58: pp. 52–56.

Nash, T.H., III, G.T. Nebeker, T.J. Moser, & T. Reeves. 1979. "Lichen vegetational gradients in relation to the Pacific coast of Baja, California: The maritime influence." *Madroño* 26: pp. 149–163.

Nash, T.H., III & L.L. Sigal. 1979. "Gross photosynthetic response of lichens to short-term ozone fumigations." *Bryologist* 82: pp. 280–285.

———. 1981. "Ecological approaches to the use of lichenized fungi as indicators of air pollution." In *The Fungal Community,* edited by D.T. Wicklow and G.C. Carroll, pp. 481–497. New York: Marcel Dekker.

Nash, T.H., III & M. Zavada. 1977. "Population studies among Sonoran Desert species of *Parmelia* subg. *Xanthoparmelia* (Parmeliaceae)." *Am. J. Bot.* 64: pp. 664–669.

National Academy of Sciences. 1975. *Air Quality and Stationary Soruce Emissions Control* Washington, D.C.: U.S. Government Printing Office, Superintendent of Documents.

Newell, S.J. & E.J. Tramer. 1978. "Reproductive strategies in herbaceous plant communities during succession." *Ecology* 59: pp. 228–234.

Nieboer, E., P. Lavoie, R.L.P. Sasserville, K.J. Puckett, & D.H.S. Richardson. 1976a. "Cation-exchange equilibrium and mass balance in the lichen *Umbilicaria muhlenbergii.*" *Can. J. Bot.* 54: pp. 720–723.

Nieboer, E., K.J. Puckett, & B. Grace. 1976b. "The uptake of nickel by *Umbilicaria muhlenbergii*: A physicochemical process." *Can. J. Bot.* 54: pp. 724–733.

Nieboer, E., K.J. Puckett, D.H.S. Richardson, F.D. Tomassini, & B. Grace. 1977a. "Ecological and physicochemical aspects of the accumulation of heavy metals and sulphur in lichens." In *International Conference on Heavy Metals in the Environment, Symp. Proc.,* Vol. 2, Pt. 1, Toronto, Ontario, Oct. 27–31, 1975.

Nieboer, E. & D.H.S. Richardson. 1981. "Lichens as monitors of atmospheric deposition." In *Atmospheric Pollutants in Natural Waters,* edited by S.J. Eisenreich, pp. 339–388. Ann Arbor, MI: Ann Arbor Science Publication.

Nieboer, E., D.H.S. Richardson, P. Lavoie, & D. Padovan. 1979. "The role of metal-ion binding in modifying the toxic effects of sulphur dioxide on the lichen *Umbilicaria muhlenbergii.* I. Potassium efflux studies." *New Phytol.* 82: pp. 621–632.

Nieboer, E., D.H.S. Richardson, K.J. Puckett, & F.D. Tomassini. 1976c. "The phytotoxicity of sulphur dioxide in relation to measurable responses in lichens." In *Effects of Air Pollutants on Plants,* edited by T.A. Mansfield, pp. 61–85. Cambridge, England: Cambridge University Press.

Nieboer, E., D.H.S. Richardson, & F.D. Tomassini. 1978. "Mineral uptake and release by lichens: An overview." *Bryologist* 81: pp. 226–246.

Nieboer, E., F.D. Tomassini, K.J. Puckett, & D.H.S. Richardson. 1977b. "A model for the relationship between gaseous and aqueous concentrations of sulphur dioxide in lichen exposure studies." *New Phytol.* 79: pp. 157–162.

Nifontova, M.G. & N.V. Kulikov. 1977. "Accumulation of strontium-90 and cesium-137 by lichens under natural conditions." *Soviet J. Ecol.* 8: pp. 270–273.

Nikonov, A.A. & T. Yu. Shebalina. 1979. "Lichenometry and earthquake age determination in central Asia." *Nature* 280: pp. 675–677.

Nishikawa, Y., K. Michishita, & G. Kurono. 1973. "Studies on water soluble constituents in lichens. I. Gas chromatographic analysis of low molecular weight carbohydrates." *Chem. Pharm. Bull.* 21: pp. 1014–1019.

Nishikawa, Y., M. Tanaka, S. Shibata, & F. Fukuoka. 1970. "Polysaccharides of lichens and fungi. IV. Antitumor active *O*-acetylated pustulan-type glucans from the lichens of *Umbilicaria* species." *Chem. Pharm. Bull.* 18: pp. 1431–1434.

Nishikawa, Y., K. Yoshimoto, R. Horiuchi, K. Michishita, M. Okabe, & F. Fukuoka. 1979. "Studies on the water-soluble constituents of lichens. III. Changes in antitumor effect caused by modifications of pustulan- and lichenan-type glucans." *Chem. Pharm. Bull.* 27: pp. 2065–2072.

Nissen, P. 1974. "Uptake mechanisms: Inorganic and organic." *Ann. Rev. Plant Physiol.* 25: pp. 53–79.

Noeske, O., A. Läuchli, O.L. Lange, G.H. Vieweg, & H. Ziegler. 1970. "Konzentration und Lokalisierung von Schwermetallen in Flechten der Erzschlackenhalden des Harzes." *Dtsch. Bot. Ges.* (Neue Folge) 4: pp. 67–79.

Nourish, R. & R.W.A. Oliver. 1976. "Chemotaxonomic studies on the *Cladonia chlorophaea-pyxidata* complex and some allied species in Britain." In *Lichenology: Progress and Problems,* edited by D.H. Brown, D.L. Hawksworth and R.H. Bailey, pp. 185–214. London: Academic Press.

Nowak, R. & S. Winkler. 1970. "Foliicole Flechten der Sierra Nevada de Santa Marta (Kolumbien) und ihre gegenseitigen Beziehungen." *Öst. Bot. Z.* 118: pp. 456–485.

———. 1972. "Foliicole Flechten von El Salvador, C.A." *Rev. Bryol. Lichénol.* 38: pp. 269–279.

———. 1975. "Foliicolous lichens of Chocó, Colombia, and their substrate abundances." *Lichenologist* 7: pp. 53–58.

Nuno, M. 1973. "Miscellaneous notes on *Cladonia* (1). *Misc. Bryol. Lichenol.* (Nichinan, Japan) 6: pp. 126–127.

Nylander, W. 1865. "Ad historiam reactionis iodi apud Lichenes et Fungos notulae." *Flora* 48: pp. 465–468.

———. 1866a. "Circa novum in studio lichenum criterium chemicum." *Flora* 49: pp. 198–201.

———. 1866b. "Quaedam addenda ad nova criteria chemica in studio lichenum ." *Flora* 49: pp. 233–234.

———. 1866c. "Les Lichens du Jardin du Luxembourg." *Bull. Soc. Bot. Fr.* 13: pp. 364–372.

Oberwinkler, F. 1970. "Die Gattungen der Basidiolichenen." *Deut. Bot. Ges. N.F.* 4: pp. 139–169.

Odum, E.P. 1969. "The strategy of ecosystem development." *Science* 164: pp. 262–270.

Olech, M., J. Kajfosz, S. Szymszyk, & P. Wodniecki. 1981. "Fluorine content in epiphytic lichens and mosses." *Zeszyty Naukowe Uniwersytetu Jagiellonskiego* 8: pp. 163–171.

Ostrofsky, A. & W.C. Denison. 1980. "Ascospore discharge and germination in

Xanthoria polycarpa." *Mycologia* 72: pp. 1171–1179.

Ott, E. 1961. "Über den Einfluss von Flechtensäuren auf die Keimung verschiedener Baumarten." *Schweiz. Z. Forstw.* 112: pp. 303–304.

Ozenda, P. 1951. "Fluorescence des lichens in lumière de Wood." *C. R. Acad. Sci.* (Paris) 233: pp. 194–195.

———. 1963. "Lichens." In *Handbuch der Pflanzenanatomie,* 2nd ed., Vol. 6, edited by W. Zimmerman and P. Ozenda, pp. 1–199. Berlin: Borntraeger.

Paran, N., Y. Ben-Shaul, & M. Galun. 1971. "Fine structure of the blue-green phycobiont and its relation to the mycobiont in two *Gonohymenia* lichens." *Arch. Mikrobiol.* 76: pp. 103–113.

Parker, B.C. & H.C. Bold. 1961. "Biotic relationships between soil algae and other microorganisms." *Am. J. Bot.* 48: pp. 185–197.

Peake, J.F. & P.W. James. 1967. "Lichens and mollusca." *Lichenologist* 3: pp. 425–428.

Pearson, L.C. 1970. "Varying environmental factors in order to grow intact lichens under laboratory conditions." *Am. J. Bot.* 57: pp. 659–664.

Pearson, L.C. & S. Benson. 1977. "Laboratory growth experiments with lichens based on distribution in nature." *Bryologist* 80: pp. 317–327.

Pearson, L.C. & E. Brammer. 1978. "Rate of photosynthesis and respiration in different lichen tissues by the Cartesian Diver technique." *Am. J. Bot.* 65: pp. 276–281.

Pearson, L.C. & E. Henriksson. 1981. "Air pollution damage to cell membranes in lichens. II. Laboratory experiments." *Bryologist* 84: pp. 515–520.

Pearson, L.C. & G.A. Rodgers. 1982. "Air pollution damage to cell membranes in lichens. III. Field experiments." *Phyton* 22: pp. 329–337.

Pearson, L.C. & E. Skye. 1965. "Air pollution affects pattern of photosynthesis in *Parmelia sulcata,* a corticolous lichen." *Science* 148: pp. 1600–1602.

Peat, A. 1968. "Fine structure of the vegetative thallus of the lichen *Peltigera polydactyla.*" *Arch. Mikrobiol.* 61: pp. 212–222.

Peet, M.M. & M.S. Adams. 1972. "Net photosynthesis and respiration of *Cladonia subtenuis* (Abb.) Evans, and comparison with a northern lichen species." *Am. Midl. Natur.* 88: pp. 446–454.

Peet, R.K. 1974. "The measurement of species diversity." *Ann. Rev. Ecol. Syst.* 5: pp. 285–307.

Pentecost, A. 1979. "Aspect and slope preferences in a saxicolous lichen community." *Lichenologist* 11: pp. 81–83.

———. 1980. "Aspects of competition in saxicolous lichen communities." *Lichenologist* 12: pp. 135–144.

Perkins, D.F., R.D. Millar, & P.E. Neep. 1980. "Accumulation of airborne fluoride by lichens in the vicinity of an aluminum reduction plant." *Environ. Pollut.* (Ser. A) 21: pp. 155–168.

Persson, B.R., E. Holm, & K. Lidén. 1974. "Radiolead (^{210}Pb) and stable lead in the lichen *Cladonia alpestris.*" *Oikos* 25: pp. 140–147.

Persson, R.B.R. 1969. "Iron-55 in northern Sweden: Relationships and annual variation from 1956 until 1967 in lichen and reindeer as well as uptake and metabolism in man." *Health Physics* 16: pp. 69–78.

Petit, P. 1982. "Phytolectins from the nitrogen-fixing lichen *Peltigera horizontalis:*

The binding pattern of primary protein extract." *New Phytol.* 91: pp. 705–710.

Pettersson, B. 1940. "Experimentelle Untersuchungen über die euanenomochore Verbreitung der Sporenpflanzen." *Acta Bot. Fenn.* 25: pp. 1–103.

Peveling, E. 1968. "Pyrenoidstrukturen in symbiontisch lebenden *Trebouxia*-Arten." *Z. Pflanzenphysiol.* 59: pp. 393–396.

———. 1969. "Elektronenoptische Untersuchungen an Flechten IV. Die Feinstruktur einigen Flechten mit Cyanophyceen-Phycobionten." *Protoplasma* 68: pp. 209–222.

———. 1970. "Die Darstellung der Oberflächenstrukturen von Flechten mit dem Raster-Elektronenmikroskop." *Ber. Deut. Bot.* Ges. (Neue Folge) 4: pp. 89–101.

———. 1973. "Fine structure." In *The Lichens,* edited by V. Ahmadjian and M.E. Hale, Jr., pp. 147–182. New York: Academic Press.

———. 1976. "Investigations into the ultrastructure of lichens." In *Lichenology: Progress and Problems,* edited by D.H. Brown, D.L. Hawksworth and R.H. Bailey, pp. 17–26. London: Academic Press.

Peveling, E. & M. Galun. 1976. "Electron-microscopical studies on the phycobiont *Coccomyxa* Schmidle." *New Phytol.* 77: pp. 713–718.

Peveling, E. & J. Poelt. 1974. "Glaszilien in der Flechtenfamilie Physciaceae: ihre Ultrastruktur und die Unterschiede gegenüber Rhizinen." *Nova Hedwigia* 25: pp. 639–649.

Peveling, E. & H. Robenek. 1980. "The plasmalemma structure in the phycobiont *Trebouxia* at different stages of humidity of a lichen thallus." *New Phytol.* 84: pp. 371–374.

Phillips, H.C. 1963. "Growth rate of *Parmelia isidiosa* (Müll. Arg.) Hale." *J. Tenn. Acad. Sci.* 38: pp. 95–96.

Pianka, E.R. 1970. "On *r* and *K* selection." *Am. Nat.* 104: pp. 592–597.

———. 1976. "Competition and niche theory." In *Theoretical Ecology: Principles and Applications,* edited by R. May, pp. 114–143. Philadelphia: Saunders.

———. 1978. *Evolutionary Ecology.* 2nd ed. New York: Harper & Row.

Pierce, W.G. & K.A. Kershaw. 1976. "Studies on lichen-dominated systems. XVII. The colonization of young raised beaches in NW Ontario." *Can. J. Bot.* 54: pp. 1672–1683.

Pike, L.H. 1978. "The importance of epiphytic lichens in mineral cycling." *Bryologist* 81: pp. 247–257.

Pike, L.H., D.M. Tracy, M. Sherwood, & D. Nielsen. 1972. "Estimates of biomass and fixed nitrogen of epiphytes from old-growth Douglas fir." In *Proceedings of Research on Coniferous Forest Ecosystems,* edited by J.F. Franklin, L.J. Dempster, and R.H. Waring. Portland, OR: USDA Forest Service.

Plessl, A. 1963. "Über die Beziehungen von Haustorientypen und Organisationshöhe bei Flechten." *Oesterr. Bot. Z.* 110: pp. 194–269.

Plummer, G.L. & B.D. Gray. 1972. "Numerical densities of algal cells and growth in the lichen genus *Cladonia.*" *Am. Midl. Natur.* 87: pp. 355–365.

Poelt, J. 1972. "Die Taxonomische Behandlung von Artenpaaren bei den Flechten." *Bot. Notiser* 125: pp. 77–81.

———. 1973. "Systematic evaluation of morphological characters." In *The Lichens,* edited by V. Ahmadjian and M.E. Hale, Jr., pp. 91–115. New York: Academic Press.

———. 1977. "Types of symbiosis with lichens." In *Proc. 2nd Intern. Mycol. Congr. Abstr. Vol. M-Z,* edited by H.E. Bigelow and E.G. Simmons, p. 526. Tampa, FL: 2nd Int. Mycological Congress.

Proctor, M.C.S. 1977. "The growth curve of the crustose lichen *Buellia canescens* (Dicks.) de Not." *New Phytol.* 79: pp. 659–663.

Puckett, K.J. 1976. "The effect of heavy metals on some aspects of lichen physiology." *Can. J. Bot.* 54: pp. 2695–2703.

Puckett, K.J. & M.A.S. Burton. 1981. "The effect of trace elements on lower plants." In *Effect of Heavy Metal Pollution on Plants,* Vol. 2, edited by N.W. Lepp, pp. 213–238. London: Applied Science Publications.

Puckett, K.J., E. Nieboer, W.P. Flora, & D.H.S. Richardson. 1973a. "Sulphur dioxide: Its effect on photosynthetic ^{14}C fixation in lichens and suggested mechanisms of phytotoxicity." *New Phytol.* 72: pp. 141–154.

Puckett, K.J., E. Nieboer, M.J. Gorzynski, & D.H.S. Richardson. 1973b. "The uptake of metal ions by lichens: A modified ion-exchange process." *New Phytol.* 72: pp. 329–342.

Puckett, K.J., D.H.S. Richardson, W.P. Flora, & E. Nieboer. 1974. "Photosynthetic ^{14}C fixation by the lichen *Umbilicaria muhlenbergii* (Ach.) Tuck. following short exposures to aqueous sulphur dioxide." *New Phytol.* 73: pp. 1183–1192.

Puckett, K.J., F.D. Tomassini, E. Nieboer, & D.H.S. Richardson. 1977. "Potassium efflux by lichen thalli following exposure to aqueous sulphur dioxide." *New Phytol.* 79: pp. 135–145.

Punz, W. 1979. "Der Einfluss isolierte und kombinierter Schadstoffe auf die Flechtenphotosynthese." *Photosynthetica* 13: pp. 428–433.

Pyatt, F.B. 1967. "The inhibitory influence of *Peltigera canina* on the germination of graminaceous seeds and the subsequent growth of the seedlings." *Bryologist* 70: pp. 328–329.

———. 1968. "Ascospore germination in *Pertusaria pertusa* (L.) Tuck." *Rev. Bryol. Lichénol.* 36: pp. 316–328.

———. 1973. "Lichen propagules." In *The Lichens,* edited by V. Ahmadjian and M.E. Hale, Jr., pp. 117–145. New York: Academic Press.

Quispel, A. 1943–1945. "The mutual relations between algae and fungi in lichens." *Rec. Trav. Bot. Neerl.* 40: pp. 413–541.

———. 1959. "Lichens." In *Handbuch der Pflanzenphysiologie,* Vol. 11, edited by W. Ruhland, pp. 577–604. Berlin: Springer-Verlag.

———. 1960. "Respiration of lichens." In *Handbuch der Pflanzenphysiologie,* Vol. 12, edited by W. Ruhland, pp. 455–460. Berlin: Springer-Verlag.

Rahola, T. & J.K. Miettinen. 1969–1970. *Radionuclides in plants and reindeer meat in Lapland—Preliminary report in radioactive foodchains in the subarctic environment,* Ann. Rep. Dept. Radiochemistry, Helsingfors University, Helsinki, Finland, 1969–1970.

Rai, A.N., P. Rowell, & W.D.P. Stewart. 1981. "Nitrogenase activity and dark CO_2 fixation in the lichen *Peltigera aphthosa* Willd." *Planta* 151: pp. 256–264.

Ramaut, J.-L. & M. Corvisier. 1975. "Effets inhibiteurs des extraits de *Cladonia impexa* Harm., *C. gracilis* (L.) Willd. et *Cornicularia muricata* (Ach.) sur la germination des graines de *Pinus sylvestris* (L.)." *Oecologia Plantarum* 10: pp. 295–299.

Ramkaer, K. 1978. "The influence of salinity on the establishment phase of rocky shore lichens." *Bot. Tidsskr.* 72: pp. 119–123.

Rao, D.N. & F. LeBlanc. 1965. "A possible role of atranorin in the lichen thallus." *Bryologist* 68: pp. 284–289.

Rao, D.N., G. Robitaille, & F. LeBlanc. 1977. "Influence of heavy metal pollution on lichens and bryophytes." *J. Hattori Bot. Lab.* 42: pp. 213–239.

Rasmussen, L. 1977. "Epiphytic bryophytes as indicators of the changes in the background levels of airborne metals from 1951–75." *Environ. Pollut.* 14: pp. 37–45.

Rathore, J.S. & S.K. Mishra. 1971. "Inhibition of root elongation by some plant extracts." *Indian J. Exp. Biol.* 9: pp. 523–524.

Reddy, P.V. & P.S. Rao. 1978. "Influence of some lichen substances on mitosis in *Allium cepa* root tips." *Indian J. Exp. Biol.* 16: pp. 1019–1021.

Renner, B. & D.J. Galloway. 1982. "Phycosymbiodemes in *Pseudocyphellaria* in New Zealand." *Mycotaxon* 16: pp. 197–231.

Rhoades, D.F. & R.G. Cates. 1976. "A general theory of plant antiherbivore chemistry." *Rec. Advances Phytochem.* 10: pp. 168–213.

Rhoades, F.M. 1977. "Growth rates of the lichen *Lobaria oregana* as determined from sequential photographs." *Can. J. Bot.* 55: pp. 2226–2233.

Rice, E.L. 1974. *Allelopathy*. New York: Academic Press.

Rice, P. 1970. "Some biological effects of volatiles emanating from wood." *Can. J. Bot.* 48: pp. 719–735.

Richardson, D.H.S. 1967. "The transplantation of lichen thalli to solve some taxonomic problems in *Xanthoria parietina* (L.) Th. Fr." *Lichenologist* 3: pp. 386–391.

———. 1971. "Lichens." In *Methods in Microbiology*, pp. 267–293. New York: Academic Press.

———. 1975. *The Vanishing Lichens*. Vancouver: David & Charles.

Richardson, D.H.S., D.J. Hill, & D.C. Smith. 1968. "Lichen physiology. XI. The role of the alga in determining the pattern of carbohydrate movement between lichen symbionts." *New Phytol.* 67: pp. 469–486.

Richardson, D.H.S. & E. Nieboer. 1980. "Surface binding and accumulation of metals in lichens." In *Cellular Interactions in Symbiosis and Parasitism*, edited by C.B. Cook, P.W. Pappas and E.D. Rudolph, pp. 75–94. Columbus: Ohio State University Press.

Richardson, D.H.S., E. Nieboer, P. Lavoie, & D. Padovan. 1979. "The role of metal-ion binding in modifying the toxic effects of sulphur dioxide on the lichen *Umbilicaria muhlenbergii*. II. ^{14}C-fixation studies." *New Phytol.* 82: pp. 633–643.

Richardson, D.H.S. & K.J. Puckett. 1973. "Sulphur dioxide and photosynthesis in lichens." In *Air Pollution and Lichens*, edited by B.W. Ferry, M.S. Baddeley, & D.L. Hawksworth, pp. 283–298. Toronto: University of Toronto Press.

Richardson, D.H.S. & D.C. Smith. 1966. "The physiology of the symbiosis in *Xanthoria aureola* (Ach.) Erichs." *Lichenologist* 3: pp. 202–206.

———. 1968. "Lichen physiology. IX. Carbohydrate movement from the *Trebouxia* symbiont of *Xanthoria aureola* to the fungus." *New Phytol.* 67: pp. 61–68.

Richardson, D.H.S. & C.M. Young. 1977. "Lichens and vertebrates." In *Lichen Ecology*, edited by M.R.D. Seaward, pp. 121–144. London: Academic Press.

Rickard, W.H., J.J. Davis, W.C. Hanson, & D.G. Watson. 1965. "Gamma-emitting radionuclides in Alaskan tundra vegetation, 1959, 1960, 1961." *Ecology* 46: pp. 352–356.

Ried, A. 1960a. "Stoffwechsel und Verbreitungsgrenzen von Flechten I. Flechtenzonierung an Bachufern und ihre Beziehungen zur jährlichen Überflutungsdauer und zum Mikroklima." *Flora* 148: pp. 612–638.

———. 1960b. "Stoffwechsel und Verbreitungsgrenzen von Flechten II. Wasser- und Assimilationshaushalt, Entquellungs- und Submersionsresistenz von Krustenflechten benachbarter Standorte." *Flora* 149: pp. 345–385.

———. 1960c. "Thallusbau und Assimilationshaushalt von Laub- und Krustenflechten." *Biol. Zentralbl.* 79: pp. 129–151.

———. 1960d. "Nachwirkungen der Entquellung auf den Gaswechsel von Krustenflechten." *Biol. Zentralbl.* 79: pp. 657–678.

Roberts, B.A. & L.K. Thompson. 1980. "Lichens as indicators of fluoride emission from a phosphorus plant, Long Harbour, Newfoundland, Canada." *Can. J. Bot.* 58: pp. 2218–2228.

Robinson, H.R. 1959. "Lichen succession in abandoned fields in the Piedmont of North Carolina." *Bryologist* 62: pp. 254–259.

———. 1975. "Considerations on the evolution of lichens." *Phytologia* 32: pp. 407–413.

Robitaille, G., F. LeBlanc, & D.N. Rao. 1977. "Acid rain: A factor contributing to the paucity of epiphytic cryptogams in the vicinity of a copper smelter." *Rev. Bryol. Lichénol.* 43: pp. 53–66.

Rogers, R.W. 1972a. "Soil surface lichens in arid and sub-arid south-eastern Australia. II. Phytosociology and geographic zonation." *Australian J. Bot.* 20: pp. 215–227.

———. 1972b. "Soil surface lichens in arid and subarid south-eastern Australia. III. The relationship between distribution and environment." *Australian J. Bot.* 20: pp. 301–316.

———. 1974. "Lichens from the T.G.B. Osborn vegetation reserve at Koonamore in arid South Australia." *Trans. Roy. Soc. South Austral.* 98: pp. 113–124.

———. 1977. "Lichens of hot arid and semi-arid lands." In *Lichen Ecology,* edited by M.R.D. Seaward, pp. 211–252. London: Academic Press.

Rogers, R.W. & R.T. Lange. 1972. "Soil surface lichens in arid and sub-arid south-eastern Australia. I. Introduction and floristics." *Australian J. Bot.* 20: pp. 197–213.

Rondon, Y. 1966. "Action inhibitrice de l'extrait du lichen *Roccella fucoides* (Dicks.) Vain. sur la germination." *Bull. Soc. Bot. France* 113: pp. 1–2.

Rose, C.I. & D.L. Hawksworth. 1981. "Lichen recolonization in London's cleaner air." *Nature* 289: pp. 289–292.

Rosentreter, R. & V. Ahmadjian. 1977. "Effect of ozone on the lichen *Cladonia arbuscula* and the *Trebouxia* phycobiont of *Cladina stellaris.*" *Bryologist* 80: pp. 600–605.

Roskin, P.A. 1970. "Ultrastructure of the host-parasite interaction in the basidiolichen *Cora pavonia* (Web.) E. Fries." *Arch. Mikrobiol.* 70: pp. 176–182.

Rudolph, E.D. 1970. "Local dissemination of plant propagules in Antarctica." In *Antarctic Ecology,* edited by M.W. Holdgate, Vol. 2: pp. 812–817. New York: Academic Press.

Rudolph, E.D. & R. M. Giesy. 1966. "Electron microscope studies of lichen reproductive structures in *Physcia aipolia.*" *Mycologia* 58: pp. 786–796.

Rühling, Å. & G. Tyler. 1968. "An ecological approach to the lead problem." *Bot. Notiser* 121: pp. 321–342.

———. 1969. "Ecology of heavy metals—A regional and historical study." *Bot. Notiser* 122: pp. 248–259.

———. 1971. "Regional differences in the deposition of heavy metals over Scandinavia." *J. Appl. Ecol.* 8: pp. 497–507.

———. 1973. "Heavy metal deposition in Scandinavia." *Water, Air & Soil Pollut.* 2: pp. 445–455.

Rundel, P.W. 1969. "Clinal variation in the production of usnic acid in *Cladonia subtenuis.*" *Bryologist* 72: pp. 40–44.

———. 1972. "CO_2 exchange in ecological races of *Cladonia subtenuis.*" *Photosynthetica* 6: pp. 13–17.

———. 1974. "Water relations and morphological variation in *Ramalina menziesii* Tayl." *Bryologist* 77: pp. 23–32.

———. 1978. "The ecological role of secondary lichen substances." *Biochem. Syst. Ecol.* 6: pp. 157–170.

Rundel, P.W., G.C. Bratt, & O.T. Lange. 1979. "Habitat ecology and physiological response of *Sticta filix* and *Pseudocyphellaria delisei* from Tasmania." *Bryologist* 82: pp. 171–180.

Rydzak, J. 1961. "Investigations on the growth rate of lichens." *Ann. Univ. Mariae Curie—Skłódowska,* (Sec. C) 16: pp. 1–15.

Saeki, M., K. Kunii, T. Seki, K. Sugiyama, T. Suzuki, & S. Shishido. 1977. "Metal burden of urban lichens." *Environ. Res.* 13: pp. 256–266.

Salo, A. & J.K. Miettinen. 1964. "Strontium-90 and cesium-137 in arctic vegetation during 1961." *Nature* 201: pp. 1177–1179.

Samuelson, D.A. & J. Bezerra. 1977. "Concentric bodies in two species of the Loculoascomycetes." *Can. J. Microbiol.* 23: pp. 1485–1488.

Santesson, J. 1969. "Chemical studies on lichens. 10. Mass spectrometry of lichens." *Ark. Kemi* 30: pp. 363–377.

———. 1973. "Identification and isolation of lichen substances." In *The Lichens,* edited by V. Ahmadjian and M.E. Hale, Jr., pp. 633–652. New York: Academic Press.

Santesson, R. 1952. "Foliicolous lichens. I. A revision of the taxonomy of the obligately foliicolous, lichenized fungi." *Symb. Bot. Upsal.* 12: pp. 1–590.

Saunders, P.J.W. 1966. "The toxicity of sulphur dioxide to *Diplocarpon rosae* Wolf causing blackspot of roses." *Ann. Appl. Biol.* 58: pp. 103–114.

———. 1970. "Air pollution in relation to lichens and fungi." *Lichenologist* 4: pp. 337–349.

Schade, A. 1970. "Über Herkunft und Vorkommen der Calciumoxalat-Exkrete in kortizikolen *Parmeliaceen.*" *Nova Hedwigia* 19: pp. 159–187.

Schade, A. & W. Seitz. 1970. "Extremes Auftreten von Calciumoxalat-Exkreten bei einer Art der Gattung *Usnea* (Lichenes)." *Ber. Deut. Bot. Ges.* 83: pp. 121–127.

Scharf, C.S. 1978. "Birds and mammals as passive transporters for algae found in lichens." *Can. Field-Naturalist* 92: pp. 70–71.

Schofield, E. & V. Ahmadjian. 1972. "Field observation and laboratory studies of some Anatarctic cold desert cryptogams." In *Antarctic Terrestrial Biology,* edited by G.A. Llano, pp. 97–142. Antarctic Res. Series, Vol. 20, Washington, D.C.: American Geophysical Union.

Schönbeck, H. 1968. "Einfluss von Luftverunreinigungen (SO_2) auf transplantierte Flechten." *Die Naturwiss.* 9: pp. 451–452.

———. 1969. "Eine Methode zur Erfassung der biologischen Wirkung von Luftverunreinigungen durch transplantierte Flechten." *Staub* 29: pp. 14–18.

Schubert, R. & E. Fritsche. 1965. "Beitrag zur Einwirkung von Luftverunreinigungen auf xerische Flechten." *Arch. Naturschutz.* 5: pp. 107–110.

Schultz, J. & K. Mosbach. 1971. "Studies on lichen enzymes, purification and properties of an orsellinate depside hydrolase obtained from *Lasallia pustulata.*" *Eur. J. Biochem.* 22: pp. 153–157.

Schulz, K. 1931. Die Flechtenvegetation der Mark Brandenburg. Inaug. -Diss., Berlin.

Schumm, F. & K.H. Kreeb. 1979. "Die Nettphotosynthese von Flechtentransplanten als mass für die Immissionsbelastung der Luft." *Angew. Bot.* 53: pp. 31–39.

Schutte, J.A. 1977. "Chromium in two corticolous lichens from Ohio and West Virginia." *Bryologist* 80: pp. 279–283.

Scott, G.D. 1964. "The lichen symbiosis." In *Advancement of Science* (London) 31: pp. 244–248.

Sell, D.K. & C.H. Schmidt. 1968. "Chelating agents suppress pupation of the cabbage looper." *J. Econ. Entomol.* 61: pp. 946–949.

Shapiro, I.A. 1977. "Urease activity in lichens." *Soviet Plant Physiol.* 24: pp. 914–918. (original Russian paper published in *Fiziologiya Rastenii* 24: pp. 1135–1139)

Sharma, P., B. Bergman, L. Hällbom, & A.v. Hofsten. 1983. "Ultrastructural changes of *Nostoc* of *Peltigera canina* in presence of SO_2." *New Phytol.* 92: pp. 573–579.

Sheard, J.W. 1968. "The zonation of lichens on three rocky shores of Inishowen, County Donegal." *Proc. R. Ir. Acad.* 66B: pp. 101–112.

Sheridan, R.P. 1979. "Impact of emissions from coal-fired electricity generating facilities on N_2-fixing lichens." *Bryologist* 82: pp. 54–58.

Shibata, S. 1965. "Biogenetical and chemotaxonomical aspects of lichen substances." In *Beitrage zur Biochemie und Physiologie von Naturstoffen, Festschrift Kurt Mothes zum 65 Geburtstag,* pp. 451–465. Jena: Gustav Fischer Verlag.

Shibata, S., T. Furuya, & H. Iizuka. 1965. "Gas-liquid chromatography of lichen substances. I. Studies on zeorin." *Chem. Pharm. Bull.* (Tokyo) 13: pp. 1254–1257.

Shibata, S. & Y. Miura. 1949. "Antibacterial effects of lichen substances. II. Studies on didymic acid and related compounds." *Jap. Med. J.* 2: pp. 22–24.

Shibata, S., Y. Nishikawa, M. Tanaka, F. Fukuoka, & M. Nakanishi. 1968. "Antitumor activities of lichen polysaccharides." *Zeitschr. Krebsforsch.* 71: pp. 102–104.

Showman, R.E. 1972. "Residual effects of sulfur dioxide on the net photosynthetic and respiratory rates of lichen thalli and cultured lichen symbionts." *Bryologist* 75: pp. 335–341.

———. 1975. "Lichens as indicators of air quality around a coal-fired power generating plant." *Bryologist* 78: pp. 1–6.

———. 1976. "Seasonal growth of *Parmelia caperata.*" *Bryologist* 79: pp. 360–363.

———. 1981. "Lichen recolonization following air quality improvement." *Bryologist* 84: pp. 492–497.

Showman, R.E. & E.D. Rudolph. 1971. "Water relations in living, dead, and cellulose models of the lichen *Umbilicaria papulosa.*" *Bryologist* 74: pp. 444–450.

Sigal, L.L. & O.C. Taylor. 1979. "Preliminary studies of the gross photosynthetic response of lichens to peroxyacetylnitrate fumigations." *Bryologist* 82: pp. 564–575.

Sigfridsson, B. 1980. "Some effects of humidity on the light reaction of photosynthesis in the lichens *Cladonia impexa* and *Collema flaccidium.*" *Physiol. Plantarum* 49: pp. 320–326.

Sigfridsson, B. & G. Öquist. 1980. "Preferential distribution of excitation energy into photosystem I of desiccated samples of the lichen *Cladonia impexa* and the isolated lichen-alga *Trebouxia pyriformis.*" *Physiol. Plantarum* 49: pp. 329–335.

Silva-Pando, F.J. & C. Ascaso. 1982. "Modificationes ultraestructurales de liquenes epipfitos transplantados a zonas urbanas de Madrid." *Collectanea Botanica* 13: pp. 351–374.

Sivori, E.M. & J.R. Jatimliansky. 1965. "Inhibidores del crecimiento de *Scenedesmus obliquus* en *Laurus nobilis* L." *Actas 1er Coloq. Latinoamer. Biol. Suelo. B. Blanca* (Argentina).

———. 1970. "Effects of xylopine on physiological activities of *Scenedesmus obliquus.*" *Plant & Cell Physiol.* 11: pp. 921–926.

Slack, N.G. 1976. "Host specificity of bryophyte epiphytes in eastern North America." *J. Hattori Bot. Lab.* 41: pp. 107–132.

Slansky, F. 1979. "Effect of the lichen chemicals atranorin and vulpinic acid upon feeding and growth of larvae of the yellow-striped armyworm, *Spodoptera ornithogalli.*" *Environ. Entomol.* 8: pp. 865–868.

Slayman, C.L. 1965. "Electrical properties of *Neurospora crassa.* Effect of external cation on the intracellular potential." *J. Gen. Physiol.* 49: pp. 69–92.

Slocum, R.D. 1977. "Effects of SO_2 and pH on the ultrastructure of the *Trebouxia* phycobiont of the pollution-sensitive lichen *Parmelia caperata* (L.) Ach." M.Sc. dissertation, Columbus: Ohio State University.

———. 1980. "Light and electron microscope investigations in the Dictyonemataceae (basidiolichens). II. *Dictyonema irpicinum.*" *Can. J. Bot.* 58: pp. 1005–1015.

Slocum, R.D., V. Ahmadjian, & K.C. Hildreth. 1980. "Zoosporogenesis in *Trebouxia gelatinosa:* Ultrastructure, potential for zoospore release and implications for the lichen association." *Lichenologist* 12: pp. 173–187.

Slocum, R.D. & G.L. Floyd. 1977. "Light and electron microscope investigations in the Dictyonemataceae (Basidiolichens)." *Can. J. Bot.* 55: pp. 2565–2573.

Smith, A.E. & I. Morris. 1980. "Synthesis of lipid during photosynthesis by phytoplankton of the southern ocean." *Science* 207: pp. 197–199.

Smith, A.L. 1921. *Lichens.* Cambridge: Cambridge University Press.

Smith, D.C. 1960a. "Studies in the physiology of lichens. I. The effects of starvation and of ammonia absorption upon the nitrogen content of *Peltigera polydactyla.*" *Ann. Bot.* 24: pp. 52–62.

———. 1960b. "Studies in the physiology of lichens. II. Absorption and utilization of some simple organic compounds by *Peltigera polydactyla.*" *Ann. Bot.* 24: pp. 172–185.

———. 1960c. "Studies in the physiology of lichens. III. Experiments with dissected discs of *Peltigera polydactyla.*" *Ann. Bot.* 24: pp. 186–199.

———. 1961. "The physiology of *Peltigera polydactyla* (Neck.) Hoffm." *Lichenologist* 1: pp. 209–226.

———. 1962. "The biology of lichen thalli." *Biol. Rev.* 37: pp. 537–570.

———. 1973. *Symbiosis of Algae with Invertebrates.* Oxford Biology Readers No. 43. Oxford University Press.

———. 1974. "Transport from symbiotic algae and symbiotic chloroplasts to host cells." *Symposia Soc. Exp. Biol.* University Press, Cambridge, England, 28: pp. 485–520.

———. 1975. "Symbiosis and the biology of lichenised fungi." *Symposia Soc. Exp. Biol.* University Press, Cambridge, England, 29: pp. 373–405.

———. 1976. "A comparison between the lichen symbiosis and other symbioses." In *Lichenology: Progress and Problems,* edited by D.H. Brown, D.L. Hawksworth and R.H. Bailey, pp. 497–513. London: Academic Press.

———. 1978. "What can lichens tell us about real fungi." *Mycologia* 70: pp. 915–934.

Smith, D.C. & S. Molesworth. 1973. "Lichen physiology. XIII. Effects of rewetting dry lichens." *New Phytol.* 72: pp. 525–533.

Smyth, E.S. 1934. "A contribution to the physiology and ecology of *Peltigera canina* and *P. polydactyla.*" *Ann. Bot.* 48: pp. 781–818.

Snelgar, W.P., T.G.A. Green, & A.L. Wilkins. 1981. "Carbon dioxide exchange in lichens: Resistances to CO_2 uptake at different thallus water contents." *New Phytol.* 88: pp. 353–361.

Søchting, U. & J. Johnsen. 1978. "Lichen transplants as biological indicators of SO_2 air pollution in Copenhagen." *Bull. Environ. Contam. Toxicol.* 19: pp. 1–7.

Solbrig, O.T. & D.J. Solbrig. 1979. *Introduction to Population Biology and Evolution.* Reading, MA: Addison-Wesley.

Stahl, E. 1877. *Beiträge zur Entwicklungsgeschichte des Flechten.* Leipzig, Germany: Felix.

Stahl, E. & P.J. Schorn. 1961. "Dünnschicht-Chromatographie hygrophiler Arzneipflanzenauszüge. VIII. Mitteilung; Cumarine, Flavonderivate, Hydroxysäuren, Gerbstoffe, Antracenderivate und Flechteninhaltosstoffe." *Hoppe-Seyler's Zeitschr. für Physiol. Chem* 325: pp. 263–274.

Stahl, G.E. 1904. *Die Schutzmittel der Flechten gegen Tierfrass. Festschrift z. 70 Geburtstage von Ernst Haekel,* pp. 357–375. Jena: G. Fischer.

Stark, A.A., B. Kobbe, K. Matsuo, G. Büchi, G.N. Wogan, & A.L. Demain. 1978. "Mollicellins: Mutagenic and antibacterial mycotoxins." *Appl. Environ. Microbiol.* 36: pp. 412–420.

Stebbing, A.R.D. 1973. "Competition for space between the epiphytes of *Fucus Serratus* L." *J. Mar. Biol. Assoc.* (U.K.) 53: pp. 247–261.

Stephenson, N.L. & P.W. Rundel. 1979. "Quantitative variation and the ecological role of vulpinic acid and atranorin in the thallus of *Letharia vulpina* (Lichenes)." *Biochem. Syst. Ecol.* 7: pp. 263–267.

Stevens, R.B. 1941. "Morphology and ontogeny of *Dermatocarpon aquaticum.*" *Am. J. Bot.* 28: pp. 59–69.

Stewart, W.D.P. 1966. *Nitrogen Fixation in Plants.* London: Athlone Press.

Stewart, W.D.P. & P. Rowell. 1977. "Modifications of nitrogen-fixing algae in lichen symbioses." *Nature* 265: pp. 371–372.

Stocker, O. 1927. "Physiologische und ökologische Untersuchungen an Lau-bund Strauchflechten." *Flora* (Jena) 21: pp. 334–415.

Stringer, P.W. & M.H.L. Stringer. 1974. "A quantitative study of corticolous bryophytes in the vicinity of Winnipeg, Manitoba." *Bryologist* 77: pp. 551–560.

Swain, T. 1963. *Chemical Plant Taxonomy.* New York: Academic Press.

———. 1977. "Secondary compounds as protective agents." *Ann. Rev. Plant Physiol.* 28: pp. 479–501.

———. 1979. "Tannins and lignins." In *Herbivores, Their Interaction with Secondary Plant Metabolites,* edited by G.A. Rosenthal and D.H. Janzen, pp. 657–682. New York: Academic Press.

Syers, J.K. 1969. "Chelating ability of fumarprotocetraric acid and *Parmelia conspersa.*" *Plant and Soil* 31: pp. 205–208.

Syers, J.K., A.C. Birnie, & B.D. Mitchell. 1967. "The calcium oxalate content of some lichens growing on limestone." *Lichenologist* 3: pp. 409–414.

Syers, J.K. & I.K. Iskandar. 1973. "Pedogenetic significance of lichens." In *The Lichens,* edited by V. Ahmadjian and M.E. Hale, Jr., pp. 225–248. New York: Academic Press.

Takai, M., Y. Uehara, & J.A. Beisler. 1979. "Usnic acid derivatives as potential antineoplastic agents." *J. Medical Chem.* 22: pp. 1380–1384.

Takala, K., P. Kauranen, & R. Fagerstén. 1979. "Fluorine content of terricolous lichens and bryophytes on exposed rapakivi bedrock." *Ann. Bot. Fenn.* 16: pp. 90–92.

Takeda, T., M. Funatsu, S. Shibata, & F. Fukuoka. 1972. "Polysaccharides of lichens and fungi. V. Antitumor active polysaccharides of lichens of *Evernia, Acarospora & Alectoria* spp." *Chem. Pharm. Bull.* 20: pp. 2445–2449.

Tamm, C.O. 1950. "Growth and plant nutrient concentrations in *Hylocomium proliferum* Lindb. in relation to tree canopy." *Oikos* 2: pp. 60–64.

Tansey, M.R. 1977. "Microbial facilitation of plant mineral nutrition." In *Microorganisms and Minerals,* edited by E.D. Weinberg, pp. 343–385. New York: Marcel Dekker.

Tapper, R. 1981a. "Direct measurement of translocation of carbohydrate in the lichen *Cladonia convoluta,* by quantitative autoradiography." *New Phytol.* 89: pp. 429–437.

———. 1981b. "Glucose uptake by *Trebouxia* and associated fungal symbiont in the lichen symbiosis." *FEMS Microbiol. Lett.* 10: pp. 103–106.

Taylor, C.J. 1967. *The Lichens of Ohio. Part I. Foliose Lichens.* Columbus: Ohio State University.

Taylor, D.L. 1970. "Chloroplasts as symbiotic organelles." *Int. Rev. Cytol.* 27: pp. 29–64.

———. 1973. "Algal symbionts of invertebrates." *Ann. Rev. Microbiol.* 27: pp. 171–184.

Taylor, I.E.P. & A.J. Wilkinson. 1977. "The occurrence of gibberellins and gibberellin-like substances in algae." *Phycologia* 16: pp. 37–42.

Taylor, O.C. 1968. "Effects of oxidant air pollutants." *J. Occup. Med.* 10: pp. 485–496.

———. 1969. "Importance of PAN (peroxyacetyl nitrate) as a phytotoxic air pollutant." *J. Air Pollut. Contr. Assoc.* 19: pp. 347–351.

Tegler, B. & K.A. Kershaw. 1981. "Physiological-environmental interactions in lichens. XII. The seasonal variation of the heat response of *Cladonia rangiferina.*" *New Phytol.* 87: pp. 395–401.

Thøgersen, P.-J. 1977. Note on cover drawing of *Int. Lichenol. Newsletter,* 10(2).

Thomas, E.A. 1939. "Über die Biologie von Flechtenbildern." *Beitr. Kryptogamenflora Schweiz.* 9: pp. 1–208.

Thomson, J.W. 1967. *The Lichen Genus* Cladonia *in North America.* Toronto: University of Toronto Press.

Tobler, F. 1925. *Biologie der Flechten. Entwicklung und Begriff der Symbiose.* Berlin: Borntraeger.

———. 1939. "Die Kultur von Flechten." In *Abderhaldens Handbuch der biologischen Arbeitsmethoden* 12: pp. 1491–1511. Berlin: Urban & Schwarzenberg.

———. 1944. "Neue Tatsachen und Klarstellungen zu Symbiosefrage." *Forsch. Fortschr. Deutsch.* 20: pp. 205–207.

Tomassini, F.D., P. Lavoie, K.J. Puckett, E. Nieboer, & D.H.S. Richardson. 1977. "The effect of time of exposure to sulphur dioxide on potassium loss from and photosynthesis in the lichen *Cladina rangiferina* (L.) Harm." *New Phytol.* 79: pp. 147–155.

Topham, P.B. 1977. "Colonization, growth, succession and competition." In *Lichen Ecology,* edited by M.R.D. Seaward, pp. 31–68. London: Academic Press.

Travé, J. 1963. "Ecologie et biologie des Oribates (Acariens) saxicoles et arboricoles." *Vie Milieu Supp.* 14: pp. 1–267.

———. 1969. "Sur le peuplement des lichens crustacés des Iles Salvages par les Oribates (Acariens)." *Rev. Ecol. Biol. Sol.* 6: pp. 239–248.

Trench, R.K. 1979. "The cell biology of plant-animal symbioses." *Ann. Rev. Plant Physiol.* 30: pp. 485–531.

Trudgill, S.T., R.W. Crabtree, & P.J.C. Walker. 1979. "The age of exposure of limestone pavements—A pilot lichenometric study in County Clare, Eire." *Trans. Brit. Cave Res. Assoc.* 6: pp. 10–14.

Tschermak-Woess, E. 1978. "*Myrmecia reticulata* as a phycobiont and free-living; free-living *Trebouxia*—the problem of *Stenocybe septata.*" *Lichenologist* 10: pp. 69–79.

Tschermak-Woess, E. & E.I. Friedmann. 1984. "*Hemichloris antarctica,* gen. et spec. nov., an endolithic chlorococcalean alga." *Phycologia* (in press).

Tschermak-Woess, E. & J. Poelt. 1976. "*Vesdaea,* a peculiar lichen genus, and its phycobiont." In *Lichenology: Progress and Problems,* edited by D.H. Brown, D.L. Hawksworth and R.H. Bailey, pp. 89–105. London: Academic Press.

Tu, J.C. & N. Colotelo. 1973. "A new structure containing cyst-like bodies in apothecia-bearing sclerotia of *Sclerotinia borealis.*" *Can. J. Bot.* 51: pp. 2249–2250.

Tuominen, Y. 1967. "Studies on the strontium uptake of the *Cladonia alpestris* thallus." *Ann. Bot. Fenn.* 4: pp. 1–28.

Tuominen, Y. & T. Jaakkola. 1973. "Absorption and accumulation of elements." In *The Lichens,* edited by V. Ahmadjian and M.E. Hale, Jr., pp. 185–223. New York: Academic Press.

Türk, R. & V. Wirth. 1975. "The pH dependence of SO_2 damage to lichens." *Oecologia*

19: pp. 285–291.

Türk, R., V. Wirth, & O.L. Lange. 1974. "CO_2-Gaswechsel-Untersuchungen zur SO_2-Resistenz von Flechten." *Oecologia* 15: pp. 33–64.

Tyler, G. 1971. "Moss analysis—Method for surveying heavy metal deposition." *Proceedings,* Washington, D.C.: Second International Clean Air Congress, 1970, pp. 129–132.

Tysiaczny, M.J. & K.A. Kershaw. 1979. "Physiological-environmental interactions in lichens. VII. The environmental control of glucose movement from alga to fungus in *Peltigera canina* v. *praetextata* Hue." *New Phytol.* 83: pp. 137–146.

Umezawa, H., N. Shibamoto, H. Naganawa, S. Ayukawa, M. Matsuzaki, T. Takeuchi, K. Kono, & T. Sakamoto. 1974. "Isolation of lecanoric acid, an inhibitor of histidine decarboxylase from a fungus." *J. Antibiotics* 27: pp. 587–596.

Väisälä, L. 1974. "Effects of terpene compounds on the growth of wood-decomposing fungi." *Ann. Bot. Fenn.* 11: pp. 275–278.

Van Sumere, C.F., J. Albrecht, A. Dedonder, H. DePooter, & I. Pé. 1975. "Plant proteins and phenolics." In *Ann. Proc. Phytochem. Soc.*, Vol. 11, edited by J.B. Harborne and C.F. Van Sumere, pp. 211–264.

Vartia, K.O. 1973. "Antibiotics in lichens." In *The Lichens,* edited by V. Ahmadjian and M.E. Hale, Jr., pp. 547–561. New York: Academic Press.

Vězda, A. 1975. "Foliikole Flechten aus Tanzania (Öst Africa)." *Folia Geobot. Phytotax., Praha* 10: pp. 383–432.

Vicente, C. & L. Xavier Filho. 1979. "Urease regulation in *Cladonia verticillaris* (Raddi) Fr." *Phyton* (Buenos Aires) 37: pp. 137–144.

Vidrich, V., C.A. Cecconi, G.G. Ristori, & P. Fusi. 1982. "Verwitterung toskanischer Gesteine unter Mitwirkung von Flechten." *Z. Pflanzenernaehr. Bodenk.* 145: pp. 384–389.

Virtanen, O.E. & O.E. Kilpiö. 1957. "On the *in vitro* fungistatic activity of an usnic acid preparation with the trade name USNO." *Suom. Kemistilehti B* 30: pp. 8–9.

Vobis, G. 1977. "Studies on the germination of lichen conidia." *Lichenologist* 9: pp. 131–136.

———. 1980. *Bau und Entwicklung der Flechten-Pycnidien und ihrer Conidien.* Vaduz: Cramer.

Vobis, G. & D.L. Hawksworth. 1981. "Conidial lichen-forming fungi." In *Biology of Conidial Fungi,* edited by G.T. Cole and B. Kendrick, pp. 245–273. New York: Academic Press.

Wachtmeister, C.A. 1952. "Studies on the chemistry of lichens. I. Separation of depside components by paper chromatography." *Acta Chem. Scand.* 6: pp. 818–825.

———. 1959. "Flechtensäuren." In *Papierchromatographie in der Botanik,* 2nd ed., edited by H.F. Linskens, pp. 135–141. Berlin: Springer-Verlag.

Waggoner, P.E. & G.R. Stephens. 1970. "Transition probabilities for a forest." *Nature* 255: pp. 1160–1161.

Walker, A.T. 1968. "Fungus-alga ultrastructure in the lichen, *Cornicularia normoerica.*" *Am. J. Bot.* 55: pp. 641–648.

Watanabe, A. & T. Kiyohara. 1963. "Symbiotic blue-green algae of lichens, liverworts and cycads." In *Studies on Microalgae and Photosynthetic Bacteria,* edited by Japanese Soc. Plant Physiologists, pp. 189–196. Tokyo: University of Tokyo Press.

Watson, D.G., W.C. Hanson, & J.J. Davis. 1964. "Strontium-90 in plants and animals of arctic Alaska, 1959–61." *Science* 144: pp. 1005–1009.

Watt, A.S. 1947. "Pattern and process in the plant community." *J. Ecol.* 35: pp. 1–22.

Weaver, J.E. & F.E. Clements. 1938. *Plant Ecology.* New York: McGraw-Hill.

Webber, M. & P. Webber. 1970. "Ultrastructure of lichen haustoria: Symbiosis in *Parmelia sulcata.*" *Can. J. Bot.* 48: pp. 1521–1524.

Webber, P.J. & J.T. Andrews. 1973. "Lichenometry: A commentary." *Arctic & Alpine Res.* 5: pp. 295–302.

Weigel, H.-J. & H.-J. Jäger. 1979. "Changes in proline concentration of the lichen *Pseudevernia furfuracea* during drought stress." *Phyton* (Austria) 19: pp. 163–167.

Werner, P.A. 1976. "Ecology of plant populations in successional environments." *Syst. Bot.* 1: pp. 246–268.

West, V. & R. Stotler. 1977. "Saxicolous bryophyte and macrolichen associations in southern Illinois. II. Panther's Den, Union County." *Bryologist* 80: pp. 612–618.

Wetherbee, R. 1969. "Population studies in the chemical species of the *Cladonia chlorophaea* group." *Mich. Bot.* 8: pp. 170–174.

Whiton, J.C. & J.D. Lawrey. 1982. "Inhibition of *Cladonia cristatella* and *Sordaria fimicola* ascospore germination by lichen acids." *Bryologist* 85: pp. 222–226.

———. 1984. "Inhibition of crustose lichen spore germination by lichen acids." *Bryologist* 87: 42–43.

Whittaker, R.H. & P.P. Feeny. 1971. "Allelochemics: Chemical interaction between species." *Science* 171: pp. 1757–1770.

Wilhelmsen, J.B. 1959. "Chlorophylls in the lichens *Peltigera, Parmelia,* and *Xanthoria.*" *Bot. Tidssk.* 55: pp. 30–36.

Williams, A.H. 1963. "Enzyme inhibition by phenolic compounds." In *The Enzyme Chemistry of Phenolic Compounds,* edited by J.B. Pridham, pp. 87–95. Oxford: Pergamon Press.

Williams, G.C. 1975. *Sex and Evolution.* Princeton, NJ: Princeton University Press.

Williams, M.E. & E.D. Rudolph. 1974. "The role of lichens and associated fungi in the chemical weathering of rock." *Mycologia* 66: pp. 648–660.

Wilson, M.J., D. Jones, & J.D. Russell. 1980. "Glushinskite, a naturally occurring magnesium oxalate." *Minerological Magazine* 43: pp. 837–840.

Winterringer, G.S. & A.G. Vestal. 1956. "Rockledge vegetation in Southern Illinois." *Ecol. Monogr.* 26: pp. 105–130.

Wirth, V. 1972. *Die Silikatflechten-Gemeinschaften im ausseralpinen Zentraleuropa. Vduz: Cramer.*

Wirth, V. & R. Türk. 1975. "Zur SO_2-Resistenz von Flechten verschiedener Wuchsform." *Flora* 164: pp. 133–143.

Woodring, J.P. & E.F. Cook. 1962. "The biology of *Ceratozetes cisalpinus* Berlese, *Scheloribates laevigatus* Koch, and *Oppia neerlandica* Oudemans (Oribatei), with a description of all stages." *Acarologia* 4: pp. 101–137.

Woolhouse, H.W. 1968. "The measurement of growth rates in lichens." *Lichenologist* 4: pp. 32–33.

Xavier Filho, L. & C. Vicente. 1978. "Exo- and endourease from *Parmelia roystonea* and their regulation by lichen acids." *Bol. Soc. Brot.* (Sér 2) 52: pp. 51–65.

Yamamoto, Y., K. Nishimura, & N. Kiriyama. 1976. "Studies on the metabolic products of *Aspergillus terreus*. I. Metabolites of the strain IFO 6123." *Chem. Pharm. Bull.* 24: pp. 1853–1859.

Yarranton, G.A. 1972. "Distribution and succession of epiphytic lichens on black spruce near Cochrane, Ontario." *Bryologist* 75: pp. 462–480.

———. 1975. "Population growth in *Cladonia stellaris* (Opiz) Pouz. and Vězda." *New Phytol.* 75: pp. 99–110.

Yarranton, G.A. & W.G.E. Green. 1966. "The distribution patterns of crustose lichens on limestone cliffs at Rattlesnake Point, Ontario." *Bryologist* 69: pp. 450–461.

Yom-Tov, Y. & M. Galun. 1971. "Note on feeding habits of the desert snails *Sphincterochila boissieri* Charpentier and *Trochoidea (Xerocrassa) seetzeni* Charpentier." *The Veliger* 14: pp. 86–88.

Yosioka, I., T. Nakanishi, & I. Kitagawa. 1969. "Lichen triterpenoids. II. The stereostructure of zeorin." *Chem. Pharm. Bull.* 17: pp. 291–295.

Zabka, G.C. & W.R. Lazo. 1962. "Reciprocal transfer of materials between algal cells and myxomycete plasmodia in intimate association." *Am. J. Bot.* 49: pp. 146–148.

Zehnder, A. 1949. "Über den Einfluss von Wuchsstoffen auf Flechtenbildner." *Ber. Schweiz. Bot. Ges.* 59: pp. 201–267.

*Zopf, W. 1896. "Zur biologischen Bedeutung der Flechtensä*uren." *Biol. Centralbl.* 16: pp. 593–610.

———. 1907. *Die Flechtenstoffe in chemischer, botanischer, pharmakologischer und technischer Beziehung*. Jena: Gustav Fischer.

Zukal, H. 1879. "Das Zusammenleben von Moos und Flechten." *Oesterr. Bot. Z.* 29: pp. 189–191.

———. 1895. "Morphologische und biologische Untersuchungen über die Flechten." *Sber. K. Böhm. Ges. Wiss. Math.-nat. Kl.* 104: pp. 1303–1395.

Taxonomic Index

Author and Subject Index

About the Author

James D. Lawrey is Associate Professor of Biology at George Mason University in Fairfax, Virginia. He received a B.S. degree in biology from Wake Forest University in 1971, an A.M. degree in biology from the University of South Dakota in 1973, and a Ph.D. in botany from The Ohio State University in 1977. His research focuses broadly on lichen biology, particularly lichen chemical ecology. He has recently been investigating the regulatory role of lichen phenolic compound production on lichen-invertebrate interactions. This work has been supported by the National Science Foundation since 1979. He has also pursued research interests in lichens as biological pollution monitors. Much of this work was done while he was a research collaborator in the Department of Botany at the Smithsonian Institution in Washington, D.C., where he worked with Dr. Mason E. Hale, Jr., Curator of Cryptogams.